Advances in Intensional Logic

APPLIED LOGIC SERIES

VOLUME 7

SCOPE OF THE SERIES

Logic is applied in an increasingly wide variety of disciplines, from the traditional subjects of philosophy and mathematics to the more recent disciplines of cognitive science, computer science, artificial intelligence, and linguistics, leading to new vigor in this ancient subject. Kluwer, through its Applied Logic Series, seeks to provide a home for outstanding books and research monographs in applied logic, and in doing so demonstrates the underlying unity and applicability of logic.

The titles published in this series are listed at the end of this volume.

Advances in Intensional Logic

edited by

MAARTEN DE RIJKE

Department of Computer Science,
University of Warwick,
Coventry, U.K.

Springer-Science+Business Media, B.V.

A C.I.P. Catalogue record for this book is available from the Library of Congress.

ISBN 978-90-481-4897-4 ISBN 978-94-015-8879-9 (eBook)
DOI 10.1007/978-94-015-8879-9

Logo design by L. Rivlin

Printed on acid-free paper

EDITORIAL PREFACE

Intensional logic has emerged, since the 1960's, as a powerful theoretical and practical tool in such diverse disciplines as computer science, artificial intelligence, linguistics, philosophy and even the foundations of mathematics. The present volume is a collection of carefully chosen papers, giving the reader a taste of the frontline state of research in intensional logics today. Most papers are representative of new ideas and/or new research themes. The collection would benefit the researcher as well as the student. This book is a most welcome addition to our series.

The Editors

CONTENTS

CONTENTS

PREFACE

Intensional logic has many faces. In this preface we identify some prominent ones without aiming at completeness. We list some of the current trends and directions as far as they pertain to the contributions to the present volume, and after that we briefly examine the contents of the individual contributions with these generalities in mind.

Intensional Concepts. Traditionally, research in intensional logic has focused on analyzing 'the logic' of a wide variety of intensional concepts such as necessity, knowledge, belief, obligation, provability and implication. Questions that one typically encounters in this area include: Is the calculus sound and complete with respect to the intended intuitive interpretation? What does the formal modeling *explain*? Similar issues are still being addressed in current 'operator oriented' interpretations of intensional languages, both in computing (e.g., in the agent based paradigm, cf. [Fagin et al., 1995]), and in more traditional areas in which intensional logic is used, such as philosophy and cognitive science, cf. [Segerberg, to appear]. Several of the papers in this volume are devoted to the study of intensional concepts. In his contribution to the volume, Seligman introduces and analyses an intensional calculus for talking about sentences that *correctly* describe situations. Zalta's paper focuses on the object calculus — an intensional system for reasoning about abstract objects.

The Advent of Semantics. While the analysis of intensional concepts has by no means vanished as a research line in intensional logic, the arrival of a decent notion of semantics for intensional logics in the early 1960s brought with it a whole new line of research questions and uses for intensional logic. The former included issues like axiomatic completeness and expressive power. As to the latter, the advent of semantics cleared the way for the use of intensional languages as *description languages* for the structures on which they are interpreted. Initially, this use was limited to the description of temporal phenomena and the structure of time; cf. [van Benthem, 1984]. But contemporary modal and temporal logic has been profoundly influenced by its applications as a description tool in other areas, particularly in theoretical computer science. Indeed, some of the most noteworthy systems in the area are due to computer scientists, not modal or temporal logicians; examples include such well-known calculi as propositional dynamic logic and the modal mu-calculus. This descriptive use is exactly the use to which modal logic is put in the paper by Blackburn and Meyer-Viol in this volume. The authors develop a modal language for describing finite ordered binary trees as they are used in computational linguistics.

A Proof-Theoretic Turn. The advent of semantics brought with it a wealth of new questions and tools (for example for distinguishing between proposed logical systems). But to be able to actually *use* intensional logics, and to be able to *compute* with them, one needs ways of syntactically manipulating formulas in non-trivial fragments of intensional logics. Since the early 1980s (see e.g., [Fitting, 1983]) the interest in developing *general* proof-theoretic formats for intensional logics has steadily increased, and recent attempts at mechanizing logics have only strengthened the interest. In some cases the peculiarities of intensional languages have forced authors to consider novel proof formats. Indeed, two of the three proof-theoretic contributions to this volume focus on such novel formats; Wansing interrelates a number of approaches in his paper, while De Queiroz and Gabbay adapt the labelled deductive approach (due to the latter author) to the setting of modal logic. In his contribution on a logic of correct description, Seligman sticks to a more traditional proof-theoretic approach; he offers a Gentzen-style proof system, and proves that it admits cut elimination.

Modalizing. A recent trend offers a new perspective on the 'operator oriented' view of intensional logic. The essence of this trend is to adopt a 'modal perspective' on existing (non-modal) calculi, and to somehow view these as modal logics after all. One of the first areas where this *modalizing* strategy has become a useful and powerful technique is algebraic logic, and more specifically, algebras of relations and cylindric algebras; cf. [Marx and Venema, 1996] for a recent monograph in this field. But the method has also been applied in other areas, including generalized quantifiers (see below) and knowledge representation (cf. [Schild, 1991]). The benefits of modalizing are perceived to be notational simplicity, and the availability of an intuitive and concrete semantics about which one can reason using very simple graph-like pictures. In this volume, modalizing is at the heart of three of the contributions. Van Benthem and Alechina adopt a modal perspective on generalized quantifiers, focusing on the interplay between (modal) quantifier principles and semantic conditions. In the multi-modal systems considered by Rybakov the modal operators $\Box_i$ simulate universal quantifiers $\forall x_i$; this modalizing is used to study schemes of theorems of first-order theories. Finally, Vakarelov modalizes directed multi-graphs by introducing an 'arrow' language for reasoning about these structures (and the associated algebras).

Logical Concerns. Some of the issues outlined above take modal and intensional logic in new and unexplored directions. But in parallel with these developments there has been, and still is, a steady flow of work on traditional logical concerns in the area. For example, a general complexity-theoretic classification of modal logics has been developed by [Spaan, 1993], and a number of recent publications have advanced the model theory of intensional languages, cf. [Barwise and van Benthem, 1996; de Rijke, to appear]. In this volume Zakharyaschev is also concerned with classical questions from modal and intuitionistic logic; among others these concern completeness and decidability. Zakharyaschev describes a general method for dealing with such questions; cf. also his forthcoming book with Chagrov, [Chagrov and Zakharyaschev, to appear].

After having discussed some recent trends and directions in intensional logic, let us now turn to the individual contributions in more detail.

Van Benthem and Alechina adopt a modal perspective on generalized quantifiers. They give up the assumption of standard first-order logic that any domain element is 'available' for a quantifier. Instead, they interpret quantifiers in a modal fashion, using a dependence relation between elements. The authors study the resulting model theory, and in particular, they develop a systematic correspondence between axioms for quantifiers and properties of the dependence relation.

Blackburn and Meyer-Viol develop a modal logic to deal with finite ordered binary trees as they are used in (computational) linguistics. The authors introduce a modal language with operators for the 'mother of,' 'first daughter of,' and 'second daughter of' relations together with their transitive reflexive closures. The relevant class of tree models is defined and three linguistic applications of this language are discussed: context-free grammars, command relations, and trees decorated with feature structures. An axiomatic proof system is given for which completeness is shown with respect to the class of finite ordered binary trees. A number of decidability results are also presented.

De Queiroz and Gabbay are concerned with proof-theoretic aspects of traditional modal logic. They argue that the lack of a natural proof-theoretic counterpart of the semantic notion of *accessibility relation* makes it difficult to use natural deduction as the appropriate framework for a general proof theory for modal logics. In Gabbay's *Labelled Deductive Systems* [Gabbay, 1996] the *label-and-formula* perspective allows for a direct proof-theoretic treatment of referents and dependency functions. De Queiroz and Gabbay apply the labelled deduction perspective to modal logic. They consider the 'necessity' operator as a universal quantifier over structured collections of labelled formulas, and explore an extension of the (Curry-Howard-Tait-like) functional interpretation to some well-behaved notions of modal 'necessity'. The emphasis is on the characterization of a proof theory for necessity which will fall out 'naturally' from the proof theory for propositional implication [Gabbay and de Queiroz, 1992].

Rybakov studies so-called schemes of theorems of first-order theories. He adopts a general perspective on such schemes and the connection to the problem of admissibility of rules (which the author himself has studied extensively, cf.

[Rybakov, 1987]). Rybakov gives a complete description for first-order theories which have an algorithm for recognizing admissibility of inference rules. As a corollary, a description of first-order theories with decidable logics of schemes is obtained. The author also considers the connection between logics of schemes and poly-modal propositional logics, and he compares the complexities of first-order theories and their logics of schemes.

Seligman formulates Austin's theory of truth in terms of a relation of *correct description* holding between a sentence and a situation. A recursive definition of correct description is provided for first-order languages that contain terms denoting situations as well as a predicate denoting correct-description. Seligman examines a very strong logic of situations, by restricting attention to *situated consequence* between descriptions of *omniscient situations*, arguing that weaker logics may be obtained using standard methods. Rules of natural deduction for the logic are introduced by way of examples of natural reasoning using spatial indexicals. Finally, a Gentzen-style sequent calculus is offered.

Vakarelov's contribution to this volume can both be seen as a case study in describing directed multi-graphs, and as a case study in 'modalizing' certain relation algebraic formalisms. The author introduces the notion of an arrow structure as an algebraic version of directed multi-graphs. By means of a representation theorem for arrow structures it is shown that all information contained in an arrow structure is also contained in the set of its arrows equipped with four binary relations describing the four possibilities for two arrows to have a point in common. This enables one to use arrow structures as a semantic base for a special poly-modal logic, called basic arrow logic. Basic arrow logic — and some of its extensions — are used for expressing different properties of arrow structures in a modal setting. Several kinds of completeness theorems are proved, both for basic arrow logic and for some other arrow logics, including completeness with respect to classes of finite models.

The starting point of Wansing's paper is a statement of Segerberg that dates back over a decade: "a deductive treatment congenial to modal logic is yet to be found" [Bull and Segerberg, 1984]. There is still no *general* proof theory of modal or intensional logic. Since Segerberg's statement was published, however, various generalizations of the standard Gentzen-style proof-theoretic formats have been developed and investigated. These include higher-level [Došen, 1985], higher-dimensional [Masini, 1992] and higher-arity proof systems [Blamey and Humberstone, 1991]. Moreover, general proof-theoretic frameworks like Belnap's display logic [Belnap, 1982] and Gabbay's labelled deductive systems [Gabbay, 1996] have been developed. Wansing's contribution to this volume brings together two of these recent approaches, applies them to simple tense logic, and shows these instances and certain axiomatic presentations to be mutually equivalent.

Zakharyaschev surveys the vast body of results he has obtained over the past 15 years on extensions of **K4** and intermediate logics. The key notion here is the notion of a canonical formula. Without doubt the introduction of canonical formulas has led to some of the most significant advances in traditional modal logic. Briefly, the most important features of these formulas are, firstly, that every modal logic extending **K4** and every intermediate logic can be represented by canonical formulas, and,secondly, that they explicitly display the structure of frames for the logic they axiomatize. Using these features, Zakharyaschev has been able to obtain numerous results on (independent) axiomatizability, modal companions, decidability, elementarity, persistence, the disjunction property, and Halldén completeness. These results are collected together in an extensive and integrated survey in which the main emphasis is on motivations, examples and formulations of the results.

In his contribution to the volume Zalta is concerned with the modal object calculus. This is an intensional logic that has some rather interesting features in and of itself, independent of the proper axioms that can be formulated in the calculus to assert the existence of abstract objects. Zalta defines the calculus, discusses its applications, and recasts the intended interpretation of the calculus in a new and interesting way; the latter yields a philosophically more correct definition of truth and logical truth. Finally, the author describes a new interpretation for the calculus.

This volume has been in the making for a long time. Many people have been helpful in bringing this book about — I would like to thank the authors for their contributions and patience, the referees for their thorough work, and Annie Kuipers of Kluwer Academic Publishers for her encouragement. I am especially grateful to Mrs. Jane Spurr of Imperial College, London, for her generous help in finalizing the camera-ready copy.

Maarten de Rijke

REFERENCES

[Barwise and van Benthem, 1996] J. Barwise and J. van Benthem. Interpolation, preservation and pebble games. Department of Mathematics, Indiana University and CSLI, Stanford, 1996.

[Belnap, 1982] N. Belnap. Display logic. *Journal of Philosophical Logic*, 11:14–25, 1982.

[van Benthem, 1984] J. van Benthem. *The Logic of Time*. Reidel, Dordrecht, 1984.

[Blamey and Humberstone, 1991] S. Blamey and I. Humberstone. A perspective on modal sequent logic. *Publ. RIMS Kyoto University*, 27:763–782, 1991.

[Bull and Segerberg, 1984] R.A. Bull and K. Segerberg. Basic modal logic. In D. Gabbay and F. Guenthner, editors, *Handbook of Philosophical Logic. Vol. 2*, pages 1–88. Reidel, Dordrecht, 1984.

[Chagrov and Zakharyaschev, to appear] A.V. Chagrov and M.V. Zakharyaschev. *Modal Logic*. Oxford Logic Guides. Oxford University Press, to appear.

[Došen, 1985] K. Došen. Sequent-systems for modal logic. *Journal of Symbolic Logic*, 50:149–168, 1985.

[Fagin et al., 1995] R. Fagin, J.Y. Halpern, Y. Moses, and M.Y. Vardi. *Reasoning about Knowledge*. The MIT Press, Cambridge, MA and London, England, 1995.

[Fitting, 1983] M.C. Fitting. *Proof Methods for Modal and Intuitionistic Logics*. Synthese Library 169. Reidel, Dordrecht, 1983.

[Gabbay, 1996] D.M. Gabbay. *Labelled Deductive Systems*. Oxford Logic Guides. Oxford University Press, Oxford, 1996.

[Gabbay and de Queiroz, 1992] D.M. Gabbay and R. de Queiroz. Extending the Curry-Howard interpretation to linear, relevant and other resource logics. *Journal of Symbolic Logic*, 57:1319–1365, 1992.

[Marx and Venema, 1996] M. Marx and Y. Venema. *Multi-Dimensional Modal Logic*. Applied Logic Series, Kluwer Academic Publishers, Dordrecht, 1996.

[Masini, 1992] A. Masini. 2-Sequent calculus: a proof theory of modalities. *Annals of Pure and Applied Logic*, 58:229–246, 1992.

[de Rijke, to appear] M. de Rijke. Modal model theory. *Annals of Pure and Applied Logic*, to appear.

[Rybakov, 1987] V.V. Rybakov. Bases of admissible rules of the modal system *grz* and of intuitionistic logic. *Math. USSR Sbornic*, 56:311–331, 1987.

[Segerberg, to appear] K. Segerberg. Belief revision from the point of view of doxastic logic. *Logic Journal of the IGPL*, to appear.

[Schild, 1991] K. Schild. A correspondence theory for terminological logics. In *Proceedings 12th IJCAI*, pages 466–471, Sydney, 1991.

[Spaan, 1993] E. Spaan. *Complexity of Modal Logics*. PhD thesis, ILLC, University of Amsterdam, 1993.

JOHAN VAN BENTHEM AND NATASHA ALECHINA

MODAL QUANTIFICATION OVER STRUCTURED DOMAINS

1 QUANTIFIERS AS MODAL OPERATORS

1.1 Motivations

The semantics for quantifiers described in this paper can be viewed both as a new semantics for generalized quantifiers and as a new look at standard first-order quantification, bringing the latter closer to modal logic.

The standard semantics for generalized quantifiers interprets a monadic generalized quantifier Q as a set of subsets of a domain. For example, the quantifier "there are precisely two" is interpreted by the set of all subsets of the domain which contain precisely two elements. A formula $Qx\varphi$ is true in a model if the set of elements satisfying φ belongs to the interpretation of the quantifier; in our example, if there are precisely two elements satisfying φ. The existential quantifier can be treated as a generalized quantifier, too: it is interpreted as the set of all non-empty subsets of the domain. The universal quantifier is interpreted by the singleton set containing the whole domain.

The quantifiers listed so far are first-order definable in the following sense: they can be defined using ordinary quantifiers and equality. Many interesting generalized quantifiers are not first-order definable. The present study is motivated by the work of [van Lambalgen, 1991] on Gentzen-style proof theory for the quantifiers "for many" (its dual is interpreted as a non-principal filter), "for uncountably many" and "for almost all" (the latter contains all subsets of the domain which have Lebesgue measure 1). All those quantifiers are not first-order definable. They have Hilbert-style axiomatizations, but until recently no one believed that they can have a reasonable Gentzen-style proof theory. In order to devise such a proof theory, van Lambalgen used a translation of generalized quantifier formulas into a first-order language enriched with a predicate R of indefinite arity. $Qx\varphi(x, y_1, \ldots, y_n)$, where Q is a universal-type generalized quantifier (distributing over conjunction), is translated as

$$\forall x(R(x, y_1, \ldots, y_n) \to \varphi(x, y_1, \ldots, y_n)),$$

and its dual Q^d as

$$\exists x(R(x, y_1, \ldots, y_n) \wedge \varphi(x, y_1, \ldots, y_n))$$

M. de Rijke (ed.), Advances in Intensional Logic, 1–27.

(all free variables displayed). Observe that this translation is reminiscent of the standard translation of modal formulas into first-order logic, with the sequence of free variables playing the role of the "actual world" and the quantifier ranging over the variables "accessible" from the given sequence. The idea behind such a translation is as follows. When generalized quantifiers are viewed as first-order operators (binding first-order variables), it becomes clear that a variable bound by a generalized quantifier cannot in general take any possible value. Its range is restricted, and this restriction can be defined using an accessibility relation. Then the elimination rule for Q with a premise $Qx\varphi(x, \bar{y})$ would introduce a variable $x_{\bar{y}}$ ranging over the set $\{x : R(x, \bar{y})\}$.

It turns out that some quantifier axioms correspond to first-order conditions on R in the following sense: any set of generalized quantifier formulas is consistent with the axiom if and only if the set of translations is consistent with the corresponding first-order condition on R. For example, $Qx\varphi \wedge Qx\psi \rightarrow Qx(\varphi \wedge \psi)$ corresponds in this sense to $R(x, \bar{y}\bar{z}) \rightarrow R(x, \bar{y})$. In case such a correspondent exists, it is easier to find side conditions for elimination and introduction rules for the generalized quantifier satisfying the axiom. In the example above, the rule becomes

$$\frac{Qx\varphi(x, \bar{y})}{\varphi(x_{\bar{z}}, \bar{y})}$$

where $\bar{y} \subseteq \bar{z}$ and $x_{\bar{z}}$ ranges over the set $\{x : R(x, \bar{z})\}$. Thus, we have a (generalized) modal logic for quantifiers here, which even exhibits some of the standard modal concerns (such as correspondence). Our aim in this paper is to investigate this general modal logic as such.

Another motivation for this enterprise is a modal-style modification of first-order quantification, aimed at obtaining a system which has some nice properties of modal logic not shared by first-order logic. In this sense, our approach relates to the one taken in [Németi, 1995] where non-standard models for first-order logic were introduced, yielding a decidable quantification theory by imposing restrictions on "accessible assignments". Let us rephrase our general idea. The Tarskian truth condition for the existential quantifier reads as follows:

$$M, [\bar{d}/\bar{y}] \models \exists x\varphi(x, \bar{y}) \Leftrightarrow \exists d \in D\colon M, [d/x, \bar{d}/\bar{y}] \models \varphi(x, \bar{y}).$$

This may be viewed as a special case of a more general schema, when the element d is required in addition to stand in some relation R to $\bar{d}$ — where R is a finitary relation structuring the individual domain D:

$$M, [\bar{d}/\bar{y}] \models \Diamond_x\varphi(x, \bar{y}) \Leftrightarrow \exists d \in D\colon R(d, \bar{d}) \;\&\; M, [d/x, \bar{d}/\bar{y}] \models \varphi(x, \bar{y}).$$

In the above-mentioned work on the generalized quantifiers "many", "uncountably many" and "almost all", R has properties which are common to different independence relations: linear independence in algebra, probabilistic independence, etc. But we can think of more general applications too, with domains being arranged in different levels of accessibility, or with procedures drawing objects in possible dependencies upon one another. One might read $R(d, \bar{e})$ as

- d can be constructed using $\bar{e}$,
- d is not "too far" from the e's,
- after you have picked up e's from the domain without replacing them, d is still available,

et cetera. Ordinary predicate logic then becomes the special case of flat individual domains admitting of "random access", whose R is the universal relation.

This semantics has some clear analogies with modal logic, with an existential generalized quantifier as an existential modality over some domain with, not a binary, but an arbitrary finitary "accessibility relation". As a consequence, we can apply standard ideas concerning modal completeness and correspondence to understand this broader concept of quantification.

Both motivations, generalized quantifiers and generalized first-order semantics, give rise to a variety of questions, including model theory, first-order completeness, canonicity, frame correspondence, definability of R-properties by quantifier axioms etc. In this paper we explore some of them.

1.2 Language and Models

The language of the logic $EL(\exists, \Diamond)$ with a generalized quantifier is the ordinary language of first-order predicate logic with equality (without functional symbols) plus an existential generalized quantifier $\Diamond$. The notion of a wff is extended as follows: if φ is a wff, then so is $\Diamond_x \varphi$. A universal dual of $\Diamond$ is defined as usual: $\Box_x \varphi =_{df} \neg \Diamond_x \neg \varphi$. We shall refer to the sublanguage without ordinary quantifiers as $EL(\Diamond)$.

$M = (D, R, V)$ is a *model* for $EL(\exists, \Diamond)$ if D and V are an ordinary domain and interpretation for first-order logic, and R is a binary relation between $d \in D$ and finite sets of elements from D. We will write $R(d, \bar{d})$ for $R(d, \{d_1, \ldots, d_n\})$, tacitly assuming that the second argument of R is invariant under permutations and repetitions.

The relation $M, v \models \varphi$ ("φ is *true* in M under assignment v") is defined as follows:

- $M, v \models P_i^n(x_{j1} \ldots x_{jn}) \Leftrightarrow \langle v(x_{j1}) \ldots v(x_{jn}) \rangle \in V(P_i^n)$;
- $M, v \models \neg\varphi \Leftrightarrow M, v \not\models \varphi$;
- $M, v \models \varphi \wedge \psi \Leftrightarrow M, v \models \varphi$ and $M, v \models \psi$;

- $M, v \models \exists x\psi(x) \Leftrightarrow$ there exists a variable assignment v' which differs from v at most in its assignment of a value to x ($v' =_x v$) such that $M, v' \models \psi(x)$;

- $M, v \models \Diamond_x \psi(x, y_1, \ldots, y_n) \Leftrightarrow$ there exists $v' =_x v$ such that $R(v'(x), v'(y_1), \ldots, v'(y_n))$ and $M, v' \models \psi(x, \bar{y})$ where $\bar{y}$ are all (and just the) free variables of $\Diamond_x \psi$.

It is easy to see that

- $M, v \models \Box_x \psi(x, \bar{y}) \Leftrightarrow$ if for all $v' =_x v$: $R(v'(x), v'(\bar{y})) \Rightarrow M, v' \models \psi(x, \bar{y})$.

We say that $M \models \varphi$ iff $M, v \models \varphi$ for all variable assignments v.

Let us define a *frame* (analogously to modal logic) $F = (D, R)$ as the underlying structure of a set of models with all possible interpretations of predicate letters. $F, v \models \varphi$ if $M, v \models \varphi$ for all models M on F. The formula φ is (globally) valid in F if, for all v, $F, v \models \varphi$ ("$F \models \varphi$").

This system resembles first-order logic in many respects, but no standard property can be taken for granted any more:

Monotonicity is restricted. Assume that for all variable assignments v $M, v \models \varphi(x_1, x_2) \rightarrow \psi(x_1, x_3)$, and that for some assignment v $M, v \models \Diamond_{x_1} \varphi(x_1, x_2)$ (there exists $v' =_{x_1} v$ such that $R(v'(x_1), v(x_2))$ and $M, v' \models \varphi(x_1, x_2)$). It does not follow that $M, v \models \Diamond_{x_1} \psi(x_1, x_3)$, because although $M, v' \models \psi(x_1, x_3)$, it is not necessary that $R(v'(x_1), v(x_3))$ holds. Indeed, the general monotonicity rule

$$\frac{\Sigma \vdash \varphi(x, \bar{y}) \rightarrow \psi(x, \bar{z})}{\Sigma \vdash \Diamond_x \varphi(x, \bar{y}) \rightarrow \Diamond_x \psi(x, \bar{z})},$$

with x not free in Σ, is invalid. We can accept only **Restricted Monotonicity**, where φ and ψ have the same free variables.

Extensionality is also restricted. Properties which hold for exactly the same objects, are no longer identical. Consider a property P which holds for a single object a: $\forall x(P(x) \equiv x = a)$. Let $R(a, \emptyset)$ and $\neg R(a, a)$. Then, $\Diamond_x P(x)$ is true and $\Diamond_x x = a$ is false.

Substitution therefore should also be restricted: only formulas with the same parameters can be substituted. We do not have in general that

$$D, R, V, v \vdash \varphi[\alpha/P] \Leftrightarrow D, R, V[P := [\alpha]_{M,v}], v \models \varphi.$$

2 AXIOMATICS AND COMPLETENESS

We shall now develop the basic deductive calculus for our modal quantifier logic.

DEFINITION 2.1 The *minimal logic* for $EL(\exists, \Diamond)$ is a calculus of *sequents* $\Sigma \vdash \varphi$ satisfying the usual rules for first-order logic, including all Boolean principles, as well as the following quantifier rules:

Restricted Monotonicity plus Distribution

$$\frac{\Sigma \vdash \varphi(x, \bar{y}) \rightarrow \bigvee_{i=1}^{i=n} \psi_i(x, \bar{y})}{\Sigma \vdash \Diamond_x \varphi(x, \bar{y}) \rightarrow \bigvee_{i=1}^{i=n} \Diamond_x \psi_i(x, \bar{y})},$$

where x is not free in Σ, and free variables are exactly those displayed (only x does not necessarily occur free in ψ_i). The convention here is that an empty disjunction is a falsum, both in the premise and the conclusion.

Alphabetic Variants

$$\vdash \Diamond_x \varphi(x, \bar{y}) \equiv \Diamond_z \varphi(z, \bar{y})$$

where z does not occur (free or bound) in $\varphi(x, \bar{y})$.

Here are some derivations in this system, corresponding to obvious validities given the above existential truth condition for the quantifier $\Diamond$:

1. $\vdash \bot \rightarrow \bot$

   $\vdash \Diamond_x \bot \rightarrow \bot$

   $\vdash \neg \Diamond_x \bot$

2. $\neg\varphi(\bar{y}) \vdash \varphi(\bar{y}) \rightarrow \bot$

   $\neg\varphi(\bar{y}) \vdash \Diamond_x \varphi(\bar{y}) \rightarrow \bot$

   $\vdash \Diamond_x \varphi(\bar{y}) \rightarrow \varphi(\bar{y})$, provided that x is not among the $\bar{y}$

3. Suppose that $\vdash \varphi \rightarrow \psi$ with x not free in ψ. Then:

   $\neg\psi \vdash \neg\varphi$

   $\neg\psi \vdash \varphi \rightarrow \bot$

   $\neg\psi \vdash \Diamond_x \varphi \rightarrow \bot$

   $\neg\psi \vdash \neg \Diamond_x \varphi$

whence $\vdash \Diamond_x \varphi \rightarrow \psi$.

4. An application of (3) is:

$\vdash \Diamond_x\varphi \to \Diamond_x\varphi$

$\vdash \Diamond_x\Diamond_x\varphi \to \Diamond_x\varphi$

5. Also,

$\vdash \varphi \to \exists x\varphi$

$\vdash \Diamond_x\varphi \to \exists x\varphi$

6. As a final illustration, we prove a useful principle for later reference, namely: $\neg\Diamond_z(\psi(z,\bar{y}) \wedge \neg\Diamond_x\psi(x,\bar{y}))$:

$\vdash \psi(z,\bar{y}) \wedge \neg\Diamond_x\psi(x,\bar{y}) \to \psi(z,\bar{y})$

$\vdash \Diamond_z(\psi(z,\bar{y}) \wedge \neg\Diamond_x\psi(x,\bar{y})) \to \Diamond_z\psi(z,\bar{y})$

$\vdash \Diamond_z(\psi(z,\bar{y}) \wedge \neg\Diamond_x\psi(x,\bar{y})) \to \Diamond_x\psi(x,\bar{y})$

and

$\vdash \psi(z,\bar{y}) \wedge \neg\Diamond_x\psi(x,\bar{y}) \to \neg\Diamond_x\psi(x,\bar{y})$

$\vdash \Diamond_z(\psi(z,\bar{y}) \wedge \neg\Diamond_x\psi(x,\bar{y})) \to \neg\Diamond_x\psi(x,\bar{y})$

(the latter step is as in example (3) above). Therefore,

$\vdash \Diamond_z(\psi(z,\bar{y}) \wedge \neg\Diamond_x\psi(x,\bar{y})) \to \bot$

$\vdash \neg\Diamond_z(\psi(z,\bar{y}) \wedge \neg\Diamond_x\psi(x,\bar{y}))$

THEOREM 2.2 *The minimal logic is complete for universal validity.*

Proof. By a standard Henkin construction. The key point, as usual, is to create a maximally consistent set of formulas Σ — this time, adding suitable witnesses (new variables) for accepted formulas $\Diamond_x\varphi$:

> If Σ_n is consistent with $\Diamond_x\varphi(x,\bar{y})$, then add a *new* individual variable z with
>
> 1. $\varphi(z,\bar{y})$,
> 2. $\{\psi(z,\bar{y}) \to \Diamond_x\psi(x,\bar{y})$: for all formulas $\psi\}$.

CLAIM. *This extension is consistent.*

Proof. Suppose it were inconsistent. Then, for some fresh variable z and some finite disjunction of formulas ψ_i:

$$\Sigma_n \vdash \varphi(z,\bar{y}) \to \bigvee_i (\psi_i(z,\bar{y}) \wedge \neg\Diamond_x\psi_i(x,\bar{y})).$$

Then also

$$\Sigma_n \vdash \Diamond_z \varphi(z,\bar{y}) \rightarrow \bigvee_i \Diamond_z(\psi_i(z,\bar{y}) \wedge \neg\Diamond_x \psi_i(x,\bar{y})).$$

Therefore, since $\vdash \Diamond_x \varphi(x,\bar{y}) \equiv \Diamond_z \varphi(z,\bar{y})$ (by Alphabetic Variants), the set $\{\Sigma_n, \Diamond_x \varphi(x,\bar{y})\}$ must be consistent with some

$$\Diamond_z(\psi_i(z,\bar{y}) \wedge \neg\Diamond_x \psi_i(x,\bar{y})).$$

But this contradicts the earlier derivability of the formula

$$\neg\Diamond_z(\psi(z,\bar{y}) \wedge \neg\Diamond_x \psi(x,\bar{y})). \quad \dashv$$

Now construct the Henkin model as usual, and set

$$R(z,y_1,\ldots,y_n) \Leftrightarrow_{df} \forall\varphi\colon \varphi(z,\bar{y}) \in \Sigma \Rightarrow \Diamond_x \varphi(x,\bar{y}) \in \Sigma$$

This definition may be compared with the usual introduction of the alternative relation R in completeness proofs for Modal Logic. To demonstrate the adequacy of the present Henkin model, all we have to prove is the following decomposition:

$$\Diamond_x \varphi(x,\bar{y}) \in \Sigma \text{ iff } \exists z\colon R(z,\bar{y}) \ \&\ \varphi(z,\bar{y}) \in \Sigma$$

From left to right this is guaranteed by the above construction of Σ (through the addition of all formulas of the second kind). From right to left, this is a trivial consequence of the definition of R. $\dashv$

If we look at the above completeness proof (and earlier examples of derivabilities), we see that no structural contraction rule or ordinary quantifier rules have been used. This observation (which is quite analogous with the situation in the minimal modal logic) motivates the conjecture the minimal logic without ordinary quantifiers is decidable. Indeed, the following theorem holds

THEOREM 2.3 *The minimal logic without ordinary quantifiers and without equality is decidable.*

Proof. This is shown in [Alechina, 1995]. $\dashv$

3 MODEL THEORY

Now, to illustrate the semantical properties of modal quantifiers, we shall consider an analogue to the basic model-theoretic invariance relation of modal logic. In what follows, we talk about the language $EL(\Diamond)$ (without ordinary quantifiers).

DEFINITION 3.1 A *bisimulation* $\mathcal{B}$ between two models $M_1 = \langle D_1, R_1, V_1\rangle$ and $M_2 = \langle D_2, R_2, V_2\rangle$ is a family of partial isomorphisms π with the following properties:

1. π is a partial bijection with $dom(\pi) \subseteq D_1$ and $ran(\pi) \subseteq D_2$;

2. If $\{d_1, \ldots, d_n\} \subseteq dom(\pi)$, then for all predicate letters

$$\langle d_1, \ldots, d_n \rangle \in V_1(P^n) \Leftrightarrow \langle \pi(d_1), \ldots, \pi(d_n) \rangle \in V_2(P^n)$$

($d_1, \ldots, d_n$ are not necessarily distinct).

3a. If $D \subseteq dom(\pi)$ and $R_1(d, D)$, then there exists an element d' in D_2 such that $R_2(d', \pi[D])$ and $\{\langle d, d' \rangle\} \cup \pi \in \mathcal{B}$.

3b. If $D' \subseteq ran(\pi)$, $D' = \pi[D]$, and $R_2(d', D')$, then there exists an element d in D_1 such that $R_1(d, D)$ and $\{\langle d, d' \rangle\} \cup \pi \in \mathcal{B}$.[1]

LEMMA 3.2 (Invariance Lemma) *If φ is a formula of $EL(\Diamond)$ with the set of free variables $VAR(\varphi) \subseteq \{y_1, \ldots, y_n\}$, M_1 and M_2 are bisimilar models, and for all y_i $(1 \leq i \leq n)$ $v_1(y_i) \in dom(\pi)$ and $v_2(y_i) = \pi(v_1(y_i))$, $\pi \in \mathcal{B}$, then*

$$M_1, v_1 \models \varphi \Leftrightarrow M_2, v_2 \models \varphi.$$

Proof. By induction on the length of φ.

- φ is a k-place predicate letter. — By clause (2) in the definition of bisimulation.

- $\varphi = (x = y)$. $M_1, v_1 \models x = y$ if and only if $v_1(x) = v_1(y)$. Since π is a function, and $v_1(x) \in dom(\pi)$, $\pi(v_1(x)) = \pi(v_1(y))$, that is, $v_2(x) = v_2(y)$ and $M_2, v_2 \models x = y$. Backwards: the same argument, using the fact that π is a bijection.

- The boolean cases are easy.

- $\varphi = \Diamond_x \psi(x, \bar{y})$. Assume $M_1, v_1 \models \Diamond_x \psi(x, \bar{y})$. By the semantic truth definition, there exists an assignment v_1' which differs from v_1 at most in its assignment of value to x, such that $R(v_1'(x), v_1'(\bar{y}))$ and $M_1, v_1' \models \psi(x, \bar{y})$. By assumption, $y_1, \ldots, y_n \subseteq dom(\pi)$. By clause 3a, there is $d' \in D_2$ with $R(d', \pi v_1'(\bar{y}))$, i.e. $R(d', v_2(\bar{y}))$ (since v_1' and v_1 agree on $\bar{y}$), and $\{\langle d, d' \rangle\} \cup \pi \in \mathcal{B}$. Put $v_2' =_x v_2$, $v_2'(x) = d'$. Then, for the $\pi' \in \mathcal{B}$ which consists of π and the pair $\langle d, d' \rangle$, $v_2'(x) = \pi'(v_1'(x))$, and for all y_i, $v_2'(y_i) = \pi(v_1'(y_i))$. By the inductive hypothesis, $M_2, v_2' \models \psi(x, \bar{y})$. But then $M_2, v_2 \models \Diamond_x \psi(x, \bar{y})$. The same argument can be applied backwards. ⊣

[1] Alternatively, we could restrict clause 3 to R-successors of the whole domain and range, while adding a further clause closing $\mathcal{B}$ under restrictions.

Continuing the analogy with modal logic, we define a translation of $EL(\Diamond)$ formulas into the appropriate first-order logic, which is our original base language enriched with a dependence predicate R. The *standard translation* ST is defined as follows:

$$\begin{array}{rcl} ST(P_i^n(t_1 \ldots t_n)) & := & P_i^n(t_1 \ldots t_n); \\ ST(t_1 = t_2) & := & (t_1 = t_2); \\ ST & & \text{commutes with classical connectives;} \\ ST(\Diamond_x \varphi(x, \bar{y})) & := & \exists x(R(x, \bar{y}) \wedge ST(\varphi(x, \bar{y}))). \end{array}$$

CLAIM 3.3 *If φ is a formula of $EL(\Diamond)$, then $M, v \models \varphi \Leftrightarrow M', v \models ST(\varphi)$, for the classical model $M' = (D, V')$, where V' extends V to interpret the predicate R as R_M.*

DEFINITION 3.4 The *modal formulas* (being those formulas which are standard translations of $EL(\Diamond)$ formulas) are the least set X of first-order formulas such that

- atomic formulas belong to X,
- if ψ_1 and ψ_2 are in X, then so are $\neg\psi_1$ and $\psi_1 \wedge \psi_2$,
- if $\varphi(x, \bar{y}) \in X$, then $\exists x(R(x, \bar{y}) \wedge \varphi(x, \bar{y}))$ is in X too.

THEOREM 3.5 *A first-order formula φ is equivalent to a modal formula if and only if it is preserved under bisimulation.*

Proof. The direction from left to right follows from the Invariance Lemma above. For the converse, let φ be a first-order formula with variables $x_1, \ldots, x_n$, preserved under bisimulation. We want to prove that it is equivalent to a modal formula.

Define the set $CONS_\Diamond(\varphi)$ as $\{\alpha : \alpha$ is a modal formula, $\varphi \models \alpha$ and the free variables of α are among $x_1, \ldots, x_n\}$. If we can prove that

$$CONS_\Diamond(\varphi) \models \varphi, \qquad (*)$$

then we are done. For, by compactness, there will be some finite subset $\alpha_1, \ldots, \alpha_m$ of $CONS_\Diamond(\varphi)$ with $\alpha_1, \ldots, \alpha_m \models \varphi$. By definition, $\varphi \models \alpha_1, \ldots, \alpha_m$. So, then φ is equivalent to $\alpha_1 \wedge \ldots \wedge \alpha_n$, which is a conjunction of standard translations of $EL(\Diamond)$ formulas, i.e. a standard translation of the conjunction of those formulas.

Now we start proving $(*)$. Assume that $M, v \models CONS_\Diamond(\varphi)$. We show that $M, v \models \varphi$. Let us denote the set of all modal formulas true in M and having free variables among $x_1, \ldots, x_n$ as X_M. This is consistent with φ: for, if it is not, there is a finite set $\psi_1, \ldots, \psi_k$ of formulas from X_M, such that $\bigwedge_i \psi_i \rightarrow \neg\varphi$. Then $\varphi \rightarrow \bigvee_i \neg\psi_i$. But $\bigvee_i \neg\psi_i$ is a modal formula (if every ψ_i is). Since it is a consequence of φ, it must be true in X_M. A contradiction. Therefore there should be a model N for $\varphi \cup X_M$: say, $N, v' \models \varphi \cup X_M$.

Let $v(x_1) = d_1, \ldots, v(x_n) = d_n$ in M and $v'(x_1) = d'_1, \ldots, v'(x_n) = d'_n$ in N. Now, take ω-saturated elementary extensions $\mathcal{M}$ and $\mathcal{N}$ of M and N. We define a relation of bisimulation between $\mathcal{M}$ and $\mathcal{N}$ as follows:

> $\mathcal{B}$ is the family of partial mappings π such that $\pi = \{(e_1, \pi(e_1)), \ldots, (e_n, \pi(e_n))\}$ if for all modal formulas ψ with at most free variables $x_1, \ldots, x_n$ and any two assignments v, v' with $v(x_i) = e_i$, $v'(x_i) = \pi(e_i)$ $(1 \leq i \leq n)$, (**)
>
> $$\mathcal{M}, v \models \psi \Leftrightarrow \mathcal{N}, v' \models \psi.$$

To prove that (**) indeed defines a bisimulation relation, we must check that the properties (1)–(3) hold for $\mathcal{B}$. Here, (1) is trivial. Case (2) is immediate, since atomic formulas are also standard translations of (atomic) formulas in $EL(\Diamond)$). Next, we check the zigzag clause 3a. Assume that $e_1, \ldots, e_k \in dom(\pi)$ and $R(e, e_1, \ldots, e_k)$. We must prove that there exists e' in $\mathcal{N}$ such that $R(e', \pi(e_1), \ldots, \pi(e_k))$ and $\{\langle e, e' \rangle\} \cup \pi \in \mathcal{B}$. Take the set Ψ of all modal formulas with variables interpreted as $e, e_1, \ldots, e_k$ which are true in $\mathcal{M}$ under variable assignment v. We need an element e' in $\mathcal{N}$ such that all formulas in Ψ are true in $\mathcal{N}$ under v' when e' is assigned to the variable which was assigned e in $\mathcal{M}$. By saturation, it suffices to find such an e' for each finite subset Ψ_0 of Ψ. But these must exist, because the modal formula $ST(\Diamond_x \bigwedge \Psi_0(x, e_1, \ldots, e_k))$ holds in $\mathcal{M}$ and hence $ST(\Diamond_x \bigwedge \Psi_0(x, \pi(e_1), \ldots, \pi(e_k)))$ holds in $\mathcal{N}$. The appropriate check for the converse direction 3b is proved analogously.

Recall that $v(x_i) = d_i$ and $v'(x_i) = d'_i$, $1 \leq i \leq n$. We must also show that $\{\langle d_1, d'_1 \rangle, \ldots, \langle d_n, d'_n \rangle\} \in \mathcal{B}$. But this is so because for all modal formulas ψ with variables interpreted as $d_1, \ldots, d_n$ in M,

$$M, v \models \psi \Leftrightarrow N, v' \models \psi$$

(by the construction of N), and hence $\mathcal{M}, v \models \psi \Leftrightarrow \mathcal{N}, v' \models \psi$.

Finally, since φ is invariant under bisimulation and

$$\{\langle d_1, d'_1 \rangle, \ldots, \langle d_n, d'_n \rangle\} \in \mathcal{B},$$

$\mathcal{N} \models \varphi(d'_1, \ldots, d'_n)$ will now imply $\mathcal{M} \models \varphi(d_1, \ldots, d_n)$. Since $\mathcal{M}$ is an elementary extension of M, $M \models \varphi(d_1, \ldots, d_n)$, that is, $M, v \models \varphi(x_1, \ldots, x_n)$, and we are done. ⊣

To conclude this section, we add some remarks about preservations properties for the full $EL(\exists, \Diamond)$ language. Since it includes the whole first-order language, bisimulation is obviously not enough to preserve all formulas:

CLAIM 3.6 *If φ does contain $\forall$ or $\exists$, bisimulation does not preserve truth.*

Proof. Let M_1 and M_2 be as follows:

$$M_1 = \langle D_1, R_1, V_1\rangle\colon D_1 = \{d, d'\},\ R_1 = \emptyset,\ V_1(P) = \{\langle d, d'\rangle\};$$

$$M_2 = \langle D_2, R_2, V_2\rangle\colon D_2 = \{e\},\ R_2 = \emptyset,\ V_2(P) = \emptyset.$$

Then $M_1, [d/x] \models \exists y P(x, y)$ and $M_2, [e/x] \not\models \exists y P(x, y)$. But at the same time, a bisimulation between these two models exists: $\mathcal{B} = \{\langle d, e\rangle\}$. ⊣

One can strengthen the above notion of bisimulation to preserve the full language, much as happens in modal logic extended with a "universal modality". The result is essentially the standard notion of "partial isomorphism" $\cong_p$ from abstract model theory.

4 FRAME CORRESPONDENCE

In this section and in the following one we extend the standard translation to $EL(\exists, \Diamond)$. An extra clause for ordinary quantifiers has to be added; as it is to be expected, ST commutes with ordinary quantifiers.

If a formula φ of $EL(\exists, \Diamond)$ is valid in a frame F (under an assignment v), then classically

$$F, v \models \forall P_i^n \ldots \forall P_l^m ST(\varphi),$$

where $P_i^n, \ldots, P_l^m$ are the predicate letters in φ. If this second-order formula has a first-order equivalent (containing only R and $=$), φ is called *first-order definable*. This means that if φ is true in all models over F, then R has the property defined by φ, and vice versa. Additional quantifier principles added to the minimal logic will now express special conditions on the relation R. One bunch of examples arises if we look at some properties of the standard existential quantifier $\exists$:

Unrestricted Distribution $\quad \Diamond_x(\varphi \vee \psi) \leftrightarrow \Diamond_x\varphi \vee \Diamond_x\psi.$

In one direction, this gives us unrestricted "Monotonicity" for $\Diamond$:

$$\Diamond_x\varphi \to \Diamond_x(\varphi \vee \psi).$$

This corresponds to the frame condition of

Upward Monotonicity $\quad R(x, \bar{y}) \to R(x, \bar{y}\bar{z})$

Proof. Suppose that $R(x, \bar{y})$. Define the following predicate:

$$P(u, \bar{v}) := u = x \wedge \bar{v} = \bar{y}.$$

We have $R(x, \bar{y}) \wedge P(x, \bar{y})$, whence $\Diamond_x P(x, \bar{y})$ holds. Therefore,

$$\Diamond_x(P(x, \bar{y}) \vee \bot(\bar{z}))$$

(where $\bot(\bar{z})$ is any contradiction involving $\bar{z}$): i.e., there exists d with $R(d, \bar{y}, \bar{z})$ and $P(d, \bar{y}) \vee \bot(\bar{z})$: the latter must be because $P(d, \bar{y})$: i.e. $d = x$, and hence $R(x, \bar{y}, \bar{z})$. $\dashv$

By a similar kind of argument, again making an appropriate substitution for the two predicates involved, the opposite direction $\Diamond_x(\varphi \vee \psi) \rightarrow \Diamond_x\varphi \vee \Diamond_x\psi$ corresponds to the frame condition of

Downward Monotonicity $\quad R(x, \bar{y}\bar{z}) \rightarrow R(x, \bar{y})$

Together, these reduce the finitary relation R to an essentially *unary* "restriction" to the subdomain of all objects d satisfying the condition $R(d)$. It would also be of interest to see whether we can stop short of this, with quantifiers merely reducing the finitary relation R to a compound of *binary* ones (as happens in the generalized modal semantics for program operators proposed in [van Benthem, 1992]).

REMARK 4.1 Classical analogies may be slightly misleading here. E.g., the implication

$$\Diamond_x(\varphi \wedge \psi) \rightarrow \Diamond_x\varphi$$

(cf. $\Diamond(\varphi \wedge \psi) \rightarrow \Diamond\varphi$) expresses Downward Monotonicity, rather than the Upward Monotonicity of

$$\Diamond_x\varphi \rightarrow \Diamond_x(\varphi \vee \psi)$$

(cf. $\Diamond\varphi \rightarrow \Diamond(\varphi \vee \psi)$), even though the latter is equivalent with it in standard modal logic. Thus, it should in fact imply unlimited distribution - as may be seen using the available distribution in our minimal logic. In the latter calculus, "limited distribution" sanctions

1. $\Diamond_x(\varphi(x, \bar{y}) \vee \psi(x, \bar{z})) \rightarrow \Diamond_x((\varphi(x, \bar{y}) \wedge \top(\bar{z})) \vee (\psi(x, \bar{z}) \wedge \top(\bar{y})))$
2. $\Diamond_x((\varphi(x, \bar{y}) \wedge \top(\bar{z})) \vee (\psi(x, \bar{z}) \wedge \top(\bar{y}))) \rightarrow$
   $\Diamond_x(\varphi(x, \bar{y}) \wedge \top(\bar{z})) \vee \Diamond_x(\psi(x, \bar{z}) \wedge \top(\bar{y}))$,
3. from which the unlimited version $\Diamond_x(\varphi \vee \psi) \rightarrow \Diamond_x\varphi \vee \Diamond_x\psi$ follows by the above implication, passing to the appropriate conjuncts.

Finally, the above unary relation gets trivialized to universality by the principle of

Instantiation $\quad \varphi \rightarrow \Diamond_x\varphi$

This corresponds to the frame condition $\forall x \forall y R(x, y)$ (provided that we assume non-empty individual domains, that is). The idea is this: let x,y be arbitrary, and let $P(x, y)$ hold of just these. We must have that $\Diamond_x P(x, y)$: i.e., some object d exists with $R(d, y)$ and $P(d, y)$: whence $R(x, y)$.

Another source of examples is the analysis of various properties of the standard quantifier $\exists$ which are all lumped together as being "valid" in ordinary predicate logic, but which now become distinguishable as different properties of dependence. To be sure, such differences also become visible in other more sensitive semantics, such as those for intuitionistic predicate logic, or the logic of polyadic generalized quantifiers. Indeed, one concrete interpretation of the above structured domains would be the following:

individuals	pairs $\langle$world, individual$\rangle$
dependence	$(w,x)R(v,y)$ iff $w \subseteq v$ & $y = x$,

inspired by standard possible worlds semantics for intuitionistic logic. We continue with one example of this kind:

Prenex operations $\quad \Diamond_x(\varphi \vee \psi) \leftrightarrow \varphi \vee \Diamond_x\psi$, where x not free in φ.

The direction $\rightarrow$ here turns out universally valid in case $\psi, \varphi \vee \psi$ have the same free variables, and hence derivable:

$$\neg\varphi \vdash (\varphi \vee \psi) \rightarrow \psi$$

$$\neg\varphi \vdash \Diamond_x(\varphi \vee \psi) \rightarrow \Diamond_x\psi$$

$$\vdash \Diamond_x(\varphi \vee \psi) \rightarrow \Diamond_x\psi \vee \varphi,$$

Otherwise, it will enforce Downward Monotonicity: $R(x, \bar{y}\bar{z}) \rightarrow R(x, \bar{y})$. The direction $\leftarrow$ corresponds to the conjunction of $R(x, \bar{y}) \rightarrow R(x, \bar{y}\bar{z})$ and $\exists x R(x, \bar{y})$.

A comparison of quantifier axioms and similar modal axioms can also provide some interesting correspondences. For example, how would one write a quantifier version of the well-known K4-axiom: as $\Diamond_x\varphi \rightarrow \Diamond_x\Diamond_x\varphi$ or with the more complex decoration

$$\Diamond_x\varphi \rightarrow \Diamond_y\Diamond_x\varphi?$$

The first one is universally valid, the second one defines $R(x, y\bar{z}) \rightarrow R(y, \bar{z})$ (in case y is free in φ). Another direction is also possible: which quantifier principles correspond to well known properties of Kripke frames? Well-known examples are the three defining properties of equivalence relations:

Reflexivity $\quad R(x, x)$

Transitivity $\quad R(y, x) \wedge R(z, y) \rightarrow R(z, x)$

Symmetry $\quad R(x, y) \rightarrow R(y, x)$

FACT 4.2 *The following principles are definable in* $EL(\exists, \Diamond)$*:*

- *Reflexivity corresponds to* $\Diamond_y x = y$;
- *Transitivity corresponds to* $\Diamond_y(\top(x) \wedge \Diamond_z(\top(y) \wedge P(z))) \rightarrow \Diamond_z(\top(x) \wedge P(z))$
- *Symmetry corresponds to* $\forall x \Box_y P(x,y) \rightarrow \forall y \Box_x P(x,y)$.

A proof of the above facts will be given in Section 5 below.

Some negative results concerning definability of first-order properties in $EL(\Diamond)$ alone can be obtained using frame constructions familiar from modal logic.

DEFINITION 4.3 Let $F = \langle D, R \rangle$ be a frame and $d_1, \ldots, d_n \in D$. A subframe $F' = \langle D', R' \rangle$ of F is *generated* by $d_1, \ldots, d_n$ if

- D' is the smallest subdomain of D containing $d_1, \ldots, d_n$ which is closed under accessibility, and
- R' is the restriction of R to D'.

THEOREM 4.4 *Let F' be a generated subframe of F, v a valuation restricted to the elements of D', and φ a formula of $EL(\Diamond)$. Then*

$$F, v \models \varphi \Leftrightarrow F', v \models \varphi$$

In other words, $EL(\Diamond)$-formulas are invariant for generated subframes.

Proof. For any pair of models $M = \langle F, V \rangle$ and $M' = \langle F', V \rangle$ the identity map from D' to D gives an obvious bisimulation, and we can apply our invariance results from Section 3. ⊣

EXAMPLES 4.5 (Modal undefinability)

- $\exists x \neg R(x,x)$ is not definable by an $EL(\Diamond)$ formula. Consider

  $$F = \langle \{d_1, d_2\}, \{\langle d_1, d_1 \rangle\} \rangle,$$

  where it holds, and the generated subframe

  $$F' = \langle \{d_1\}, \{\langle d_1, d_1 \rangle\} \rangle,$$

  where it fails.
- $\forall x \forall y (x \neq y \Rightarrow R(x,y))$ is not definable in $EL(\Diamond)$. Consider the same two frames, but now in the opposite direction.

The language of $EL(\exists, \Diamond)$ with ordinary quantifiers added is much more powerful. For instance, $\exists x \neg R(x,x)$ is definable as $\exists x \neg \Diamond_y x = y$, and

$$\forall x \forall y (x \neq y \Rightarrow R(x,y))$$

as $\exists y(x \neq y \wedge P(x,y)) \rightarrow \Diamond_y P(x,y)$. Of course, a great deal of expressive power is due to the presence of identity in this language. Here is a more general result demonstrating this.

THEOREM 4.6 *Every purely universal R-condition is* $\Diamond, \exists$*-definable.*

Proof. (Cf. Proposition 2.4 in [de Rijke, 1992a]; cf. also [de Rijke, 1992b].) Consider any R-condition of the following form:

$$\forall y_1 \ldots \forall y_n \; BOOL(R, =, y_1, \ldots, y_n),$$

where "*BOOL*" is a purely Boolean condition. Introduce a predicate P_{y_i} for every universally quantified variable y_i, which holds exactly for y_i: $\exists! x P_{y_i}(x)$. Define a translation * of first-order formulas with the relation symbol R into $EL(\exists, \Diamond)$, such that * commutes with Boolean connectives and $=$, where

$$(R(y, \bar{z}))^* = \Diamond_u (P_y(u) \wedge \top(\bar{z})).$$

Then the $EL(\exists, \Diamond)$ equivalent of the R-property will be

$$\exists! x P_{y_1}(x) \wedge \ldots \wedge \exists! x P_{y_n}(x) \rightarrow (BOOL(R, =, y_1, \ldots, y_n))^*. \quad \dashv$$

OPEN PROBLEM. *Are all first-order properties of R* $\exists, \Diamond$*-definable?*

We conjecture that the answer to this question is negative. A possible counterexample is the first-order formula $\exists x \forall y R(x,y)$.

There is a more general theory behind these various observations. Each of the above axioms whose frame correspondence was analysed had a "Sahlqvist form" in a suitably general sense, and the proof method depends on finding suitable "minimal substitutions." In the next section, we make this precise.

For now, we conclude with another aspect of modal frame correspondence. It is known that the question whether a modal axiom corresponds to a first-order condition on frames is undecidable [Chagrova, 1991]. One would expect that the same holds for relational generalized quantifiers. And indeed we have this

PROPOSITION 4.7 *First-order correspondence for* $EL(\Diamond)$ *formulas is undecidable.*

Proof. The idea is as follows. Let φ be a modal formula. It defines a first-order condition on frames if and only if $\forall P_1 \ldots \forall P_n ST(\varphi)$ has a first-order equivalent, where $P_1, \ldots, P_n$ are all predicate symbols in $ST(\varphi)$ and $ST(\varphi)$ is the standard translation of φ in the first-order language. Analogously for the generalized quantifier formulas. We find an effective fragment of the language $EL(\Diamond)$ whose first-order translations are effectively equivalent to the standard translations of modal formulas. Thus, a modal formula is first-order definable iff its $EL(\Diamond)$-counterpart is, and hence the correspondence problem for generalized quantifiers (in the latter language) is undecidable.

Consider the following translation $(\ )^i$ taking modal formulas to formulas of $EL(\Diamond)$ with one free variable w_i:

- $(p_n)^i = P_n(w_i)$
- commute with the Booleans
- $(\Box\varphi)^i = \Box_{w_{i+1}}(\top(w_i) \wedge (\varphi)^{i+1})$

The only thing to prove is that for every modal formula φ, $ST(\varphi)$ is provably equivalent in first-order logic to the standard translation of φ^0. (There is a minor difference which does not influence the result: modal R is the converse of our R.)

This may be shown by induction on complexity of φ. Let φ be a propositional variable. Then $ST(\varphi)$ is an atomic formula which is a standard translation of $(\varphi)^0$. In case φ is a negation or a conjunction, apply the inductive hypothesis (both standard translations commute with the Booleans). Let $\varphi = \Box\psi$. Then

$$\begin{aligned} ST(\Box\psi)[w] &= \forall w'(R(w,w') \rightarrow ST(\psi)[w']) \\ &= \forall w'(R(w,w') \rightarrow \top(w) \wedge ST(\psi)[w']) \\ &= ST(\Box_{w'}(\top(w) \wedge ST(\psi)[w']) \\ &\quad \text{(by the inductive hypothesis).} \quad \dashv \end{aligned}$$

5 A SAHLQVIST THEOREM

THEOREM 5.1 *All formulas of the "Sahlqvist form"* $\bigwedge_i Qu_1 \ldots Qu_k(\varphi \rightarrow \psi)$, *where* Qu_j *is either* $\forall u_j$ *or* $\Box_{u_j}$, *and*

1. φ *is constructed from*
   - *atomic formulas, possibly prefixed by* $\Box_x$, $\forall$;
   - *formulas in which predicate letters occur only negatively*

   using $\wedge, \vee, \Diamond_x, \exists$
2. *in* ψ *all predicate letters (except =) occur only positively*

are first-order definable.

Proof. If every conjunct is first-order definable, the whole conjunction is. So, without loss of generality we can concentrate on a formula of the form $Qu_1 \ldots Qu_k(\varphi \to \psi)$. First we translate it into second-order logic:

$$\forall P_1^n \ldots \forall P_l^m \forall u_1 \ldots \forall u_k(\mathcal{R} \wedge ST(\varphi) \to ST(\psi)),$$

where $P_1^n \ldots P_l^m$ are all the predicates in $\varphi \to \psi$ and $\mathcal{R}$ is a conjunction of R-statements corresponding to the $\Box$-quantifiers in the prefix. Then we remove all "empty" quantifiers (those binding variables not occurring in their scope), and rename bound individual variables in such a way that every quantifier gets its own variable which is distinct from any free variable occurring in the formula. Now it is possible to move all existential quantifiers occurring in positive subformulas of $ST(\varphi)$ to a prefix, using the following equivalences:

$$\begin{aligned} \exists x A(x) \vee \exists y B(y) &\equiv \exists x \exists y (A(x) \vee B(y)) \\ \exists x A(x) \wedge B &\equiv \exists x (A \wedge B) \end{aligned}$$

with the usual provisos on freedom and bondage. $ST(\varphi)$ has now been rewritten as

$$\exists y_1 \ldots \exists y_m \varphi'.$$

Since ψ does not contain $y_1, \ldots, y_m$ free, $\forall u_1 \ldots \forall u_k(\mathcal{R} \wedge ST(\varphi) \to ST(\psi)$ is equivalent to

$$\forall x_1 \ldots \forall x_n(\mathcal{R} \wedge \varphi' \to ST(\psi)),$$

where $x_1, \ldots, x_n$ include $\bar{u}$ and $\bar{y}$.

Next, it would be convenient to get rid of the disjunctions in φ'. Let $\varphi' \equiv \phi_1 \vee \phi_2$.

$$\forall x_1 \ldots \forall x_n((\mathcal{R} \wedge \phi_1) \vee (\mathcal{R} \wedge \phi_2) \to ST(\psi))$$

is equivalent to

$$\forall x_1 \ldots \forall x_n(\mathcal{R} \wedge \phi_1 \to ST(\psi)) \wedge \forall x_1 \ldots \forall x_n(\mathcal{R} \wedge \phi_2 \to ST(\psi)).$$

We can restrict attention to one of these conjuncts (if both components have a first-order equivalent, then so has their conjunction). So, assume that there are no disjunctions in the antecedent. Thus, we have a formula

$$\forall P_1^n \ldots \forall P_l^m \forall x_1 \ldots \forall x_n(\varphi' \to ST(\psi)),$$

where $P_1^n \ldots P_l^m$ are all the predicates in $\varphi' \to ST(\psi)$, and φ' is a conjunction of "blocks" which are of one of the following forms:

1. standard translations of atomic formulas possibly preceded by universal and $\Box$-quantifiers,

2. R-statements,

3. formulas in which all predicate letters occur only negatively.

Next we rule out the use of negative formulas. The point is that $\varphi' \to ST(\psi)$ can always be rewritten as an implication whose antecedent does not contain negative formulas. Let $\varphi' = \phi_1 \wedge \phi_2$, where ϕ_2 is a negative formula. Then $\phi_1 \wedge \phi_2 \to ST(\psi)$ is equivalent to $\phi_1 \to \neg\phi_2 \vee ST(\psi)$, whose consequent contains only positive occurrences of predicate letters.

Let us denote the antecedent thus obtained (without negative formulas) φ^*. We shall now define the notion of *a minimal substitution* for every predicate letter in φ^*. A predicate letter P_i^n can occur in φ^* more than once. Consider an occurrence $\bar{P}_i^n$ of P_i^n in φ^*. First we have to classify the variables of this occurrence (this is the only part where the present proof becomes different from the modal case). Let us assume that

- the variables which stand at the places $i_1, \ldots, i_m$ *in this occurrence* are existentially bound or free; let us denote them $x_1, \ldots, x_m$;
- the variables at the places $j_1, \ldots, j_k$ are universally bound by quantifiers which correspond to $\Box$-quantifiers in the original formula; let us call them $z_1, \ldots, z_k$;
- the rest of the variables is bound by ordinary universal quantifiers; let us call them $v_1, \ldots v_l$.

Before defining a minimal substitution we have to define the notion of an "R-condition" corresponding to the variable z_i:

1. Let $\Box_{z_1}$ be the first (leftmost) generalized quantifier in the sequence of quantifiers preceding $\bar{P}_i^n$, and before $\Box z_1$ the ordinary universal quantifiers $\forall v_1, \ldots, \forall v_s$ occur. Then the R-condition corresponding to z_1 will be $R(z_1, v_1, \ldots, v_s, \bar{x})$,
2. Let $\Box_{z_i}$ be the generalized quantifier following $\Box_{z_{i-1}}$ in our sequence (with some $\forall v_p, \ldots, \forall v_r$ possibly standing in between):

$$\ldots \Box_{z_{i-1}} \forall v_p \ldots \forall v_r \Box_{z_i} \ldots \bar{P}_i^n.$$

   If the condition corresponding to z_{i-1} was $R(z_{i-1}, \bar{y})$, then the condition corresponding to z_i is $R(z_i, v_p, \ldots, v_r, z_{i-1}, \bar{y})$.

The minimal substitution $Sb(\bar{P}_i^n)$ for the occurrence of P_i^n in φ^* described above will be:

$P_i^n(u_1, \ldots u_n)$ is the conjunction of

1. $u_{i1} = x_1, \ldots, u_{im} = x_m$;
2. $\top(v_1), \ldots, \top(v_l)$,

3. $R(u_{\alpha_1}, \ldots, u_{\alpha_f})$ where $u_{\alpha_1}, \ldots, u_{\alpha_f}$ are the variables standing at the places $\alpha_1, \ldots, \alpha_f$, and in φ^* for these variables some R-condition (corresponding to one of the variables $z_1, \ldots, z_k$) holds.

Finally, we define

$$Sb(P_i^n, \varphi^*) = \bigvee Sb(\bar{P}_i^n)$$

for all occurrences of P_i^n in φ^*.[2]

The result of substituting $Sb(P_i^n, \varphi^*)$ in $\forall x_1 \ldots \forall x_m(\varphi^* \rightarrow \psi')$, which we shall denote as

$$\forall x_1 \ldots \forall x_m(s(\varphi) \rightarrow s(\psi))$$

is our intended first-order equivalent, which contains no predica'- symbols other than R and $=$. It is easy to see that it follows from the original ὑahlqvist axiom, being an instantiation of a universal second-order formula

$$\forall P_i^n \ldots \forall P_l^m \forall x_1 \ldots \forall x_m(\varphi^* \rightarrow \psi').$$

We must prove the other direction to have an equivalence.

Assume that $\forall x_1 \ldots \forall x_m(s(\varphi) \rightarrow s(\psi))$ holds in some frame F under a variable assignment v. Assume, for some interpretation function V, that φ^* holds in $M = \langle F, V \rangle$. To show that ψ' holds in the same model, we need the following two assertions:

LEMMA 5.2 *For all* M, v: $M, v \models \varphi^* \;\Rightarrow\; M, v \models s(\varphi)$.

LEMMA 5.3 *Let* $M, v \models \varphi^*$, *and let* $v(x_1) = d_1, \ldots, v(x_m) = d_m$. *Define* $V^*(P_i^n)$ *as the set of all* n*-tuples which satisfy* $Sb(P_i^n, \varphi^*)$ *under* v *(that is, with* $d_1, \ldots, d_m$ *assigned to* $x_1, \ldots, x_m$*). Then* $V^*(P_i) \subseteq V(P_i)$.

From the first lemma it follows that $s(\varphi)$ also holds for V and v; and hence $s(\psi)$ holds. Since ψ' is pos:tive, Lemma 5.3 (with the Monotonicity Lemma for classical logic) implies that $M, v \models \psi'$, as was to be shown. $\dashv$

Proof of Lemma 5.2. φ^* has the form $\Psi \wedge \Gamma \wedge \Theta$, where Ψ is a conjunction of R-statements corresponding to the translations of $\Diamond$-quantifiers, Γ is a conjunction of atomic formulas, and Θ a conjunction of universally bound implications. It is easy to check that the two latter conjuncts turn into tautologies after substituting $Sb(P_i, \varphi^*)$ for every P_i in φ^*. It means that $\vdash s(\varphi) \equiv \Psi$, so it follows from any conjunction including Ψ. $\dashv$

[2] Note that we do not need existential quantifiers here to deal with iterations of $\Box$, as in modal logic; instead of $R^n(x, y)$, which is short for $\exists y_1(R(x, y_1) \wedge \ldots \wedge \exists y_{n-1} R(y_{n-1}, y))$, we have, for iterated modalities, $R(y_1, x) \wedge \ldots \wedge R(y, y_{n-1}, \ldots, y_1, x)$.

Proof of Lemma 5.3. Consider the case when the occurrence of P_i is in Γ. Every V which makes the formula true under v should include at least one tuple which satisfies the conditions from Ψ. Then it contains the tuple which satisfies $Sb(\bar{P}_i)$. (b.) Let $\bar{P}_i$ be in Θ. Then it is of the form

$$\forall y_1 \ldots \forall y_{k+l}(\mathcal{R}_1 \wedge \ldots \wedge \mathcal{R}_k \to P_i(\bar{y}, \bar{x})),$$

where $\mathcal{R}_1 \ldots \mathcal{R}_k$ are the R-conditions corresponding to the generalized quantifiers. If φ^* is true under V and v, then this subformula is true, too, which means that $V(P_i)$ includes at least all tuples $\langle d_1, \ldots, d_n \rangle$ for which the relation R holds between $\alpha_1, \ldots, \alpha_f$th members, for each of the k R-conditions. So, again it contains all tuples which satisfy $Sb(\bar{P}_i, \varphi^*)$. But if for every occurrence of P_i, the set of tuples satisfying $Sb(\bar{P}_i, \varphi^*)$ is a subset of $V(P_i)$, then so is their union. Thus, $V^*(P_i) \subseteq V(P_i)$. $\dashv$

EXAMPLES 5.4 Here is how the above Sahlqvist algorithm works on the earlier examples of reflexivity, transitivity and symmetry.

Reflexivity. Consider $\Diamond_y x = y$. Its standard translation is

$$\exists y(R(y,x) \wedge x = y),$$

which is equivalent to $R(x,x)$.

Transitivity. The standard translation of

$$\Diamond_y(\top(x) \wedge \Diamond_z(\top(y) \wedge P(z))) \to \Diamond_z(\top(x) \wedge P(z))$$

gives us

$$\begin{aligned}&\forall P[\exists y(R(y,x) \wedge \top(x) \wedge \exists z(R(z,y) \wedge \top(y) \wedge P(z)))\\&\to \quad \exists u(R(u,x) \wedge \top(x) \wedge P(u))]\end{aligned}$$

which can be rewritten in accordance with the Sahlqvist algorithm as

$$\forall P \forall y \forall z(R(y,x) \wedge R(z,y) \wedge P(z)) \to \exists u(R(u,x) \wedge P(u)))$$

The minimal substitution for $P(u)$ is $u = z$, so we obtain

$$\forall y \forall z(R(y,x) \wedge R(z,y) \wedge z = z \to \exists u(R(u,x) \wedge u = z),$$

which is a first-order equivalent of transitivity:

$$\forall y \forall z(R(y,x) \wedge R(z,y) \to R(z,x))$$

Symmetry. The formula $\forall x \Box_y P(x,y) \to \forall y \Box_x P(x,y)$ is translated as

$$\forall P(\forall x \forall y(R(y,x) \to P(x,y)) \to \forall y \forall x(R(x,y) \to P(x,y)))$$

The minimal substitution for $P(u, v)$ is $\top(u) \wedge R(v, u)$:

$$\forall x \forall y(R(y, x) \rightarrow \top(x) \wedge R(y, x)) \rightarrow \forall x \forall y(R(x, y) \rightarrow \top(x) \wedge R(y, x))$$

The antecedent becomes trivial:

$$\top \rightarrow \forall x \forall y(R(x, y) \rightarrow R(y, x))$$

which can again be written more elegantly as $\forall x \forall y(R(x, y) \rightarrow R(y, x))$.

So far our illustations concerned finding first-order equivalents of modal formulas. Recall the reverse problem of defining first-order properties of R by modal formulas.

FACT 5.5 *All purely universal R-conditions can be defined using Sahlqvist formulas only.*

Proof. We show that the algorithm for defining R-properties in $EL(\exists, \Diamond)$ described in the Theorem 4.6 produces Sahlqvist formulas. First, the formula

$$\bigwedge_i \exists! x P_{y_i}(x)$$

is a Sahlqvist antecedent: it can be rewritten as

$$\bigwedge_i \exists y_i P_{y_i}(y_i) \wedge \bigwedge_i \forall x \forall z(P_{y_i}(x) \wedge P_{y_i}(z) \rightarrow x = z).$$

In the second conjunct, all predicate letters occur negatively (but when it is moved to the consequent in accordance with the Sahlqvist algorithm, those occurrences become positive).

Next, in the consequent wc have $(BOOL(R, =, y_i))^*$, where some predicate letters again can occur negatively. Rewrite it as a conjunction of disjunctions of "atomic" statements $(\Diamond_u(P_{y_i}(u) \wedge \top(\bar{z})))$ and their negations:

$$\Phi \rightarrow \Psi_1 \wedge \ldots \wedge \Psi_n.$$

The above expression is equivalent to the following conjunction:

$$(\Phi \rightarrow \Psi_1) \wedge \ldots \wedge (\Phi \rightarrow \Psi_n),$$

where each of Ψ_j's is a disjunction of atomic statements and their negations. Now move negations of atomic statements to the antecedents:

$$\Phi \rightarrow \neg \Diamond_u(P_y(u) \wedge \top(\bar{z})) \vee \Psi \text{ becomes } \Phi \wedge \Diamond_u(P_y(u) \wedge \top(\bar{z})) \rightarrow \Psi.$$

It follows that there are no negative occurrences of predicate letters in the consequents. ⊣

EXAMPLE 5.6 (Symmetry revisited) Here is one more illustration of the preceding technique. Symmetry can be also defined "locally" using

$$\begin{aligned}&P(x) \wedge \neg\exists x'(x' \neq x \wedge P(x')) \wedge Q(y) \wedge \neg\exists y'(y' \neq y \wedge Q(y'))\\&\quad\rightarrow \neg\Diamond_u(P(u) \wedge \top(y)) \vee \Diamond_u(Q(u) \wedge \top(x)).\end{aligned}$$

The latter formula becomes

$$\begin{aligned}&P(x) \wedge Q(y) \wedge \exists u(R(u,y) \wedge P(u))\\&\quad\rightarrow \exists x'(x' \neq x \wedge P(x') \vee \exists y'(y' \neq y \wedge Q(y')) \vee \exists v(R(v,x) \wedge Q(v)),\end{aligned}$$

or

$$\begin{aligned}&\forall u[P(x) \wedge Q(y) \wedge R(u,y) \wedge P(u)\\&\quad\rightarrow \exists x'(x' \neq x \wedge P(x') \vee \exists y'(y' \neq y \wedge Q(y')) \vee \exists v(R(v,x) \wedge Q(v))].\end{aligned}$$

The minimal substitutions are as follows:

$$\begin{aligned}P(z) &:= (z = x \vee z = u)\\Q(z) &:= z = y.\end{aligned}$$

The resulting formula will then be

$$\begin{aligned}&\forall u((x = x \vee x = u) \wedge y = y \wedge R(u,y) \wedge (u = x \vee u = u)\\&\quad\rightarrow \exists x'(x' \neq x \wedge (x' = x \vee x' = u) \vee\\&\qquad\qquad \exists y'(y' \neq y \wedge y' = y) \vee\\&\qquad\qquad \exists v(R(v,x) \wedge v = y));\end{aligned}$$

applying predicate logic gives

$$\forall u(R(u,y) \wedge \forall x'(x' = u \rightarrow x' = x) \rightarrow R(y,x),$$

which is equivalent to $\forall u(R(u,y) \rightarrow R(y,u))$.

6 LIMITATIVE RESULTS

If a formula does not have the form described in our Sahlqvist Theorem, it may lack a first-order equivalent. The proof that a combination $\Box(\ldots \vee \ldots)$ in the antecedent can be fatal, is adapted from the analogous proof for modal logic (see [van Benthem, 1985, Lemma 10.6], and [van Benthem, 1984]).

LEMMA 6.1 $\Box_x(\Box_y(P(y) \wedge \top(x,z)) \vee P(x)) \rightarrow \Diamond_x(\Diamond_y(P(y) \wedge \top(x,z)) \wedge P(x))$ *is not first-order definable.*

Proof. Define a class of frames F_n as follows:

$$\begin{aligned} D_n &= \{0, 1, \ldots, 2n+1\} \\ R_n &= \{\langle i, 0\rangle : 1 \leq i \leq 2n+1\} \cup \{\langle i+1, i, 0\rangle : 1 \leq i \leq 2n, \} \cup \\ &\qquad \{\langle 1, 2n+1, 0\rangle\}. \end{aligned}$$

Here is a picture illustrating this with $R(j, i, 0)$ represented as "there is a line from 0 to i and an arrow from i to j":

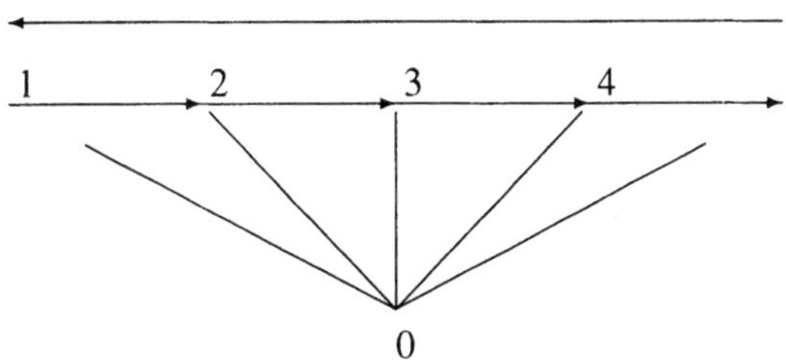

For every n and V,

$$F_n, V, [z/0] \models \Box_x(\Box_y(P(y) \wedge \top(x)) \vee P(x)) \rightarrow \Diamond_x(\Diamond_y(P(y) \wedge \top(x)) \wedge P(x)).$$

Indeed, the antecedent is true if

$$\forall x(R(x, z) \rightarrow \forall y(R(y, x, z) \rightarrow P(y)) \vee P(x));$$

that is, if for every i with $R(i, 0)$ $P(i)$ is true or P holds for each j with $R(j, i, 0)$. Each such i has exactly one "successor" j with $R(j, i, 0)$ and "predecessor" k with $R(i, k, 0)$. They form a chain which has by definition an odd number of members. That is why, if the antecedent is true, then P should hold for some pair of neighbours in this chain. But then the consequent is also true:

$$\exists x(R(x, z) \wedge \exists y(R(y, x, z) \wedge P(y)) \wedge P(x)).$$

Now, assume that our formula had a first-order equivalent. For arbitrary large n, it is consistent with the following set of first-order sentences describing the frames F_n:

- $\forall x \forall y(R(x, y) \rightarrow \neg R(y, x))$
- $\forall x \forall y \forall z(R(x, y, z) \rightarrow \neg R(y, x, z))$
- $\exists! z \forall y R(y, z)$
- $\forall y(\exists! x R(x, y, z) \wedge \exists! u R(y, u, z))$

- $\neg\exists x_1 \ldots \exists x_{2n} \exists y(R(x_2, x_1 y) \wedge \ldots R(x_{2n}, x_{2n-1}, y) \wedge R(x_1, x_{2n}, y))$.

The latter formula forbids "loops" of length less than $2n+1$; that is why it is true in F_k for all $k \geq n$. By compactness, since each finite set of these formulas has a model for suitably large n, they also have a countable model simultaneously. But in all countable models with the above properties (which are isomorphic copies of $\mathbf{Z}$ with ternary R interpreted as $R(j, i, 0) := S(j, i)$ and 0 being a fixed element preceding all other elements: $R(i, 0)$ for all $i \neq 0$) the formula can easily be refuted by putting $P(i)$ iff $\neg P(i-1)$ and $\neg P(i+1)$. ⊣

The same result holds for the combination $\Box_x \ldots \Diamond_y$ in the antecedent (the proof is analogous to the proof of Lemma 10.2 in [van Benthem, 1985] for the McKinsey axiom):

LEMMA 6.2 $\Box_x \Diamond_y (P(y) \wedge \top(x, z)) \rightarrow \Diamond_x \Box_y (P(y) \wedge \top(x, z))$ *does not have a first-order equivalent.*

Proof. Consider the following class of models:

$$\begin{aligned} D &= \{0\} \cup \{y_n : n \in N\} \cup \{y_{n_i} : n \in N, i \in \{0,1\}\} \cup \\ &\quad \{z_f : f : N \rightarrow \{0,1\}\} \\ R &= \{\langle y_n, 0\rangle : n \in N\} \cup \{\langle y_{n_i}, y_n, 0\rangle : n \in N, i \in \{0,1\}\} \cup \\ &\quad \{\langle z_f, 0\rangle : f : N \rightarrow \{0,1\}\} \cup \\ &\quad \{\langle y_{n_{f(n)}}, z_f, 0\rangle : n \in N, f : N \rightarrow \{0,1\}\}. \end{aligned}$$

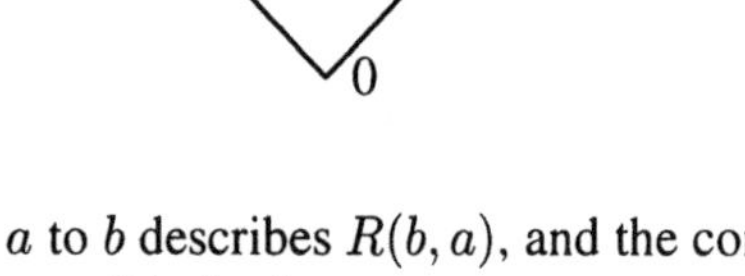

Here an arrow from a to b describes $R(b, a)$, and the combination of arrows from a to b and from b to c — $R(c, b, a)$.

Any model of this class validates the formula in question: assume

$$M, v = [z/0] \models \Box_x \Diamond_y (P(y) \wedge \top(x, z)).$$

This means that $\forall x(R(x, 0) \rightarrow \exists y(R(y, x, 0) \wedge P(y) \wedge \top(x, 0))$ is true, which implies that $\forall n \exists i P(y_{n_i})$ holds. Since for every n either y_{n_0} or y_{n_1} satisfies P, we can choose f such that $P(y_{f(n)})$ for every n. Then the consequent is also true: $\exists x(R(x,0) \wedge \forall y(R(y,x,0) \to (P(y) \wedge \top(x,0))))$ (via $x = z_f$), whence

$$F, v = [z/0] \models \Box_x \Diamond_y (P(y) \wedge \top(x,z)) \to \Diamond_x \Box_y (P(y) \wedge \top(x,z)).$$

M is obviously uncountable. Consider any countable elementary submodel M' of F which includes 0, y_n, y_{n_0}, y_{n_1} for all n. If our formula had a first-order equivalent, it would be true in M'. But it can be refuted there: since M' is countable, it does not contain some z_f. Put $y_{n_i} \in V(P)$ iff $i = f(n)$. Then the antecedent is still true (all elements which had a successor in P, still have it), but the consequent is false. ⊣

Another limitation to the above result emerges when we try to obtain its natural generalization towards *completeness* of Sahlqvist logics. Here is a striking problem, due to Michiel van Lambalgen.

EXAMPLE 6.3 (Sahlqvist incompleteness) Consider the following three axioms:

A1. $\Diamond_x x = x$;

A2. $\neg \Diamond_y x = y$;

A3. $\Diamond_x \varphi(x, \bar{y}) \to \Diamond_x (\varphi(x, \bar{y}) \vee \psi(x, \bar{z}))$

These properties are consistent (think of an interpretation for $\Diamond$ like "there exist at least two"). According to the Sahlqvist theorem, these axioms define the following properties of R:

R1. $\exists x R(x)$;

R2. $\neg R(x,x)$;

R3. $R(x, \bar{y}) \to R(x, \bar{y}, \bar{z})$;

But together R1–R3 imply $\bot$:

1. $R(x)$ – R1
2. $R(x) \to R(x,x)$ – R3
3. $R(x,x)$ – 1,2
4. $\neg R(x,x)$ – R2
5. $\bot$

The above example shows that the match between correspondence and completeness is not as good for modal quantifiers as it is for ordinary modal logic. A natural question arises, whether an analogue of the Sahlqvist's theorem can be proved for *correspondence for completeness*. The answer is given in [Alechina and van Lambalgen, 1995]. In fact, the class of *weak Sahlqvist formulas* which have correspondents in the sense of completeness turns out to be a proper subclass of Sahlqvist formulas: namely, $\Diamond$ and $\exists$ quantifiers are not allowed in the antecedent.

7 FURTHER DIRECTIONS

In this paper we studied a number of properties of "modal" quantifiers, mostly their model theory and frame correspondence theory. Correspondence in the sense of completeness and its connection to the proof theory of generalized quantifiers is the main topic of [Alechina and van Lambalgen, 1995].

Another line of research is experimenting further with the truth definition so that to make the quantifier behave even more like a modal operator (and escape the incompleteness phenomenon described above). This can be achieved by replacing our accessibility relation between elements with one between assignments [van Benthem, 1994; Venema, 1991].

There are interesting connections between both lines of research outlined above and developments in algebraic logic (cf. [Németi, 1995]), namely in cylindric algebras. "Logically," cylindric algebras correspond to first order models with restricted sets of possible assignments. In [van Benthem, 1994] a set of conditions is found under which "abstract" frames for modal quantifiers (with the new truth definition) can be represented as "assignment frames," i.e., frames of such models. These and similar connections with algebraic logic form one of our directions of further research.

Johan van Benthem
ILLC
University of Amsterdam

Natasha Alechina
School of Computer Science
University of Birmingham

REFERENCES

[Alechina, 1995] A. Alechina. On a decidable generalized quantifier logic corresponding to a decidable fragment of first-order logic. *Journal of Logic, Language and Information*, 4:177–189, 1995.

[Alechina and van Lambalgen, 1995] A. Alechina and M. van Lambalgen. Correspondence and completeness for generalized quantifiers. *Bulletin of the IGPL*, 3:167–190, 1995.

[van Benthem, 1984] J. van Benthem. Correspondence theory. In D.M. Gabbay and F. Guenthner, editors, *Handbook of Philosophical Logic*, volume 2, pages 167–247. Reidel, Dordrecht, 1984.

[van Benthem, 1985] J. van Benthem. *Modal Logic and Classical Logic*. Bibliopolis, 1985.

[van Benthem, 1992] J. van Benthem. Logic as programming. *Fundamentae Informaticae*, 17:285–317, 1992.

[van Benthem, 1994] J. van Benthem. Modal foundations for predicate logic. Technical report, CSLI, Stanford University, 1994. To appear in E. Orlowska, editor, *Memorial Volume for Elena Rasiowa*, Studia Logica Library, Kluwer, Dordrecht.

[Chagrova, 1991] L.A. Chagrova. An undecidable problem in correspondence theory. *Journal of Symbolic Logic*, 56:1261–1272, 1991.

[van Lambalgen, 1991] M. van Lambalgen. Natural deduction for generalized quantifiers. In J. van der Does and J. van Eijck, editors, *Generalized Quantifier Theory and Applications*, pages 143–154. Dutch Network for Language, Logic and Information, 1991.

[Németi, 1995] I. Németi. Decidable versions of first-order logic and cylindric-relativized set algebras. In L. Csirmaz, D.M. Gabbay and M. de Rijke, editors, *Logic Colloquium '92*, pages 177–241, Studies in Logic, Language and Information, CSLI Publications, Stanford, 1995.

[de Rijke, 1992a] M. de Rijke. How not to generalize Sahlqvist's theorem, 1992. Manuscript, ILLC, University of Amsterdam.

[de Rijke, 1992b] M. de Rijke. The modal logic of inequality. *Journal of Symbolic Logic*, 57:566–584, 1992.

[Venema, 1991] Y. Venema. *Many-Dimensional Modal Logic*. PhD thesis, Department of Mathematics and Computer Science, University of Amsterdam, 1991.

PATRICK BLACKBURN AND WILFRIED MEYER-VIOL

MODAL LOGIC AND MODEL-THEORETIC SYNTAX

Many computational linguists use ideas from logic to analyse and develop syntactical frameworks. Their interest is not confined to the (fairly obviously applicable) tools offered by proof and complexity theory: there is a growing perception that the mathematical ontologies underlying linguistic theorising are interesting in their own right, and that the grammatical formalisms that deal with them should have an explicitly formulated semantics. In short, model theory is increasingly seen as valuable.

Nonetheless, this model theoretic turn has been curiously one sided. It has primarily been directed to the analysis of the role played by *feature structures* in unification formalisms, and has tended to ignore other important aspects of linguistic ontologies. For example, Generalised Phrase Structure Grammar (GPSG) and Lexical Functional Grammar (LFG) are commonly regarded as unification formalisms, yet they both make important use of *finite trees* in addition to feature structures. However most model theoretic work in feature logic says nothing about this additional structure.

Such one-sidedness is unfortunate. For a start, one of the most interesting aspects of GPSG and LFG is precisely the fact that they deal with hybrid ontologies. How do the different components fit together? What sort of expressive power is needed for talking about composite structures? To ignore the presence of trees in such systems is to ignore some of the most interesting issues. Moreover, many syntactic theories (notably Chomsky's Government and Binding (GB) theory, probably the dominant contemporary syntactic paradigm) are firmly tree based. If the insights of such frameworks are to be reconciled with the insights of unification tradition, it seems important to extend the model theoretic perspective to the role played by finite trees.

Some recent papers have made a start in this direction. [Rogers and Vijay-Shankar, 1992] and [Vijay-Shankar, 1992] propose various languages for describing trees. Their motivation is to combine insights from the Tree Adjoining Grammar and Unification traditions. [Kracht, 1995] uses a modal 'orientation language' over the parse trees of context free grammars to relate GB and GPSG. Finally, [Blackburn *et al.*, 1993] introduce a modal language for talking about trees, and, by 'layering' this language across a feature logic, give an account of some of the leading ideas of GPSG.

The present paper builds on [Blackburn *et al.*, 1993]. We discuss two linguistic ontologies, namely *finite ordered binary trees*, and *finite ordered binary trees fibred*

M. de Rijke (ed.), Advances in Intensional Logic, 29–60.

over feature structures, formulate languages for talking about them, and prove a number of results. We proceed as follows. In the first section we introduce finite ordered binary trees and a simple modal language $\mathcal{L}$ for talking about them. In the second section we outline three linguistic applications of $\mathcal{L}$: talking about the parse trees of context free languages, talking about GB command relations, and, by 'layering' $\mathcal{L}$ across a feature logic, modeling aspects of GPSG. In the third section the technical work begins: we introduce and discuss an axiomatisation of the logic of finite ordered binary trees, and in the fourth section prove it to be complete. In the fifth section we discuss some of the consequences of this completeness result, notably the decidability of the logic, and, building on the work of [Finger and Gabbay, 1992], the decidability of the 'layered language'. We close the paper by noting a number of directions for future work.

1 THE LANGUAGE $\mathcal{L}$ AND ITS SEMANTICS

Trees can be found on the pages of most syntax textbooks; they are an important part of the ontologies posited in most grammatical theories. In this section we isolate a linguistically important class of trees, namely the *finite ordered binary trees* together with a *collection of unary relations*, and define a simple modal language $\mathcal{L}$ for talking about it.

First some terminology. We assume that the reader knows what a *finite tree* is, and that such locutions as 'a node w_1 immediately dominates node w_2', or equivalently 'w_1 is the mother of w_2' or equivalently 'w_2 is a daughter of w_1' are understood. Next, certain special tree nodes are important. Every tree has a unique node called the *root* which has the property that it is not the daughter of any node. Moreover, a non-empty subset Θ of the nodes of any finite tree are *terminal nodes*, that is, nodes without daughters. A tree is a *binary* tree if no node in the tree immediately dominates more than two nodes.

In linguistics trees are typically thought of as *ordered*. For binary trees this means the following: the daughters of any node (of which there are at most two) are uniquely classified as being either the *first daughter* or the *second daughter*. We think of the first daughter as preceding the second daughter (if in fact there is a second daughter) and when finite ordered binary trees are drawn this is conventionally represented by placing the first daughter to the left of the second.

Such finite ordered binary trees lie at the heart of much syntactic analysis. We will usually present them as tuples of the form

$$\langle W, \succ_1, \succ_2, root, \Theta \rangle.$$

Here W is a finite, non-empty set, the set of tree nodes; Θ ($\subseteq W$) contains all and only the tree's terminal nodes; and *root* is the (unique) root node of the tree. As for $\succ_1$ and $\succ_2$, these are binary relations defined as follows: for all $w, w' \in W$, $w \succ_1 w'$ iff w' is the first daughter of w, and $w \succ_2 w'$ iff w' is the second daughter of w. Note that both $\succ_1$ and $\succ_2$ are partial functions, for any node in an ordered binary tree has at most one first daughter and at most one second daughter. Further note that if $w \succ_2 w'$ then there exists a unique w'' such that $w \succ_1 w''$. Moreover $w'' \neq w'$.

This presentation of finite ordered binary trees enables us to recover four other binary relations we will make heavy use of, namely $\succ$, $\prec$, $\succ_*$ and $\prec_*$. We define $\succ$ to be $\succ_1 \cup \succ_2$, thus $w \succ w'$ means that w' is a daughter of w. We define $\prec$ to be the converse relation of $\succ$, thus $w \prec w'$ means that w' is the mother of w. Note that $\prec$ is a partial function that is defined on all nodes except *root*. As to $\succ_*$ and $\prec_*$, they are the reflexive transitive closures of $\succ$ and $\prec$ respectively. Sometimes it is convenient to make some of these defined relations explicit in our presentations. For example, when defining satisfaction it will be natural to present finite ordered binary trees **O** as tuples of the form

$$\langle W, \succ_1, \succ_2, \prec, \succ_*, \prec_*, root, \Theta\rangle.$$

Finite ordered binary trees are clearly Kripke *frames*, but how does syntactic theory give rise to Kripke *models*? The answer is not hard to find. Linguists annotate trees with further information. For example, nodes may be marked by the symbols S, NP or VP, indicating that they are sentential nodes, noun phrase nodes, or verb phrase nodes respectively. Indeed nodes often bear multiple annotations. For instance, a noun phrase node might be marked +N, BAR-2 and CASE-GENITIVE. In short, the structures underlying much linguistic theorising can be seen as finite ordered binary trees *together with a collection of unary relations on the tree nodes*, and such structures are Kripke models. Incidentally, while trees are used by a wide range of grammatical theories, different theories typically use different unary relations. Thus the frames we have isolated are common to many theories, but the models built over them will vary.

Now that we know the structures we're interested in, how shall we talk about them? In this paper we explore the use of propositional modal language called $\mathcal{L}$.[1] The primitive alphabet of $\mathcal{L}$(**Prop**) consists of following items: a non-empty set of propositional symbols **Prop**, a truth functionally adequate collection of Boolean operators, five primitive unary 'diamond' modalities $\downarrow^1$, $\downarrow^2$, $\uparrow$, $\downarrow^*$ and $\uparrow^*$, and the punctuation symbols (and). For any choice of **Prop**, the wffs of $\mathcal{L}$(**Prop**) are built up in the usual way: all elements of **Prop** are wffs; if ϕ and ψ wffs, then so are $\neg\phi$, $(\phi \wedge \psi)$, $\downarrow^1\phi$, $\downarrow^2\phi$, $\uparrow\phi$, $\downarrow^*\phi$ and $\uparrow^*\phi$; and nothing else is a wff. We normally suppose that some choice of **Prop** has been fixed and speak simply of the language $\mathcal{L}$.

We use a number of defined modalities. Firstly, we define a new unary diamond by stipulating that $\downarrow\phi$ is to be $\downarrow^1\phi \vee \downarrow^2\phi$. Second, we define the dual 'box' operators

[1] Modal languages seem a natural choice for linguistic applications: although extremely simple, they offer significant expressive power. For further discussion see [Blackburn *et al.*, 1993].

to our diamonds in the usual fashion: $\Downarrow^1\phi$ is defined to be $\neg\downarrow^1\neg\phi$, $\Downarrow^2\phi$ is defined to be $\neg\downarrow^2\neg\phi$, $\Downarrow\phi$ is defined to be $\neg\downarrow\neg\phi$, $\Uparrow\phi$ is defined to be $\neg\uparrow\neg\phi$, $\Downarrow^*\phi$ is defined to be $\neg\downarrow^*\neg\phi$, and $\Uparrow^*\phi$ is defined to be $\neg\uparrow^*\neg\phi$. Third, we define two nullary modalities (or constants) as follows: s is defined to be $\Uparrow\bot$ and t is defined to be $\Downarrow\bot$. We sometimes call $\downarrow^1$, $\downarrow^2$, $\downarrow$ and $\uparrow$ *basic operators.*

This language can talk about trees in linguistically interesting ways. The atomic symbols of **Prop** will enable us to talk about node labels. Intuitively, an atomic symbol (say, NP) will be true at a node iff that node is labeled with the corresponding property (that is, iff it is a noun phrase node). The modal operators are to 'move us round' the trees in a natural manner: $\downarrow^1$ and $\downarrow^2$ will look for information at first and second daughters respectively, $\uparrow$ will look for information at mothers, and $\downarrow^*$ and $\uparrow^*$ will explore the 'dominates' and 'is dominated by' relations respectively.

Let's make this precise. The standard interpretation for our language is in terms of Kripke models built over arbitrary Kripke frames of the form

$$\langle W, R_{\succ_1}, R_{\succ_2}, R_{\prec}, R_{\succ_*}, R_{\prec_*}\rangle,$$

that is, frames with a transition relation for each modality. However, because we are interested in linguistic applications, we are only interested in models built over frames

$$\mathbf{O} = \langle W, \succ_1, \succ_2, \prec, \succ_*, \prec_*, root, \Theta\rangle$$

that are presentations of finite ordered binary trees. That is, our intended semantics is in terms of those pairs $\langle \mathbf{O}, V\rangle$ where $\mathbf{O}$ is such a presentation and V is a map from **Prop** to $Pow(W)$. In what follows we reserve the terms 'Kripke model' and 'model' for our intended structures.

Now for the satisfaction definition. For any model $\mathbf{M}$ $(= \langle \mathbf{O}, V\rangle)$, for any node w of $\mathbf{M}$, and for any wff ϕ we define:

$$\begin{array}{rcl}
\mathbf{M}, w \models p & \text{iff} & w \in V(p), \text{ for all } p \in \mathbf{Prop} \\
\mathbf{M}, w \models \neg\phi & \text{iff} & \mathbf{M}, w \not\models \phi \\
\mathbf{M}, w \models \phi \wedge \psi & \text{iff} & \mathbf{M}, w \models \phi \text{ and } \mathbf{M}, w \models \phi \\
\mathbf{M}, w \models \downarrow^1\phi & \text{iff} & \exists w'(w \succ_1 w' \text{ and } \mathbf{M}, w' \models \phi) \\
\mathbf{M}, w \models \downarrow^2\phi & \text{iff} & \exists w'(w \succ_2 w' \text{ and } \mathbf{M}, w' \models \phi) \\
\mathbf{M}, w \models \uparrow\phi & \text{iff} & \exists w'(w \prec w' \text{ and } \mathbf{M}, w' \models \phi) \\
\mathbf{M}, w \models \downarrow^*\phi & \text{iff} & \exists w'(w \succ_* w' \text{ and } \mathbf{M}, w' \models \phi) \\
\mathbf{M}, w \models \uparrow^*\phi & \text{iff} & \exists w'(w \prec_* w' \text{ and } \mathbf{M}, w' \models \phi).
\end{array}$$

If $\mathbf{M}, w \models \phi$ then we say ϕ is *true* in $\mathbf{M}$ at w, or ϕ is *satisfied* in $\mathbf{M}$ at w. For any wff ϕ, if there is a model $\mathbf{M}$ and a node w in $\mathbf{M}$ such that $\mathbf{M}, w \models \phi$, then we say that ϕ is *satisfiable*. If ϕ is true at all nodes in a model $\mathbf{M}$ then we say it is *valid in the model* $\mathbf{M}$. The notion of validity in a model has an important role to play for us. As discussed in the next section, we think of grammars G as $\mathcal{L}$ wffs ϕ^G. The trees admitted by the grammar are precisely those models in which ϕ^G is valid. Another important concept is validity: if a wff ϕ is valid in all models then we say it is *valid* and write $\models \phi$.

The satisfaction definition clearly captures the intended interpretation of $\mathcal{L}$. Also, note what the defined operators mean. In particular, note that:

$$\begin{array}{rll}\mathbf{M}, w \models s & \text{iff} & w = \mathit{root} \\ \mathbf{M}, w \models t & \text{iff} & w \in \Theta \\ \mathbf{M}, w \models \downarrow\phi & \text{iff} & \exists w'(w \succ w' \text{ and } \mathbf{M}, w' \models \phi).\end{array}$$

That is, s and t are constants true at only the root node and terminal nodes respectively, while $\downarrow$ looks for information at daughter nodes. Other useful defined operators abound, for $\mathcal{L}$ is very expressive over its intended models. For example, we can define the *universal modality*: $[U]\phi =_{def} \uparrow^*(s \wedge \Downarrow^*\phi)$. This says that ϕ is true at all points in a model, thus the modality allows universal constraints on grammatical well-formedness to be stated in the object language; see [Blackburn and Spaan, 1993] for further discussion.

Before considering applications, there is an aspect of $\mathcal{L}$'s semantics that is worth discussing, namely the way we have restricted our discussion to *binary* trees. Actually, from a logical point of view the important point is not that our trees are binary, but that there is a fixed upper bound on their branch factor. We could easily extend $\mathcal{L}$ to permit third daughters, fourth daughters, ..., n-th daughters to be talked about using unary operators $\downarrow^3$, $\downarrow^4$, ..., $\downarrow^n$, and the results of this paper generalise straightforwardly to such extensions. However if we do *not* have a fixed upper bound, matters are different. We briefly discuss the issue in the paper's conclusion.

Linguistically the restriction to binary trees is reasonable, though not uncontroversial. In GB (see [Chomsky, 1981]) binary branching trees are widely considered to be fundamental. On the other hand, most versions of GPSG (see [Gazdar *et al.*, 1985]) regard co-ordinations as 'flat'. For example, 'John and Sue and Bill and Lou and Butch and Peggy-Sue' would be represented by a node with six daughters. As no upper bound can be placed on the number of conjuncts, such versions of GPSG place no upper bound on the branch factor.

2 THREE LINGUISTIC APPLICATIONS

The aim of this section is to give the reader a taste of $\mathcal{L}$ in action. We give three examples. First we show that $\mathcal{L}$ can pick out the parse trees of any context free phrase structure grammar. Second we show that $\mathcal{L}$ can express many of the 'command relations' used in GB. Third, we combine $\mathcal{L}$ with a 'feature logic' in a particularly simple fashion. The resulting 'layered language' enables some of the more important ideas of GPSG to be captured.

2.1 *Context Free Grammars*

We now show how to construct a $\mathcal{L}$ formula that distinguishes the parse trees of a given context free grammar from all other models. To keep notation simple we will work with grammars in which no non-terminal symbol occurs on the left-hand side of two distinct productions, and which rewrites no symbol to more than two symbols. The extension to arbitrary grammars is straightforward; see [Blackburn *et al.*, 1993] for further details.

So suppose $\mathcal{G} = \langle S, \mathcal{N}, \mathcal{T}, \mathcal{P} \rangle$ is a context free grammar with start symbol S, set of non-terminal symbols $\mathcal{N}$, set of terminal symbols $\mathcal{T}$, and set of productions $\mathcal{P}$ satisfying the restrictions just mentioned. We can assume that all these sets are finite and that $\mathcal{N}$, $\mathcal{T}$ and $\{S\}$ are pairwise disjoint. Our goal is to define a wff $\phi^{\mathcal{G}}$ that is valid on all and only the parse trees of the grammar $\mathcal{G}$.

The first step is to specify our modal language; that is, to find a suitable choice of **Prop**. For any context free grammar $\mathcal{G}$ there is an obvious choice. If the grammar contains no epsilon productions (that is, the grammar rewrites no symbol to the null string), take **Prop** to be $\mathcal{N} \cup \mathcal{T} \cup \{S\}$. On the other hand, if the grammar contains epsilon productions choose some distinct new symbol ϵ and take **Prop** to be $\mathcal{N} \cup \mathcal{T} \cup \{S\} \cup \{\epsilon\}$.

The next step is to capture the effect of the productions in $\mathcal{G}$. As we assume that $\mathcal{G}$ rewrites no symbol to more than two symbols, every production has one of the following three forms:

$$X \longrightarrow YZ \quad \text{or} \quad X \longrightarrow Y \quad \text{or} \quad X \longrightarrow \epsilon,$$

where $X \in \mathcal{N} \cup \{S\}$ and $Y, Z \in \mathcal{N} \cup \mathcal{T} \cup \{S\}$. For each production $P \in \mathcal{P}$ we form the corresponding wff ϕ^P, namely:

$$X \rightarrow (\downarrow^1 Y \wedge \downarrow^2 Z) \quad \text{or} \quad X \rightarrow \downarrow^1 Y \wedge \neg\downarrow^2 \top \quad \text{or} \quad X \rightarrow \downarrow^1 \epsilon \wedge \neg\downarrow^2 \top.$$

Let $\phi^{\mathcal{P}}$ be $\bigwedge_{P \in \mathcal{P}} \phi^P$. Clearly $\phi^{\mathcal{P}}$ is valid on every parse tree for $\mathcal{G}$.

But we want to define an $\mathcal{L}$ formula that is valid on all *and only* the parse trees for $\mathcal{G}$, so we have a little more work to do: we must express in $\mathcal{L}$ certain general facts about parse trees. First, we insist that each node of a parse tree must be labeled by at least one element of **Prop**:

$$\bigvee_{p \in \mathbf{Prop}} p$$

achieves this. Second, we insist that each node of a parse tree must be labeled by at most one element of **Prop**:

$$p \rightarrow \bigwedge_{q \in \mathbf{Prop} \setminus \{p\}} \neg q$$

achieves this. Third, we insist that the root node of a parse tree must be decorated by the start symbol S of the grammar: $s \rightarrow S$ achieves this. Fourth, we insist that non-terminal symbols label only nonterminal nodes of parse trees: $\bigwedge_{p \in \mathcal{N}}(p \rightarrow \neg t)$ achieves this. Fifth, we insist that terminal symbols label only terminal nodes of parse trees: $\bigwedge_{p \in \mathcal{T}}(p \rightarrow t)$ achieves this. Finally, we insist that ϵ labels only terminal nodes: $\epsilon \rightarrow t$ achieves this. Call the conjunction of these wffs $\phi^{\mathcal{U}}$.

Now we can define the required formula: let $\phi^{\mathcal{G}}$ be $\phi^{\mathcal{P}} \wedge \phi^{\mathcal{U}}$. Clearly for any model $\mathbf{M}$, $\mathbf{M} \models \phi^{\mathcal{G}}$ iff $\mathbf{M}$ is (isomorphic to) a parse tree for $\mathcal{G}$.

The most important point to note about this logical reconstruction of $\mathcal{G}$ is the way that something essentially *procedural* (namely rewrite rules on strings) have been turned into something *declarative* (namely a collection of axioms regulating the local structure of parse trees). This is not a new idea in linguistics. It is standard in both GB and GPSG to insist that phrase structure 'rewrite rules' should *not* be thought of as operations on strings (or indeed, as operations at all) but rather as *node admissibility conditions on trees*. In GB this is a consequence of the idea that representations, not rules, are fundamental. In GPSG it is a consequence of an explicitly model theoretic stance towards syntactic theorising: the primary task facing the linguist is to construct a logical theory characterising precisely those structures exhibited by grammatical sentences. How these structures are to be constructed is an interesting, but separate, concern.

It's also worth remarking that even the idea of turning rewrite rules into *modal* axioms is not novel: essentially the same thing was done by [McCawley, 1968]. McCawley doesn't use the terminology of modal logic, nonetheless the heart of his insight is that the rewrite arrow of formal language theory can be regarded as a modal operator interpreted over tree based Kripke models. Thus our choice of a modal language for talking about trees was not capricious. Modal operators arise quite naturally in the transition from the 'rule based' perspective on linguistic structure to the declarative model theoretic perspective, a theme that recurs in contemporary work on feature logic.

2.2 *Command Relations*

The tree relations we have been considering are rather conventional. The 'mother of' and 'daughter of' relations, together with their transitive closures, are fundamental relations that occur in practically any research area concerned with trees. However many linguistic frameworks, and in particular GB, work with a much richer collection of tree relations. Perhaps the prime examples of such relations are the *command relations* of GB. Intuitively, command relations deal with the sphere of influence of syntactic mechanisms, and they play an essential role in various components of GB theory. Here we will investigate how $\mathcal{L}$ copes with command relations. Our point of departure is the definition given by [Barker and Pullum, 1990].

DEFINITION 2.1 (Command Relations on Tree Models) Let $\mathbf{M}$ be a model and P be a unary relation on the nodes of $\mathbf{M}$. The *P-command relation* C_P on $\mathbf{M}$ is defined as

$$\{\langle w, v\rangle : \forall k(k \neq w \,\&\, k \succ_* w \,\&\, k \in P \Rightarrow k \succ_* v)\}.$$

P is called the *generating property* of the command relation C_P. The set $\{v \in W : C_P(w, v)\}$ is called the *P-command domain* of w in $\mathbf{M}$. Note that because the P-command relations are defined on *trees*, when $C_P(w, v)$ holds in $\mathbf{M}$, then all P-nodes properly dominating w dominate v and, consequently, the (unique) *lowest* such P-node dominates v. Conversely if the lowest P-node dominating w dominates v, then all P-nodes dominating w do. So the P-command domain of w is determined by the lowest P-node properly dominating w. By checking the definition we see that, if P is the empty set or if P is the singleton set $\{root\}$, then the P command domain of w is the entire tree.

Given command relations C_P, are the obvious modalities corresponding to these relations expressible in $\mathcal{L}$? To be more precise, it would be pleasant to have at our disposal for each P of interest a modality $\langle P\rangle$ with the following semantics:

$$\mathbf{M}, w \models \langle P\rangle\psi \text{ iff } \exists v(C_P(w, v) \,\&\, \mathbf{M}, v \models \psi).$$

Are such modalities definable in $\mathcal{L}$? The answer is 'yes' whenever there is an $\mathcal{L}$ formula ϕ^P which holds in $\mathbf{M}$ on all and only all nodes in P. For suppose we have such a formula ϕ^P. Then it is clear that defining

$$\langle P\rangle\psi =_{def} \Uparrow\Uparrow^*((s \vee \phi^P) \rightarrow \downarrow^*\psi)$$

has the required effect. (The presence of s ensures that if no ϕ^P node dominates the point of evaluation then the command domain will be the entire tree; this is the approach adopted by Pullum and Barker.) The dual operator $[P]$ can also be defined:

$$[P]\psi =_{def} \uparrow\uparrow^*((s \vee \phi^P) \wedge \Downarrow^*\psi).$$

Because of the correspondence between the relations P and the wffs ϕ^P which pick them out, it is natural to use the $\mathcal{L}$ formulas themselves to label the modalities. That is, if ϕ is the wff corresponding to P we will write $\langle\phi\rangle\psi$ rather than $\langle P\rangle\psi$. Using this 'command modality' notation, here are some examples of expressible command relations:

1. the formula $\langle\top\rangle\phi$ insists that ϕ holds in the *smallest* command domain of w, for $C_\top(w, v)$ iff v is dominated by the mother of m.

2. the formula $\langle\bot\rangle\phi$ insists that ϕ holds in the *largest* command domain of w, for $C_\bot(w, v)$ iff v is dominated by *root*.

3. the formula $\langle\downarrow^2\top\rangle\phi$ insists that ϕ holds somewhere in what is known as the *c-command* domain of w. Note that $C_{\downarrow^2\top}(w, v)$ iff v is dominated by the first *branching* node dominating w.

4. the formula $\langle \mathrm{S}\rangle\phi$ insists that ϕ holds in the S-command domain of w. Note that the generating property of this relation consists of all the nodes in the tree decorated with S for 'Sentence'.

The last two examples are typical of the command relations used in GB. The terminology c-command is short for 'in construction with' and plays an important role in the GB account of anaphoric binding; the classic account is [Reinhart, 1981]. For further discussion of S-command, *Sentence* command, see [Barker and Pullum, 1990].

A full discussion of command relations would take us too far afield (for a thorough investigation see [Kracht, 1993]) but it is worth noting that a number of the characterizing properties of command relations discussed by Barker and Pullum follow straightforwardly from the semantics of our language. For example, consider the *intersection property*, which can be formulated as $C_{\phi\vee\psi} = C_\phi \cap C_\psi$: the intersection of command relations corresponds to union over their generating properties. This immediately follows from the fact that

$$\Uparrow\Uparrow^*((s \vee \phi \vee \psi) \rightarrow \downarrow^*\chi)$$

is logically equivalent to

$$\Uparrow\Uparrow^*((s \vee \phi) \rightarrow \downarrow^*\chi) \wedge \Uparrow\Uparrow^*((s \vee \psi) \rightarrow \downarrow^*\chi).$$

When written in command modality notation this boils down to the fact that $\langle\phi \vee \psi\rangle\chi \leftrightarrow \langle\phi\rangle\chi \wedge \langle\psi\rangle\chi$ is logically valid.

2.3 Trees Decorated with Feature Structures

When linguists decorate trees they do not usually do so in the manner familiar from formal language theory. For example, when a linguist decorates a node with the information NP this is usually a way of insisting that it possesses a whole bundle of properties. Sometimes it is possible to represent this information bundle using only Boolean combinations of propositional variables, but for theories such as GPSG this would be most unnatural. GPSG envisages 'atomic level information' as a structured entity, a so-called feature structure. The purpose of the present section is to show how the intuition concerning structured atomic information that underlies GPSG can be captured. First we define a standard notion of feature structure and present a language $\mathcal{L}^F$ for talking about them. We then show how to 'layer' $\mathcal{L}$ over $\mathcal{L}^F$, yielding a language $\mathcal{L}(\mathcal{L}^F)$ capable of formulating the central ideas of GPSG in a natural way.

A *feature structure* is a labeled decorated directed graph. The elements of feature structures will be called *points*, and to say that feature structures are *labeled directed* graphs simply means that one can envisage the points being linked by arrows bearing a label. The most important constraint on feature structures concerns these labeled arrows: there is no point in any feature structure from which two distinct arrows bearing the same label emerges. In effect, labeled arrows are representations of partial functions; these partial functions are usually called *features*. As to the *decorations*, we envisage the points of feature structures being adorned with pieces of information of linguistic interest.

Let's make this precise. We assume that the linguistic theory we are working with tells us what features and decorations may be used. That is, we assume that our linguistic theorising gives us a signature $\langle \mathcal{F}, \mathcal{D} \rangle$ where both $\mathcal{F}$ and $\mathcal{D}$ are non-empty denumerable sets, the set of *features* and the set of *decorations* respectively. Typical elements of $\mathcal{F}$ might be CASE, NUMBER, PERSON and AGREEMENT; while typical elements of $\mathcal{D}$ might be *genitive*, *singular*, *plural*, *1st*, *2nd*, and *3rd*. We now define:

DEFINITION 2.2 (Feature structures) By a *feature structure* of signature $\langle \mathcal{F}, \mathcal{D} \rangle$ we mean a triple $\mathbf{F}$ of the form

$$\langle U, \{R_f\}_{f \in \mathcal{F}}, \{Q_d\}_{d \in \mathcal{D}} \rangle,$$

where U is a non-empty set; for all $f \in \mathcal{F}$, R_f is a binary relation on U that is a partial function; and for each $d \in \mathcal{D}$, Q_d is a unary relation on U.

As we have defined them feature structures are multimodal Kripke models; and indeed the language $\mathcal{L}^F$ we now introduce will be the obvious modal language for talking about them, with the R_f serving to interpret its modalities, and the Q_d interpreting its propositional symbols. However we shall continue to call these entities feature structures, reserving the words 'model' and 'Kripke model' for the objects that interpret our language of trees $\mathcal{L}$.

The language $\mathcal{L}^F$ (of signature $\langle \mathcal{F}, \mathcal{D} \rangle$) contains the following items: all the elements of $\mathcal{D}$ (which we will regard as propositional symbols), a truth functionally adequate collection of Boolean connectives, and all the elements of $\mathcal{F}$ (which we will regard as one place modal operators).[2] The set of wffs of $\mathcal{L}^F$ is the smallest set containing all the propositional symbols (that is, all the elements of $\mathcal{D}$) closed under the Boolean and modal operators (that is, the elements of $\mathcal{F}$). Thus a typical wff of $\mathcal{L}^F$ might be the following:

$$\langle \text{AGREEMENT} \rangle \langle \text{PERSON} \rangle \mathit{3rd} \wedge \langle \text{CASE} \rangle \mathit{genitive}.$$

[2]Our discussion of $\mathcal{L}^F$ and its semantics is of necessity somewhat brief. For a more leisurely account see [Blackburn and Spaan, 1993].

In passing, a computational linguist would probably write this wff as follows:

$$\begin{bmatrix} \text{AGREEMENT} & [\text{PERSON } \mathit{3rd}] \\ \text{CASE} & \mathit{genitive} \end{bmatrix}$$

Such two dimensional wffs are called Attribute Value Matrices, and they are essentially a perspicuous notation for $\mathcal{L}^F$ wffs.

$\mathcal{L}^F$ wffs are interpreted on feature structures as follows. For any feature structure $\mathbf{F}$ (that is, $\langle U, \{R_f\}_{f\in\mathcal{F}}, \{Q_d\}_{d\in\mathcal{D}}, \rangle$) and any point $u \in U$:

$$\begin{array}{rll} \mathbf{F}, u \models d & \text{iff} & u \in Q_d, \text{ it for all } d \in \mathcal{D} \\ \mathbf{F}, u \models \neg\phi & \text{iff} & \text{not } \mathbf{F}, u \models \phi \\ \mathbf{F}, u \models \phi \wedge \psi & \text{iff} & \mathbf{F}, u \models \phi \text{ and } \mathbf{F}, u \models \psi \\ \mathbf{F}, u \models \langle f \rangle\phi & \text{iff} & \exists u'(uR_f u' \text{ and } \mathbf{F}, u' \models \phi). \end{array}$$

If $\mathbf{F}, u \models \phi$ then we say that ϕ is *satisfied* (or *true*) in $\mathbf{F}$ at u. The most obvious definition of validity in $\mathcal{L}^F$ is as follows: a wff ϕ is valid iff it is satisfied at all points in all feature structures. However, many feature structures are linguistically uninteresting, so we will confine our attention to *finite point-generated feature structures* (or finite *rooted* feature structures). These are finite feature structures $\mathbf{F}$ that contain a point u_0 such that any other point u of $\mathbf{F}$ can be reached by making a finite number of feature transitions from u_0. (Such a point u_0 is said to *generate* $\mathbf{F}$.) Thus an alternative definition suggests itself: an $\mathcal{L}^F$ wff is valid iff it is satisfied at all points in all finite, point generated feature structures. Actually, as we shall later see, both definitions yield the same set of validities.

With $\mathcal{L}^F$ and its semantics defined, we are ready to define a language for talking about the structures underlying GPSG: trees decorated with feature structures. The language is called $\mathcal{L}(\mathcal{L}^F)$, that is, the language $\mathcal{L}$ *layered over* the language $\mathcal{L}^F$, and it is defined in a very simple way. We simply choose **Prop** to be $\mathcal{L}^F$ and then construct the $\mathcal{L}$ wffs over this base in the usual way. As a result, we've given an 'internal structure' (namely, a modal structure) to the propositional symbols of $\mathcal{L}$. For further discussion see [Blackburn *et al.*, 1993].

Syntactically that's all there is to it; what about the semantics? There is a straightforward interpretation for $\mathcal{L}(\mathcal{L}^F)$ in terms of the following entities:

DEFINITION 2.3 (Feature decorated trees) By a (finite, ordered, binary) *feature structure decorated tree* (of signature $\langle \mathcal{F}, \mathcal{D} \rangle$) is meant a triple $\langle \mathbf{O}, Z, z \rangle$ where $\mathbf{O}$ is the presentation of a finite ordered binary tree, Z is a function that assigns to each node u of $\mathbf{O}$ a finite, point-generated feature structure (of signature $\langle \mathcal{F}, \mathcal{D} \rangle$), and z is a function that assigns to each node u of $\mathbf{O}$ a point $z(u) \in Z(u)$ that generates $Z(u)$.

Two comments about feature decorated trees are in order. First of all, they seem to do justice to the ideas of GPSG. An examination of [Gazdar *et al.*, 1985] suggests that they are a natural mathematical embodiment of the ontology underlying

GPSG (modulo the fact that we are working only with binary trees). Second, in a number of recent talks Dov Gabbay has emphasized the importance of 'fibred semantics' for combined logics. By a combined logic he means a layered language of the sort exemplified by $\mathcal{L}(\mathcal{L}^F)$; and feature decorated trees are a nice example of what he means by a fibred semantics — the definition fibres a tree over a collection of feature structures. [Finger and Gabbay, 1992] gives a very clear and detailed account of such systems and proves a number of useful results which we shall make use of later.

To interpret $\mathcal{L}(\mathcal{L}^F)$ wffs on feature structure decorated trees, all we have to do is alter the base clause of the satisfaction definition for $\mathcal{L}$. Let $\mathbf{M} = \langle \mathbf{O}, Z, z \rangle$ be a feature structure decorated tree, and w any node in $\mathbf{O}$. Then for all $\mathcal{L}^F$ wffs ϕ, $\mathbf{M}, w \models \phi$ iff $Z(w), z(w) \models \phi$. If $\mathbf{M}$ (= $\langle \mathbf{O}, Z, z \rangle$) is a feature structure decorated tree, w is a node in $\mathbf{O}$, ϕ is an $\mathcal{L}(\mathcal{L})^F$ wff and $\mathbf{M}, w \models \phi$ then we say that ϕ is *satisfied* (or *true*) in $\mathbf{M}$ at w. If ϕ is satisfied at all nodes w of all feature structure decorated trees, then we say that ϕ is *valid*.

To sum up: when in the course of evaluating an $\mathcal{L}(\mathcal{L}^F)$ wff at a node w we encounter an $\mathcal{L}^F$ wff (that is, when we reach what used to be the 'atomic' level) we jump into the feature structure associated with w (that is, $Z(w)$), and start evaluating the $\mathcal{L}^F$ wff at $z(w)$. This atomic level change is the only change needed: the remaining clauses are those that were given in the satisfaction definition for $\mathcal{L}$. In short, the 'top' layer of language $\mathcal{L}$ moves us round trees in the familiar way, while the 'bottom' layer of language $\mathcal{L}^F$ moves us round feature structures.

Having assembled the machinery, let's briefly discuss how GPSG puts it to work. One of the central insights of GPSG is that by making systematic use of the structured information embodied in feature structures, it is possible to give elegant accounts of many troublesome grammatical phenomena. First, context free rules in GPSG makes use of feature information. The GPSG analogue of $VP \longrightarrow V\ NP$ (*'A verb phrase is made up of a verb followed by a noun phrase'*) might be something like:

$$\begin{aligned} \neg noun \wedge verb \wedge \langle \text{BAR} \rangle two \quad \rightarrow \quad & \downarrow^1(\neg noun \wedge verb \wedge \langle \text{BAR} \rangle zero) \wedge \\ & \downarrow^2(noun \wedge \neg verb \wedge \langle \text{BAR} \rangle two). \end{aligned}$$

For the purposes of the present discussion it is the *form* of this wff that is important: the 'outer' (or $\mathcal{L}$) level of this wff has the same form as the wffs we used to capture the parse trees of context free grammars. But because of the 'inner' (or $\mathcal{L}^F$) level it reaches inside the feature structures and insists that certain conditions must hold there.

Secondly, and more interestingly, GPSG imposes a number of global restrictions on the way feature structures can be distributed over trees. The simplest of these is the *foot feature principle.* When a natural language is analysed GPSG style, certain information is classified as 'foot information'. For a feature decorated tree to be acceptable, foot information must be 'passed up' to the feature structure associated with the mother node.[3] This is essentially to demand the validity of the following $\mathcal{L}(\mathcal{L}^F)$ wff:

$$\langle \text{FOOT} \rangle \phi \rightarrow \Uparrow \langle \text{FOOT} \rangle \phi.$$

Now, every word in a GPSG lexicon is associated with feature information, that is, with an $\mathcal{L}^F$ wff. If a certain word bears foot information then the foot feature principle forces this information to trickle up the tree from any terminal node where the word is inserted. It is quite possible that this foot information is incompatible with other information present in the structure. If this happens the structure is judged 'bad' that is, as not being the representation of a grammatical sentence. It is in this manner that 'feature passing' allows more refined accounts of grammaticality to be developed.

In short, GPSG is essentially a collection of axioms stipulating which feature decorated trees correspond to grammatical structures and which do not. Some of the theory's constraints are essentially a generalisation of the idea of phrase structure grammars to the richer setting of feature decorated trees; but in addition there is a set of global constraints, such as the foot feature convention, which act as a further filter on ungrammaticality. The way these various principles interact enables GPSG to give a neat account of a variety of phenomena in a wide range of languages. Many of these principles can be formalised in $\mathcal{L}(\mathcal{L}^F)$. For further discussion of GPSG from the present perspective see [Blackburn *et al.*, 1993].

We close this section with a general warning. The reader should *not* conclude from our discussion that layered modal languages or feature structure decorated trees are all there is to modelling grammatical frameworks that make use of feature structures. This is simply false. For a start, different theories use feature structures in different ways. For example, Lexical Functional Grammar (LFG) (see [Kaplan and Bresnan, 1982]) uses them to model grammatical relations such as subject and object, and Head Driven Phrase Structure Grammar (HPSG) (see [Pollard and Sag, 1987]) uses them for a wide variety of tasks. Moreover, while both GPSG and LFG have composite ontologies made up of trees and feature structures, the theories 'glue' these building blocks together differently. Like GPSG, substantial parts of the LFG formalism can be viewed as a certain combination of modal logics; but the combination in question is not the simple layering + fibering idea embodied in $\mathcal{L}(\mathcal{L}^F)$, and the resulting systems have very different logical properties.

3 THE AXIOMATISATION

In this section we present an axiomatisation of the logic of finite ordered binary trees. We will later prove that this axiomatisation (which we call **Lot**) is complete.

[3] As far as later versions of GPSG are concerned this is something of a simplification; in such systems feature passing involves the use of defaults. For a discussion of some of the logical issues raised see [Evans, 1987].

As axioms we take any suitable axiomatisation of propositional calculus, together with all instances of the schemas B1–B10, E1–E4, D1–D9 and F1–F3 below. As rules of inference we take modus ponens (if ϕ and $\phi \rightarrow \psi$ are provable then so is ψ) and the rule of necessitation in $\Downarrow^1$, $\Downarrow^2$, $\Uparrow$, $\Downarrow^*$ and $\Uparrow^*$. That is, if ϕ is provable then so are $\Downarrow^1\phi$, $\Downarrow^2\phi$, $\Uparrow\phi$, $\Downarrow^*\phi$ and $\Uparrow^*\phi$. Formal proofs are finite sequences of wffs built using the axioms and rules of inference in the usual way. If a wff ϕ is provable then we write $\vdash \phi$ and say that ϕ is a theorem.

With these generalities to hand, let us examine the details. The axioms split naturally into four groups. First of all, there are the axioms for the basic operators.

B1 $\Downarrow^1(\phi \rightarrow \psi) \rightarrow (\Downarrow^1\phi \rightarrow \Downarrow^1\psi)$

B2 $\Downarrow^2(\phi \rightarrow \psi) \rightarrow (\Downarrow^2\phi \rightarrow \Downarrow^2\psi)$

B3 $\Uparrow(\phi \rightarrow \psi) \rightarrow (\Uparrow\phi \rightarrow \Uparrow\psi)$

B4 $\downarrow^1\phi \rightarrow \Downarrow^1\phi$

B5 $\downarrow^2\phi \rightarrow \Downarrow^2\phi$

B6 $\uparrow\phi \rightarrow \Uparrow\phi$

B7 $\phi \rightarrow \Downarrow^1\uparrow\phi$

B8 $\phi \rightarrow \Downarrow^2\uparrow\phi$

B9 $\phi \rightarrow \Uparrow\downarrow\phi$

B10 $\downarrow^2\top \rightarrow \downarrow^1\top$

Most of this is familiar. B1–B3 are universally valid modal principles, while B4–B6 reflect the partial functional (or 'deterministic') nature of the $\succ_1$, $\succ_2$ and $\prec$ relations. B7 and B8 are familiar from tense logic: they record the fact that both the converse of $\succ_1$ and the converse of $\succ_2$ are contained in $\prec$. B9 is closely related and says that the converse of $\prec$ is contained in $\succ_1 \cup \succ_2$ (to see this, recall that $\downarrow\phi$ is shorthand for $\downarrow^1\phi \vee \downarrow^2\phi$). Finally, B10 takes account of the fact that in ordered binary trees, the existence of a $\succ_2$ successor to some node entails the existence of a $\succ_1$ successor to that same node.

The next group of axioms deals with the transitive closure operators and their interactions with the basic operators.

E1 $\Downarrow^*\phi \leftrightarrow (\phi \wedge \Downarrow\Downarrow^*\phi)$

E2 $\Uparrow^*\phi \leftrightarrow (\phi \wedge \Uparrow\Uparrow^*\phi)$

E3 $\Downarrow^*(\phi \rightarrow \Downarrow\phi) \rightarrow (\phi \rightarrow \Downarrow^*\phi)$

E4 $\Uparrow^*(\phi \rightarrow \Uparrow\phi) \rightarrow (\phi \rightarrow \Uparrow^*\phi)$

These are familiar from temporal logic and Propositional Dynamic Logic; see for example [Goldblatt, 1987]. They reflect the fact that $\succ_*$ and $\prec_*$ are the reflexive transitive closures of $\succ$ and $\prec$ respectively.

The intended meaning of the defined symbols was discussed in the introduction; the next group of axiom pins these down:

D1 $s \leftrightarrow \Uparrow\bot$

D2 $t \leftrightarrow \Downarrow\bot$

D3 $\downarrow\phi \leftrightarrow \downarrow^1\phi \vee \downarrow^2\phi$

D4 $\Downarrow^1\phi \leftrightarrow \neg\downarrow^1\neg\phi$

D5 $\Downarrow^2\phi \leftrightarrow \neg\downarrow^2\neg\phi$

D6 $\Uparrow\phi \leftrightarrow \neg\uparrow\neg\phi$

D7 $\Downarrow\phi \leftrightarrow \neg\downarrow\neg\phi$

D8 $\Downarrow^*\phi \leftrightarrow \neg\downarrow^*\neg\phi$

D9 $\Uparrow^*\phi \leftrightarrow \neg\uparrow^*\neg\phi$

Finally we turn to the axioms that give the system its flavour, namely those that reflect the fact that our intended models are all *finite*.

F1 $\downarrow^* t$

F2 $\uparrow^* s$

F3 $\phi \rightarrow \downarrow^*(\phi \wedge \Downarrow\Downarrow^*\neg\phi)$

F1 and F2 are straightforward: no matter where we are in a finite tree we are only a finite number of $\prec$ steps away from the root node (which is what F1 says) and a finite number of $\succ$ steps away from at least one terminal node (which is what F2 says). More interesting is F3. Roughly speaking, it says that if ϕ holds at any node in a tree, then this node dominates a ϕ node not dominating any other ϕ nodes. It is this axiom that will enable us to maintain the finiteness of the tree constructed in the completeness proof. As it plays such a crucial role, let's look at it more closely.

All instances of F3 are valid in our intended semantics. For suppose some wff ϕ is true at a point w_1 in a model $\mathbf{M}$. Now, either w_1 dominates a distinct node w_2 such that $\mathbf{M}, w_2 \models \phi$, or this is not the case. If this is *not* the case then we are through: for it is immediate that $\mathbf{M}, w_1 \models \phi \wedge \Downarrow\Downarrow^*\neg\phi$, and as $w_1 \succ_* w_1$ it follows that $\mathbf{M}, w_1 \models \downarrow^*(\phi \wedge \Downarrow\Downarrow^*\neg\phi)$, and we have verified the consequent of the axiom. So suppose that there *is* a point w_2 such that $w_1 \neq w_2$, $w_1 \succ_* w_2$, and $\mathbf{M}, w_2 \models \phi$. Now we ask: does w_2 dominate a distinct point w_3 such that $\mathbf{M}, w_3 \models \phi$? If the answer is 'no' then it follows that $\mathbf{M}, w_1 \models \downarrow^*(\phi \wedge \Downarrow\Downarrow^* \neg\phi)$ and we have verified the consequent of the axiom. On the other hand, if the answer is 'yes', then we repeat the question, asking whether there is a distinct w_4 ... in short, we keep asking whether or not there is a lower node that satisfies ϕ, and as soon as we get the answer 'no' we have our desired result. And we *must* eventually get the answer 'no', for as we are working with *finite* trees, our original point w_1 dominates only finitely many nodes.

Note that all instances of $\phi \rightarrow \uparrow^*(\phi \wedge \Uparrow\Uparrow^* \neg\phi)$, the mirror image of F3, are also valid. However we don't need them as axioms: its an easy exercise to show that they are all derivable in **Lot**. This derivability of the mirror image reflects a fairly obvious fact about our models. When we look downwards in a tree we may see complex branching structure, and a special axiom (namely F3) is needed to cope with this. However when we look upwards we see a nice regular linear structure, and the deductive power we already have copes successfully.[4]

The discussion of this section has established that **Lot** is sound: the axioms are valid, and the rules of inference clearly preserve validity. We are ready to turn to the question of its completeness.

4 THE COMPLETENESS PROOF

We begin with the following observation: a completeness result for **Lot** must be a weak completeness result, for as we are working only over *finite* trees there is an obvious compactness failure. For example, let Φ be $\{p, \uparrow p, \uparrow\uparrow p, \uparrow\uparrow\uparrow p, \ldots\}$. Any finite subset of Φ has a model, but it is impossible to satisfy all the wffs of Φ in the same model.

In fact **Lot** is weakly complete and the proof falls into two parts. In the first part ('Preliminaries') we define the basic entities we use to build our model, prove a number of results about them, and finally state and prove the Truth Lemma that we shall use. Much of this material is familiar from the literature on temporal logics for programs and Propositional Dynamic Logic. We have given fairly complete proof details, but occasionally the reader may find it useful to consult [Goldblatt, 1987] or [van Benthem and Meyer-Viol, forthcoming].

In the subsequent part ('Building the model') we turn to the heart of the proof. The problem is this: we need to build a model, but this model must be based on a *finite* tree. An inductive construction suggests itself, but can it be shown to terminate after a finite number of steps? By making use of axiom F3 it is possible to guarantee this.

[4] It's also interesting to note (as Jan van Eijck pointed out to us) that F1 can be derived from the other axioms. F3 plays a key rôle in its derivation.

4.1 Preliminaries

The first notion we need is that of a *closure* of a set of sentences. Recall that a set of sentences Σ is said to be closed under subformulas iff for all $\phi \in \Sigma$, if ψ is a subformula of ϕ then $\psi \in \Sigma$. We need to work with closures that offer more structure than just closure under subformulahood, thus, following [Fischer and Ladner, 1979] we define:

DEFINITION 4.1 (Closures) If Σ is a set of formulas, $Cl(\Sigma)$ is defined to be the smallest set of sentences containing Σ that is closed under subformulas and satisfies the following additional properties:

1. $\uparrow^*\phi \in Cl(\Sigma)$ implies $\uparrow\uparrow^*\phi \in Cl(\Sigma)$.
2. $\downarrow^*\phi \in Cl(\Sigma)$ implies $\downarrow\downarrow^*\phi \in Cl(\Sigma)$.
3. $\downarrow^1\top, \downarrow^2\top, \uparrow\top, \uparrow^* s$ and $\downarrow^* t \in Cl(\Sigma)$.
4. If $\phi \in Cl(\Sigma)$ and ϕ is not of the form $\neg\psi$ then $\neg\phi \in Cl(\Sigma)$.

$Cl(\Sigma)$ is called the *closure* of Σ. Note that if Σ is finite then $Cl(\Sigma)$ is finite.

In fact, because of the failure of compactness already noted (and because our ultimate goal is to build a *finite* tree) we shall only be interested in finite closures. The next step is to pick out the subsets of $Cl(\Sigma)$ needed for building models. Following van [Benthem and Meyer-Viol, forthcoming] we define:

DEFINITION 4.2 (Atoms) If Σ is a set of formulas, then $At(\Sigma)$ consists of all the maximal consistent subsets of $Cl(\Sigma)$. That is, $A \in At(\Sigma)$ means that A is consistent, and, if B is a consistent subset of $Cl(\Sigma)$ such that $A \subseteq B$, then $A = B$. The elements of $At(\Sigma)$ are called *atoms*.

For example, if Σ is the set of all formulas then elements of $At(\Sigma)$ are just the usual maximal consistent sets of sentences. More generally, the following relationship holds between atoms and the familiar notion of maximal consistent set of sentences:

LEMMA 4.3 *Let $\mathcal{M}$ be the set of all maximal consistent sets of sentences, and Σ be some set of sentences. Then* $At(\Sigma) = \{M \cap Cl(\Sigma) : M \in \mathcal{M}\}$.

Proof. Straightforward. $\dashv$

The next two lemmas guarantee that we have a plentiful supply of atoms, and that atoms have nice properties. Direct proofs of these facts are easy, or one can observe that analogues of these results are familiar properties of maximal consistent sets of sentences and use the previous lemma to transfer them to atoms.

LEMMA 4.4 *If* $\phi \in Cl(\Sigma)$ *and* ϕ *is consistent then there is an* $A \in At(\Sigma)$ *such that* $\phi \in A$.

LEMMA 4.5 *For any set of sentences* Σ, *and any* $A \in At(\Sigma)$:

1. *If* $\phi \in Cl(\Sigma)$ *then* $\phi \in A$ *iff* $\neg\phi \notin A$.
2. *If* $\phi \wedge \psi \in Cl(\Sigma)$ *then* $\phi \wedge \psi \in A$ *iff* $\phi \in A$ *and* $\psi \in A$
3. *If* $\phi \to \psi$ *and* $\phi \in Cl(\Sigma)$, *then* $\phi \to \psi$ *and* $\phi \in A$ *implies* $\psi \in A$.
4. *If* $\uparrow^*\phi \in Cl(\Sigma)$ *then* $\uparrow^*\phi \in A$ *iff* $\phi \in A$ *or* $\uparrow\uparrow^* A \in A$.
5. *If* $\downarrow^*\phi \in Cl(\Sigma)$ *then* $\downarrow^*\phi \in A$ *iff* $\phi \in A$ *or* $\downarrow\downarrow^* A \in A$.
6. $\uparrow^* s$, $\downarrow^* t$ *and* $\top \in A$.

In the completeness proof that follows we shall work with closures of *finite* sets of sentences Σ. Let us assume from now on that Σ always denotes a finite set of sentences.

In the finite case $At(\Sigma)$ has a very pleasant structure. We can enumerate all the singly negated formulas in $Cl(\Sigma)$ as one list $\neg\sigma_1, \ldots, \neg\sigma_n$ (we call this the *negative enumeration*) and all the non-negated formulas in $Cl(\Sigma)$ as another list $\sigma_1, \ldots, \sigma_n$ (we call this the *positive enumeration*), in such a way that the i-th item on the negative enumeration is the negation of the i-th item on the positive enumeration. Note that for any formula ϕ in $Cl(\Sigma)$ there is a formula ψ_i ($1 \leq i \leq n$) such that ϕ is logically equivalent to ψ_i and ψ_i occurs on either the negative or positive enumerations at the i-th place. Define a *pointwise selection* from the negative and positive enumerations to be the result of choosing, for each i ($1 \leq i \leq n$), a wff from one of the enumerations. Further, note that the conjunction of all the wffs in a consistent pointwise selection is logically equivalent to the conjunction of all the wffs in some atom, and conversely.

LEMMA 4.6 *Suppose* $At(\Sigma) = \{A_1, \ldots, A_n\}$. *Then* $\vdash \bigwedge A_1 \vee \cdots \vee \bigwedge A_n$.

Proof. Simple propositional logic: use the tautology $\alpha \leftrightarrow ((\alpha \wedge \beta) \vee (\alpha \wedge \neg\beta))$ repeatedly. $\dashv$

We now define the finite analogs of the canonical model:

DEFINITION 4.7 For all A, B in $At(\Sigma)$ define:

1. $A >_1 B$ iff $\bigwedge A \wedge \downarrow^1 \bigwedge B$ is consistent.
2. $A >_2 B$ iff $\bigwedge A \wedge \downarrow^2 \bigwedge B$ is consistent.
3. $A > B$ iff $A >_1 B$ or $A >_2 B$.

4. $A < B$ iff $\bigwedge A \wedge \uparrow \bigwedge B$ is consistent.

5. $A >_* B$ iff $\bigwedge A \wedge \downarrow^* \bigwedge B$ is consistent.

6. $A <_* B$ iff $\bigwedge A \wedge \uparrow^* \bigwedge B$ is consistent.

Let $\mathcal{C}^\Sigma$ be $At(\Sigma)$ together with these six relations.

$\mathcal{C}^\Sigma$ is not generally a tree; nonetheless, as the following sequence of lemmas shows, it does have a number of useful properties.[5]

LEMMA 4.8 *For all A, B in* $At(\Sigma)$*:*

1. $A >_1 B$ *implies* $B < A$.

2. $A >_2 B$ *implies* $B < A$.

3. $A < B$ *implies* $B >_1 A$ *or* $B >_2 A$.

4. $A < B$ *iff* $B > A$.

5. $A >_* B$ *implies that there is a finite sequence of atoms* $A = A_1 > \cdots > A_n = B$.

6. $A <_* B$ *implies that there is a finite sequence of atoms* $A = A_1 < \cdots < A_n = B$.

Proof. These results are standard, or simple variations on standard results. *1* and *2* are proved using B7 and B8 respectively in the manner familiar from tense logic; *3* is a minor variation on this theme, making use of B9 and D3; while *4* follows immediately from *1–3*. The proofs of *5* and *6* make use of E1–E4 in the standard fashion; see [Goldblatt, 1987] or [van Benthem and Meyer-Viol, forthcoming] for further details. $\dashv$

LEMMA 4.9 *For any atom A,* $s \in A$ *iff no formula of the form* $\uparrow\phi$ *is in A; and* $t \in A$ *iff no formula of the form* $\downarrow^1\phi$ *or* $\downarrow^2\phi$ *is in A.*

Proof. We prove the second equivalence. Suppose that $t \in A$ and further suppose for the sake of a contradiction that a formula of the form $\downarrow^1\phi$ or $\downarrow^2\phi$ is in A; without loss of generality we suppose that it is a formula of the form $\downarrow^1\phi$. As $t \in A$ we have $\Downarrow\bot \in A$ by D2, but as $\downarrow^1\phi \in A$ we have that $\downarrow^1\bot \in A$, which contradicts the consistency of A. This proves the left to right direction.

[5] It is perhaps worth making an aside for readers familiar with the approach of [Goldblatt, 1987]. One can also regard $\mathcal{C}^\Sigma$ as arising by *filtrating the canonical model*; this is Goldblatt's approach and it is probably the standard one. We have a slight preference for the present approach (developed in detail in [van Benthem and Meyer-Viol, forthcoming]) because it deals with finite structures right from the start.

For the right to left direction, suppose no formula of the form $\downarrow^1\phi$ or $\downarrow^2\phi$ is in A, and further suppose for the sake of a contradiction that $t \notin A$. Now $\downarrow^* t \in A$ (by Lemma 4.5 clause 6) and by Lemma 4.5 clause 5 this means that either $t \in A$ or $\downarrow\downarrow^* t \in A$. But $t \notin A$, thus $\downarrow\downarrow^* t \in A$. That is, either $\downarrow^1\downarrow^* t$ or $\downarrow^2\downarrow^* t$ is in A, which contradicts our original assumption. We conclude that t must be in A after all, the required result. $\dashv$

LEMMA 4.10 *Suppose* $\diamond\phi \in Cl(\Sigma)$, *where* $\diamond \in \{\downarrow^1, \downarrow^2, \downarrow, \uparrow, \downarrow^*, \uparrow^*\}$. *Then for all atoms A and B we have that* $AR^\diamond B$ *and* $\phi \in B$ *implies* $\diamond\phi \in A$, *where* $R^\diamond$ *denotes the relation on the canonical graph corresponding to the operator* $\diamond$.

Proof. We treat the case for $\downarrow^*$; the others are similar. Suppose $\downarrow^*\phi \in Cl(\Sigma)$ and further suppose that $A >_* B$ and $\phi \in B$. Thus we have that $\bigwedge A \wedge \downarrow^* \bigwedge B$ is consistent, and as $\phi \in B$ we have $\bigwedge A \wedge \downarrow^*\phi$ is consistent. As $\downarrow^*\phi \in Cl(\Sigma)$ and A is an atom (which means it is *maximal* consistent in $Cl(\Sigma)$), $\downarrow^*\phi \in A$. $\dashv$

LEMMA 4.11 *For all atoms A and B:*

1. *If* $A >_1 B$ *and* $\downarrow^1\phi \in A$ *then* $\phi \in B$.

2. *If* $A >_2 B$ *and* $\downarrow^2\phi \in A$ *then* $\phi \in B$.

3. *If* $A < B$ *and* $\uparrow\phi \in A$ *then* $\phi \in B$.

Proof. We treat the third case. Suppose $A < B$ and $\uparrow\phi \in A$. Thus $\bigwedge A \wedge \uparrow \bigwedge B$ is consistent, and as $\uparrow\phi \in A$ we have that $\uparrow\phi \wedge \uparrow \bigwedge B$ is consistent. It is an easy consequence of B6, the functionality axiom for $\uparrow$, that $\vdash \uparrow\gamma \wedge \uparrow\lambda \rightarrow \uparrow(\gamma \wedge \lambda)$; thus it follows that $\uparrow(\phi \wedge \bigwedge B)$ is consistent, which in turn means that $\phi \wedge \bigwedge B$ is consistent. But $\phi \in Cl(\Sigma)$ (as $\uparrow\phi \in A$), so as B is an atom, $\phi \in B$.

The other two cases are similar, and make use of B4 and B5. $\dashv$

DEFINITION 4.12 Let $\mathbf{O}$ $(= \langle W, \succ_1, \succ_2, \succ, root, \Theta \rangle)$ be the presentation of a finite ordered binary tree and Σ a finite set of sentences. By a *decoration of* $\mathbf{O}$ *by* $At(\Sigma)$ is meant a mapping $h : W \longrightarrow At(\Sigma)$. Suppose h is a decoration with the following properties:

1. For all $w, w' \in T$, if $w \succ_1 w'$ then $h(w) >_1 h(w')$,

2. For all $w, w' \in T$, if $w \succ_2 w'$ then $h(w) >_2 h(w')$,

3. $s \in h(root)$,

4. $t \in h(w)$, for all $w \in \Theta$.

Then h is a *sensible decoration*.

LEMMA 4.13 *Let h be a sensible decoration of* $\mathbf{O}$ *by* $At(\Sigma)$. *Then, for all nodes* w *and* w' *of* $\mathbf{O}$, $w \succ w'$ *implies* $h(w) > h(w')$ *and* $w \prec w'$ *implies* $h(w) < h(w')$.

Proof. Suppose $w \succ w'$. Thus $w' \succ_1 w$ or $w' \succ_2 w$. But h is a sensible decoration, thus $h(w') >_1 h(w)$ or $h(w') >_2 h(w)$. Thus, by definition, $h(w) > h(w')$, and the first implication is proved. As for the second implication, suppose that $w \prec w'$. Thus $w' \succ w$. Thus, making use of the first implication, $h(w') > h(w)$. Thus, by Lemma 4.8 clause 4, $h(w) < h(w')$, the required result. ⊣

DEFINITION 4.14 (Models induced by decorations) Let h be a decoration of $\mathbf{O}$ by $At(\Sigma)$. Let V be the valuation on $\mathbf{O}$ defined by $w \in V(p)$ iff $p \in h(w)$, for all nodes w in $\mathbf{O}$. The *model induced by* h is the pair $\langle \mathbf{O}, V \rangle$.

LEMMA 4.15 (Truth Lemma) *Let h be a sensible decoration of* $\mathbf{O}$ *by* $At(\Sigma)$, *and let* $\mathbf{M}$ $(= \langle \mathbf{O}, V \rangle)$ *be the model induced by h. Then, for all* $\phi \in Cl(\Sigma)$, *and all nodes* w *in* $\mathbf{M}$: $\mathbf{M}, w \models \phi$ *iff* $\phi \in h(w)$.

Proof. By induction on the structure of ϕ. For all atoms p we have that $\mathbf{M}, w \models p$ iff $w \in V(p)$ iff $p \in h(w)$ so the base case is clear. Assume the desired result holds for all wffs of degree less than n. Suppose that ϕ has degree n. If ϕ is of the form $\neg\psi$ or $\psi \vee \gamma$ then the desired result follows easily using Lemma 4.5, so let us consider the cases involving modalities.

Suppose ϕ has the form $\uparrow\psi$ and $\mathbf{M}, w \models \uparrow\psi$. Thus there is a w' such that $w \prec w'$ and $\mathbf{M}, w' \models \psi$. But as h is a sensible decoration, we get $h(w) < h(w')$ by Lemma 4.13, and by the inductive hypothesis $\psi \in h(w')$. So by Lemma 4.10 $\uparrow\psi \in h(w)$, as desired.

So assume $\mathbf{M}, w \not\models \uparrow\psi$. Then *either* $w = root$ *or* there is a w' such that $w \prec w'$ but $\mathbf{M}, w' \not\models \psi$. Now if $w = root$ then as h is a sensible decoration we have that $s \in h(w)$, thus by Lemma 4.9 we have $\uparrow\psi \notin h(w)$, the required result. On the other hand, suppose there is a w' such that $w \prec w'$ and $\mathbf{M}, w' \not\models \psi$. Thus $h(w) < h(w')$ (by Lemma 4.13) and by the inductive hypothesis $\psi \notin h(w')$. So $\uparrow\psi \notin h(w)$, the required result.

If ϕ is of the form $\downarrow^1\psi$ or $\downarrow^2\psi$, the argument is essentially the same as that just given, so let us consider the transitive closure operators.

Suppose $\mathbf{M}, w \models \downarrow^*\psi$. Then there is a node w' such that $w \succ_* w'$ and $\mathbf{M}, w' \models \psi$. That is, there is a finite sequence of nodes $w = w_1 \succ \cdots \succ w_k = w'$ such that $\mathbf{M}, w' \models \psi$. But as h is a sensible decoration we thus have (by Lemma 4.13) that $h(w) = h(w_1) > \cdots > h(w_k) = h(w')$. Moreover, by the inductive hypothesis $\psi \in h(w')$. It thus follows from Lemma 4.10 that $\downarrow^*\psi \in h(w)$, as required.

So suppose $\mathbf{M}, w \not\models \downarrow^*\psi$. Then for all w' such that $w \succ_* w'$ we have $\mathbf{M}, w' \not\models \psi$. Thus, by the inductive hypothesis, for all w' such that $w \succ_* w'$ we have $\psi \notin h(w')$. Now if we suppose that $\downarrow^*\psi \in h(w)$, then by Lemma 4.4 we have that either $\psi \in h(w)$ or $\downarrow\downarrow^*\psi \in h(w)$. Taken together with the fact that $\mathbf{O}$ is a *finite* tree, this swiftly leads to contradiction (we leave the argument to the reader); we conclude that $\downarrow^*\psi \notin h(w)$ as required.

The arguments for $\uparrow^*\psi$ are similar. $\dashv$

4.2 *Building the Model*

With these preliminaries to hand we turn to the heart of the proof. Our task is to show that any consistent sentence ϕ is satisfiable. This is equivalent to showing that $s \wedge \downarrow^*\phi$ is satisfiable. Thus a natural strategy suggests itself. Given a consistent sentence ϕ, form $At(\{s \wedge \downarrow^*\phi\})$, and inductively define a tree decorated by atoms from this set. First create a root node for the tree, and decorate it with any atom containing $s \wedge \downarrow^*\phi$. If there are no formulas of the form $\downarrow^1\psi$ or $\downarrow^2\psi$ in the decorating atom, stop. Otherwise, create the needed daughter nodes, extend the decoration in the obvious way, and so on.

Now this is the essence of what we'll do — but there is an obvious difficulty to be overcome. Our task is to make a *finite* decorated tree, but how can we guarantee that the inductive procedure just sketched produces only finitely many nodes? It is here that axiom F3 comes into play. F3 will allow us to assign each atom a unique 'level' measuring its distance from the atoms that contain the terminal nodes. Crucially, we will be able to show that if an atom A contains $\downarrow^1\psi$ (or $\downarrow^2\psi$) then there is an atom B of lower level than A containing ψ such that $A >_1 B$ (respectively, $A >_2 B$). This means that when inductively building our decorated tree we can always choose decorations of lower level, and doing this ensures that the construction halts after finitely many steps. The following sequence of lemmas shows that F3 really does allow us to impose such a level structure on the canonical graph.

LEMMA 4.16 *If ϕ is consistent then $\phi \wedge \Downarrow\Downarrow^*\neg\phi$ is consistent.*

Proof. Suppose ϕ is consistent. As $\vdash \phi \rightarrow \downarrow^*(\phi \wedge \Downarrow\Downarrow^*\neg\phi)$ (this is F3), $\downarrow^*(\phi \wedge \Downarrow\Downarrow^*\neg\phi)$ is consistent, whence $\phi \wedge \Downarrow\Downarrow^*\neg\phi$ is consistent. $\dashv$

LEMMA 4.17 *Let $\mathcal{A}$ $(= \{A_1, \ldots, A_n\})$ and $\mathcal{B}$ $(= \{B_1, \ldots, B_m\})$ be non-empty subsets of $At(\Sigma)$ that partition $At(\Sigma)$. (That is, $\mathcal{A} \cap \mathcal{B} = \emptyset$, and $\mathcal{A} \cup \mathcal{B} = At(\Sigma)$.) Then for some A_i in $\mathcal{A}$, $\bigwedge A_i \wedge \Downarrow\Downarrow^*(\bigwedge B_1 \vee \cdots \vee \bigwedge B_m)$ is consistent.*

Proof. As all atoms are consistent, any disjunction of atoms is consistent. Thus $\bigwedge A_1 \vee \cdots \vee \bigwedge A_n$ is consistent and thus by the previous lemma

$$(\bigwedge A_1 \vee \cdots \vee \bigwedge A_n) \wedge \Downarrow\Downarrow^*\neg(\bigwedge A_1 \vee \cdots \vee \bigwedge A_n)$$

is consistent. Now by Lemma 4.6 we have that

$$\vdash \bigwedge A_1 \vee \cdots \vee \bigwedge A_n \vee \bigwedge B_1 \vee \cdots \vee \bigwedge B_m,$$

or equivalently

$$\vdash \neg(\bigwedge A_1 \vee \cdots \vee \bigwedge A_n) \rightarrow \bigwedge B_1 \vee \ldots \vee \bigwedge B_m,$$

thus it follows that

$$(\bigwedge A_1 \vee \cdots \vee \bigwedge A_n) \wedge \Downarrow\Downarrow^*(\bigwedge B_1 \vee \cdots \vee \bigwedge B_m)$$

is consistent. But this in turn means that for some A_i in $\mathcal{A}$,

$$\bigwedge A_i \wedge \Downarrow\Downarrow^*(\bigwedge B_1 \vee \cdots \vee \bigwedge B_m)$$

is consistent, the required result. $\dashv$

We now inductively define the levels on $At(\Sigma)$. The 0-th level, L_0, is defined to be $\{A \in At(\Sigma) : t \in A\}$. Next, suppose the i-th level, L_i, is defined. First we define:

$$S_i = \bigcup_{0 \leq j \leq i} L_j.$$

(That is, S_i is the 'sum' of all levels up to and including level i. Note that S_0 is just L_0.) Next, if $At(\Sigma) \setminus S_i$ is non-empty then the $i+1$-th level L_{i+1} exists and is defined as follows:

$$L_{i+1} = \{A \in At(\Sigma) : A \notin S_i \text{ and } \bigwedge A \wedge \Downarrow\Downarrow^* \bigvee_{B \in S_i} \bigwedge B \text{ is consistent}\}.$$

On the other hand, if $At(\Sigma) \setminus S_i$ is empty then there is no $i+1$-th level.

LEMMA 4.18 *Every atom A in $At(\Sigma)$ belongs to exactly one level.*

Proof. It is clear from the inductive definition that every atom A belongs to at most one level, thus it remains to show that each atom belongs to some level.

We first show by induction that there are no empty levels. For the base case, note that as $\downarrow^* t$ is an axiom, t is consistent, thus there are atoms that contain t and so L_0 is non-empty. Further note that as S_0 is just L_0 we have that S_0 is also non-empty. For the inductive step we show that if S_i is non-empty and L_{i+1} exists, then L_{i+1} is non-empty. To see this note that for L_{i+1} to exist it must be the case that $At(\Sigma) \setminus S_i$ is non-empty. But then $At(\Sigma) \setminus S_i$ and S_i are a pair of non-empty sets that partition $At(\Sigma)$; thus applying Lemma 4.17 we deduce that there is an atom A in $At(\Sigma) \setminus S_i$ such that $\bigwedge A \wedge \Downarrow\Downarrow^* \bigvee_{B \in S_i} \bigwedge B$ is consistent. But then A is in L_{i+1}, thus the $i+1$-th level is non-empty. It follows by induction that no level is empty.

But now it is easy to see that every atom A belongs to at least one level. For, as every level is non-empty, and as every atom belongs to at most one level, then as there are only finitely many atoms there must be a maximum level. Call this level L_{max}. Suppose for the sake of a contradiction that there is some atom A that does not belong to any level. This means, $A \notin S_{max}$. But this means that $At(\Sigma) \setminus S_{max}$ is non-empty, thus L_{max+1} exists and is non-empty: a contradiction. So every atom belongs to some level. ⊣

Now for the vital lemma:

LEMMA 4.19 *Suppose A is an atom belonging to L_i and that there is a wff of the form $\downarrow^1\phi \in A$ (respectively, $\downarrow^2\phi \in A$). Then there is an atom B belonging to L_m where $m < i$ (respectively, there is an atom C belonging to L_n where $n < i$) such that $A >_1 B$ (respectively, such that $A >_2 C$).*

Proof. Suppose A is an atom in L_i and that there is a wff of the form $\downarrow^1\phi \in A$. Note that as $\downarrow^1\phi \in A$ we have by Lemma 4.9 that $t \notin A$, thus $i > 0$. This means that S_{i-1} exists and is non-empty. We will show that there is an atom B in S_{i-1} such that $A >_1 B$.

Suppose for the sake of a contradiction that this is not the case. That is, suppose that for all atoms B in S_{i-1} we have that $\bigwedge A \wedge \downarrow^1 \bigwedge B$ is inconsistent. This means that for all B in S_{i-1} we have $\vdash \bigwedge A \rightarrow \Downarrow^1\neg \bigwedge B$. Enumerate all the atoms in S_{i-1} as $B_1, \ldots, B_n$. It follows by simple modal reasoning that

$$\vdash \bigwedge A \rightarrow \Downarrow^1\neg(\bigwedge B_1 \vee \cdots \vee \bigwedge B_n).$$

Abbreviate $\bigwedge B_1 \vee \cdots \vee \bigwedge B_n$ to $\mathbf{B}$. Thus $\vdash \bigwedge A \rightarrow \Downarrow^1\neg\mathbf{B}$.

As A is in L_i and $i > 1$, by our definition of levels we have that $\bigwedge A \wedge \Downarrow\mathbf{B}$ is consistent. Writing $\Downarrow$ explicitly in terms of $\Downarrow^1$ and $\Downarrow^2$, we thus have that $\bigwedge A \wedge \Downarrow^1\mathbf{B} \wedge \Downarrow^2\mathbf{B}$ is consistent. But we have $\vdash \bigwedge A \rightarrow \Downarrow^1\neg\mathbf{B}$ by our previous argument, thus

$$\bigwedge A \wedge \Downarrow^1\mathbf{B} \wedge \Downarrow^2\mathbf{B} \wedge \Downarrow^1\neg\mathbf{B}$$

is consistent. This means that $\bigwedge A \wedge \Downarrow^1\mathbf{B} \wedge \Downarrow^1\neg\mathbf{B}$ is consistent, which means that $\bigwedge A \wedge \Downarrow^1\bot$ is consistent. As $\downarrow^1\phi \in A$ we deduce that $\downarrow^1\bot$ is consistent, which is impossible. So our original supposition was mistaken and we conclude that there is an atom B in S_{i-1} such that $A >_1 B$. This means that for some level L_m, where $m < i$, $B \in L_m$, the required result.

The result for the operator $\downarrow^2$ is proved analogously. ⊣

We are now ready for the inductive construction of our desired model. Suppose ϕ is consistent. Form $At(\{s \wedge \downarrow^*\phi\})$. Let $\mathcal{W}$ be some denumerably infinite set; we shall use (finitely many) of its elements as the tree nodes.

Stage 0. Choose some $w_0 \in \mathcal{W}$, and some atom A_0 in $At(\{s \wedge \downarrow^*\phi\})$ such that $s \wedge \downarrow^*\phi \in A_0$. As ϕ is consistent, so is $s \wedge \downarrow^*\phi$, thus by Lemma 4.4 such an A_0 exists. Define W_0 to be $\{w_0\}$; $\succ_1^0$ to be $\emptyset$; $\succ_2^0$ to be $\emptyset$; and h_0 to be $\{\langle w_0, A_0\rangle\}$.

Stage $n+1$. Suppose n stages of the inductive construction have been performed. We call a pair $\langle w, k\rangle$ (where $w \in W_n$ and $k \in \{1,2\}$) an *unsatisfied demand* iff $\downarrow^k \phi \in h(w)$ but there is no $w' \in W_n$ such that $w \succ_k w'$.

If there are no unsatisfied demands the construction is complete.

Otherwise, choose an unsatisfied demand $\langle w, k\rangle$. Note that as $\downarrow^k\phi \in h(w)$, $h(w)$ cannot belong to level zero, for otherwise we would plainly contradict Lemma 4.9. Let $w' \in \mathcal{W} \setminus W_n$. Let A_{n+1} be an atom such that $h(w) \succ_k A_{n+1}$ and A_{n+1} belongs to a *strictly lower* level than $h(w)$; Lemma 4.19 guarantees that such an A_{n+1} exists. Let j be 1 if k is 2, and 2 if k is 1. Define:

$$\begin{aligned} W_{n+1} &= W_n \cup \{w'\} \\ \succ_k^{n+1} &= \succ_k \cup\{\langle w, w'\rangle\} \\ \succ_j^{n+1} &= \succ_j \\ h_{n+1} &= h_n \cup \{\langle w', A_{n+1}\rangle\}. \end{aligned}$$

While adjoining a new node w' to w as described in the inductive step may result in new unsatisfied demands $\langle w', k\rangle$, where $k \in \{1,2\}$, we were careful to choose $h(w')$ from a strictly lower level than $h(w)$. This means that in the course of the construction we will be forced to map the newly adjoined node w' to an atom of level zero; but doing so cannot give rise to an unsatisfied demand. Thus the construction process terminates after finitely many steps.

Let m be the stage at which the construction is completed. Define h to be h_m. Define W to W_m; define $\succ_1$ to be $\succ_1^m$; define $\succ_2$ to be $\succ_2^m$; define *root* to be w_0; and define Θ to be $\{w' : w' \not\succ_1 w \textit{ and } w' \not\succ_1 w, \textit{ for all } w \in W\}$. Define $\mathbf{O}$ to be

$$\langle W, \succ_1, \succ_2, \succ, \prec, node, \Theta\rangle.$$

LEMMA 4.20 $\mathbf{O}$ *is the presentation of a finite tree. Moreover, h is a sensible decoration of* $\mathbf{O}$.

Proof. That $\mathbf{O}$ is the presentation of a finite tree follows straightforwardly from its inductive definition. There is only one subtlety: if $w \succ_2 w'$ then how do we know that there is a w'' such that $w \succ_1 w''$ (and $w' \neq w''$)? In fact, by using our last remaining unused axiom, namely B10, we can guarantee this. For suppose $w \succ_2 w'$. The inductive construction would only have adjoined w' if at some stage $\langle w, 2\rangle$ was an unsatisfied demand. But this means that some formula $\downarrow^2\phi \in h(w)$. But then, by B10, $\downarrow^1\top \in h(w)$. So if there is no w'' in W such that $w \succ_1 w''$ then $\langle w, 1\rangle$ is an unsatisfied demand — but there are no unsatisfied demands.

Showing that h is a sensible decoration of $\mathbf{O}$ is straightforward. ⊣

THEOREM 4.21 (Completeness) *Every consistent sentence has a model.*

Proof. Given a consistent sentence ϕ, we use the inductive construction to build a tree decorated by $At(\{s \wedge \downarrow^* \phi\})$. By the previous lemma the decoration so constructed is sensible, thus by the Truth Lemma the induced model satisfies $s \wedge \downarrow^* \phi$ at its root node, hence ϕ is true somewhere in this model. ⊣

5 CONSEQUENCES OF COMPLETENESS

The most important consequence of completeness is that validity is decidable.

THEOREM 5.1 *The set of valid wffs is recursive.*

Proof. The set of wffs that are *not* valid is recursively enumerable (r.e.). For, if a wff ϕ is not valid then it is falsifiable on some model. As all our models are based on *finite* ordered binary trees it is possible to write a procedure that systematically generates models and tests for the validity of ϕ on the models so produced. Any such procedure will eventually tell us that ϕ is not valid.

On the other hand, the set of valid wffs is also r.e.. For the set of **Lot** proofs is obviously an r.e. set (systematically generate finite sequences of wffs, discarding those that are not **Lot** proofs) and by the Completeness Theorem a wff is valid iff it is provable. As both the set of validities and its complement are r.e., the set of validities is recursive. ⊣

Next, recall that in Section 2 we defined a language $\mathcal{L}(\mathcal{L}^F)$, the result of layering $\mathcal{L}$ over $\mathcal{L}^F$. By making use of results proved by [Finger and Gabbay, 1992] we will be able to see that the set of valid $\mathcal{L}(\mathcal{L}^F)$ formulas is both recursively axiomatisable and decidable.

Finger and Gabbay's results are fairly general. Essentially they show how completeness and decidability results enjoyed by both languages participating in the layering process can be combined to obtain completeness and decidability results for the layered language. Now, we have just established such results for $\mathcal{L}$, so the next step is to ascertain that similar results hold for $\mathcal{L}^F$. As these results are well known (and rather simple) we only sketch the required proofs.

Suppose we have fixed some signature $\langle \mathcal{F}, \mathcal{A} \rangle$. The following axiomatisation suffices to capture the $\mathcal{L}^F$ wffs (over this signature) that are valid on all finite, point generated feature structures. As axioms take all the wffs of $\mathcal{L}^F$ (over this signature) that are instances of the following two schemas:

Feature 1 $[f](\phi \rightarrow \psi) \rightarrow ([f]\phi \rightarrow [f]\psi)$

Feature 2 $\langle f \rangle \phi \rightarrow [f]\phi$.

As rules of inference take modus ponens and the rule of necessitation for each 'box' modality (that is, if ϕ is provable, so is $[f]\phi$, for all $f \in \mathcal{F}$). A formal proof in this system is a sequence of wffs regulated by the axioms and rules of inference in the usual way, and $\vdash_F \phi$ means that ϕ is formally provable.

To prove completeness proceed as follows. Given a consistent $\mathcal{L}^F$ wff ϕ, use Lindenbaum's Lemma to form a maximal consistent set of sentences Φ that contains ϕ. The **Feature 2** axioms guarantee that the relations in the canonical model $\mathbf{M}^H$ for this system are partial functions, and by the usual argument $\mathbf{M}, \Phi \models \phi$; thus $\mathbf{M}^H$ is a feature structure that satisfies ϕ. Thus we have a (strong) completeness result for the axiomatisation with respect to the class of all feature structures. The next step is to transform $\mathbf{M}^H$ into a finite point-generated feature structure that satisfies ϕ and thus prove the relevant (weak) completeness result. This is routine. Take the submodel of $\mathbf{M}^F$ generated by Φ, but only generate out m steps, where m is the maximal depth of nesting of modalities in ϕ, and only generate on the relations corresponding to modalities actually occurring in ϕ. This establishes the result. It also tells us a little more. First, as the axiomatisation is complete with respect to both the class of all feature structures and the class of all finite point-generated feature structures, the two definitions of $\mathcal{L}^F$ validity given in Section 2 must coincide. A second consequence is the decidability of $\mathcal{L}^F$ validity, for the generation process yields an upper bound on the size of (finite) point generated feature structures that need to be inspected to determine ϕ's validity.

We now have all the information about $\mathcal{L}^F$ and $\mathcal{L}$ that we need to apply the results of Finger and Gabbay to $\mathcal{L}(\mathcal{L}^F)$. Following their account, here is an axiomatisation of the $\mathcal{L}(\mathcal{L}^F)$ validities:

Tree 1 All $\mathcal{L}(\mathcal{L}^F)$ instances of the **Lot** axioms.

Tree 2 All the **Lot** rules of inference.

Preserve For all wffs ϕ in $\mathcal{L}^F$, if $\vdash_F \phi$ then ϕ is provable.

It's worth being explicit about what this means. A proof in this system is any finite sequence of $\mathcal{L}(\mathcal{L}^F)$ wffs such that for any wff ϕ in the sequence *either* ϕ is an instance in $\mathcal{L}(\mathcal{L}^F)$ of our tree axioms or follows from earlier items in the sequence by making use of the rules of inference in our proof system for finite trees (this is the content of the **Tree 1** and **Tree 2** clauses); *or* ϕ is an $\mathcal{L}^F$ wff such that $\vdash_F \phi$ (this is the content of the **Preserve** rule of inference). If ϕ is the last item in some such sequence then we say that ϕ is provable. Thus the axiomatisation reflects the layered nature of $\mathcal{L}(\mathcal{L}^F)$ very clearly. The **Preserve** rule tells us that in addition we can 'lift' formulas provable in the feature logic to the layered system.

This axiomatisation is complete for the intended semantics of $\mathcal{L}(\mathcal{L}^F)$, that is, finite trees trees fibred over finite point-generated feature structures. We won't prove this here; it is a straightforward application of Finger and Gabbay's methods requiring no new ideas. Instead we will prove a simple corollary of completeness: our layered logic is decidable. First a lemma:

LEMMA 5.2 *The set of validities of $\mathcal{L}(\mathcal{L}^F)$ is recursively enumerable.*

Proof. Clearly we can write a procedure that systematically generates the finite sequences of $\mathcal{L}(\mathcal{L}^F)$ wffs; but can we 'weed out' those sequences that are *not* formal $\mathcal{L}(\mathcal{L}^F)$ proofs? It is clear that we can write a procedure for determining whether or not a wff in a putative proof sequence is licenced by the **Tree 1** or **Tree 2** clauses, thus it only remains to check that we can write a procedure for determining whether a wff in a putative proof sequence is licenced by the **Preserve** rule. But, by the decidability result for $\mathcal{L}^F$, this must be the case. In short, we can recursively enumerate $\mathcal{L}(\mathcal{L}^F)$ proof sequences. As this axiomatisation is complete, validity and formal provability coincide, and we have the result. ⊣

By systematically generating finite trees fibred over finite point-gene-rated feature structures we can recursively enumerate *non*-valid $\mathcal{L}(\mathcal{L})^F$ wffs. This fact together with the previous lemma yields:

COROLLARY 5.3 *The set of valid $\mathcal{L}(\mathcal{L}^F)$ wffs is recursive.*

To close this section we briefly discuss how to generalise these results to cover the case of models constructed over finite trees when there is a fixed upper bound on the branch factor greater than 2. Recall that in Section 1 we introduced suitable languages for such models: for any fixed $n > 2$ enrich $\mathcal{L}$ by adding operators $\downarrow^3, \downarrow^4, \ldots, \downarrow^n$ that look for information at 3rd, 4th, ..., n-th daughters respectively.

A complete axiomatisation of the validities in any such language is obtained from our axiomatisation for binary branching models as follows. First, we need a new defining axiom for the $\downarrow$ modality:

D3n $\downarrow\phi \leftrightarrow \downarrow^1 \vee \cdots \vee \downarrow^n \phi$.

Second, we generalise axiom B10 by adding, all instances of the following schema, where $2 \leq i \leq n$:

B10n $\downarrow^i\top \to \downarrow^{i-1}\top$.

It should be clear that the required completeness proof involves only trivial modification to that given for $\mathcal{L}$. With this completeness result established, the decidability result for the pure tree language follows by the argument used in Theorem 5.1, and the Finger and Gabbay methods yield completeness and decidability results for the layered languages in the manner just discussed.

6 CONCLUDING REMARKS

In this paper we established completeness and decidability results for a simple modal language $\mathcal{L}$ interpreted over finite trees. As we saw, $\mathcal{L}$ is strong enough to be linguistically interesting, but it is natural to enquire whether similar results can be proved for stronger, or different, systems. To close this paper we mention some extensions and variations it seems worthwhile investigating, and discuss the long term goals of this research.

First, in the literature on temporal logics of programs, it is common to dispense with $\downarrow^*$ in favour of stronger operators, namely various generalisations of Kamp's *Until* operator; see [Goldblatt, 1987] for definitions and discussion. We think it would be interesting to strengthen $\mathcal{L}$ in this fashion, and in addition to replace $\uparrow^*$ the stronger *Since* operator.

A second direction worth exploring is what happens when the *unbounded semantics* is employed, that is, when no finite upper bound is placed on the branch factor of the allowed models. This is important if one wants to give an 'unbounded' GPSG style analysis of co-ordination.

An obvious language for dealing with the unbounded semantics is the following. For every natural number n add a unary operator $\downarrow^n$ for looking for information at the n-th daughter. Actually, this extension isn't quite good enough. Because there is no upper bound on the branch factor, $\downarrow$ is no longer a definable operator. If we want to be able to assert that information holds at some unspecified daughter (and in most linguistic applications we probably do) we need to add $\downarrow$ as a primitive unary modality. Let us call the language with the operators $\uparrow$, $\downarrow$, $\uparrow^*$, $\downarrow^*$ and $\downarrow^n$ (for all natural numbers n) $\mathcal{L}^\infty$. Note that $\mathcal{L}^\infty$ has a 'horizontal' failure of compactness in addition to the 'vertical' failure already noted for $\mathcal{L}$. For while every finite subset of

$$\{\downarrow p, \neg\downarrow^1 p, \neg\downarrow^2 p, \neg\downarrow^3 p, \neg\downarrow^4 p, \ldots\}$$

is satisfiable, the whole set cannot be satisfied at any single node in any model.[6]

But to conclude the paper, let's change tack slightly. While we feel it is important to investigate extensions and variations of the type just mentioned, in certain respects they give a misleading picture of the long term goals of this work. Ultimately we hope to give a logical analysis of the leading grammar formalisms, but many aspects of this investigation are not covered by the extensions just mentioned. To close the paper we will indicate why this is so and which questions we think are worth pursuing.

As was already mentioned, the languages discussed in this paper make use of fairly orthodox aspects of tree structure. The operators they employ quantify

[6] In fact, after the present paper was completed, the unbounded semantics was succesfully axiomatised; see [Blackburn *et al.*, 1996]. This paper does not use $\mathcal{L}^\infty$ but a language containing only eight modalities, namely four 'single step' modalities (which explore the *mother-of*, *daughter-of*, *immediate-left-sister-of* and *immediate-right-sister-of* relations) and four additional modalities which explore the transitive closure of these relations. It is easy to see that all the modalities in $\mathcal{L}^\infty$ are definable using these eight modalities. In broad terms, the completeness proof follows that of the present paper, but it uses a 'double layering' argument: layering is needed in both the vertical and horizontal directions, and the chief difficulty of the proof is showing that the two layerings fit together well. The EXPTIME-completeness of the resulting logic is also proved. Incidentally, it follows from the proof of this result that **Lot** is EXPTIME-complete as well.

across relations on trees (such as mother, daughter and their transitive reflexive closures) that are familiar from applications in mathematics and computer science. In linguistics, when trees are used together with a notion of feature structure, these may well be the only relations that need to be considered: this seems to be the case as far as GPSG is concerned. However it is probably not the case in general. In GB, for example, the feature component is minimal. Instead, various more complex relations on trees are employed. The command relations discussed earlier are one example of the relations of interest, but these are merely the tip of the iceberg. In order to do justice to GB ideas on X-bar syntax and bounding, for example, it seems that one must explore a variety of other, less familiar, relations on trees and their interactions.

We feel that a systematic investigation of these richer structures could be rewarding. For a start, there are the more-or-less obvious logical questions: what languages are appropriate for describing these richer structures and what are their properties? But, more importantly, a precise account of (say) the GB ontology would make it possible to address more substantial issues. What, precisely, are the trade-offs between imposing more structure on trees and adopting various notions of feature structure? Can the 'dynamic' notion of movement in GB be reduced to a 'static' one? Is there a canonical way of turning a GB account into (say) an LFG account and vice-versa? Different linguistic theories usually start from different pre-theoretic notions, and model these notions using very different mathematical ontologies. Nonetheless, mainstream syntactic theories are usually concerned with more or less the same data, and in spite of the (often vitriolic) inter-theoretic disputes, common themes are discernible. We believe that model theoretic investigations may provide a perspective from which the commonalities (and differences) emerge *clearly*.

It hardly needs stressing that these are difficult issues, and much of the groundwork remains to be done. Nonetheless, already interesting work exists (we draw the reader's attention to [Backofen *et al.*, 1995; Kracht, 1993, 1995; Rogers, 1994], and we hope that the present paper has given the reader a taste of why we feel optimistic about the model-theoretic approach to linguistic formalisms.

Patrick Blackburn
Computerlinguistik
Universität des Saarlandes

Wilfried Meyer-Viol
Department of Computing
Imperial College

Acknowledgements. We are very grateful to Claire Gardent, Jan van Eijck and Yde Venema for their comments on earlier drafts, and to Maarten de Rijke for editorial help over and above the call of duty. We would also like to thank Marc Pauly and the referees for their comments on the penultimate draft.

Patrick Blackburn would like to acknowledge the (partial) financial support of the Netherlands Organisation for the Advancement of Research (project NF 102/62-356, 'Structural and Semantic Parallels in Natural Languages and Programming Languages').

REFERENCES

[Backofen *et al.*, 1995] R. Backofen, J. Rogers, and K. Vijay-Shanker. A first-order axiomatization of the theory of finite trees. *Journal of Logic, Language and Information*, 4:5–39, 1995.

[Barker and Pullum, 1990] C. Barker and G. Pullum. A theory of command relations. *Linguistics and Philosophy*, 13:1–34, 1990.

[van Benthem and Meyer-Viol, forthcoming] J. van Benthem and W. Meyer-Viol. *Logical Semantics of Programming*. Manuscript. Forthcoming.

[Blackburn *et al.*, 1996] P. Blackburn, W. Meyer-Viol, and M. de Rijke. A proof system for finite trees. In H. Kleine Büning, editor, *Computer Science Logic. 9th International Workshop, CSL '95. Annual Conference of the EACSL, Paderborn, Germany, September 1995. Selected Papers.*, pages 86–105. Lecture Notes in Computer Science 1092, Springer, 1996.

[Blackburn *et al.*, 1993] P. Blackburn, C. Gardent, and W. Meyer-Viol. Talking about trees. In *Proceedings of the 6th Conference of the European Chapter of the Association for Computational Linguistics*, pages 21–29, 1993.

[Blackburn and Spaan, 1993] P. Blackburn and E. Spaan. A modal perspective on the computational complexity of attribute value grammar. *Journal of Logic, Language and Information*, 2:129–169, 1993.

[Carpenter, 1992] B. Carpenter. *The Logic of Typed Feature Structures*. Tracts in Theoretical Computer Science. Cambridge University Press, 1992.

[Chomsky, 1981] N. Chomsky. *Lecture on Goverment and Binding*. Foris, Dordrecht, 1981.

[Evans, 1987] R. Evans. Towards a formal specification for defaults in GPSG. In *Categories, Polymorphism and Unification*. Centre for Cognitive Science/TLI, Edinburgh/Amsterdam, 1987.

[Finger and Gabbay, 1992] M. Finger and D.M. Gabbay. Adding a temporal dimension to a logic system. *Journal of Logic, Language and Information*, 1:203–233, 1992.

[Fischer and Ladner, 1979] M. Fischer and R. Ladner. Propositional dynamic logic of regular programs. *Journal of Computer and System Sciences*, 18:194–211, 1979.

[Gazdar *et al.*, 1985] G. Gazdar, E. Klein, G. Pullum, and I. Sag. *Generalised Phrase Structure Grammar*. Basil Blackwell, 1985.

[Goldblatt, 1987] R. Goldblatt. *Logics of Time and Computation*, volume 7 of *Lecture Notes*. CSLI Publications, Stanford, 1987.

[Kaplan and Bresnan, 1982] R. Kaplan and J. Bresnan. Lexical-functional grammar: a formal system for grammatical representation. In J. Bresnan, editor, *The Mental Representation of Grammatical Relations*, pages 173–281. MIT Press, 1982.

[Kracht, 1993] M. Kracht. Mathematical aspects of command relations. In *Proceedings of the 6th Conference of the European Chapter of the Association for Computational Linguistics*, pages 40–249, 1993.

[Kracht, 1995] M. Kracht. Syntactic codes and grammar refinement. *Journal of Logic, Language and Information*, 4:41–60, 1995.

[McCawley, 1968] J. McCawley. Concerning the base component of transformational grammar. *Foundations of Language*, 4:55–81, 1968.

[Pollard and Sag, 1987] C. Pollard and I. Sag. *Information-Based Syntax and Semantics: Volume 1, Fundamentals*. CSLI Publications, 1987.

[Reinhart, 1981] T. Reinhart. Definite np-anaphora and c-command domains. *Linguistic Inquiry*, 12:605–635, 1981.

[Rogers, 1994] J. Rogers. *Studies in the Logic of Trees with Applications to Grammar Formalisms*. PhD thesis, University of Delaware, 1994. To appear in Studies in Logic, Language and Information, CSLI Publications.

[Rogers and Vijay-Shankar, 1992] J. Rogers and K. Vijay-Shankar. Reasoning with descriptions of trees. In *Proceedings of the 25th Annual Meeting of the Association for Computational Linguistics*, pages 2–80, 1992.

[Vijay-Shankar, 1992] K. Vijay-Shankar. Using descriptions of trees in a tree adjoining grammar. *Computational Linguistics*, 18:481–517, 1992.

RUY J.G.B. DE QUEIROZ AND DOV M. GABBAY

THE FUNCTIONAL INTERPRETATION OF MODAL NECESSITY

1 MOTIVATION

Since the early days of Kripke-style possible-worlds semantics for modalities, there has been a significant amount of research into the development of mechanisms for handling and characterising modal logics by means of 'naming' possible worlds, either directly by introducing identifiers, or indirectly by some other means (such as, e.g., using formulas to identify the possible world). To list a few, we have [Gabbay, forthcoming]'s 'labelling' formulas with names for worlds in the framework of *Labelled Deductive Systems*; [Fitch, 1966b]'s tree-proof deduction procedures for modal logics; [Thomason and Stalnaker, 1968]'s device of *predicate abstraction* introduced to handle Skolemisation across worlds and characterise non-rigid designators; [Fitting, 1972, 1975]'s ε-calculus based axiom systems for modal logics, as well as his tableaux systems for modal logics with *predicate abstraction* [Fitting, 1981, 1989]; the *irreflexivity rule* of [Gabbay and Hodkison, 1990]'s axiomatic systems of temporal logic; the explicit reference to possible worlds in the deterministic modal logics of [Fariñas del Cerro and Herzig, 1990]; [Ohlbach, 1991]'s semantics-based translation methods for modal logics and its functional representation of possible worlds structures. In first-order predicate logics the individuals over which one quantifies are naturally assumed to have names. The main connectives of modal logics are such that they quantify over (higher-order) objects which are not usually given names, in some cases for methodological reasons.

The Curry-Howard functional interpretation of logics [Curry, 1934; Howard 1980] appears to be particularly suitable to deal with 'arbitrary' objects within a mathematical theory of deduction. Once names for arbitrary objects are introduced, the mechanism of *abstraction*, which goes back at least to Frege [1893, 1903]'s *Grundgesetze* mechanism of binding free-variables via so-called 'value-range' function-terms, serves to 'discharge' assumptions such as that the particular name denoted an arbitrary object.

The present paper reports on the conceptual apparatus as a preparation for the framework of labelled natural deduction for modal necessity. It is part of ongoing research, and the results about various metamathematical properties of the logical system will be left to a further publication.

M. de Rijke (ed.), Advances in Intensional Logic, 61–91.

2 THE FUNCTIONAL INTERPRETATION

In the functional interpretation (with a system of 'labelled natural deduction', as described in [de Queiroz and Gabbay, forthcoming]) the rule of universal generalisation — ∀-*introduction* —, discharges the assumption that an arbitrary element inhabits the domain over which one wants to quantify by Λ-abstracting over the particular variable in the term which shows why that arbitrary individual does satisfy the property which is being asserted for all individuals from that particular domain:

$$\frac{\begin{array}{c}[x : \mathbf{D}]\\ f(x) : \mathbf{P}(x)\end{array}}{\Lambda x.f(x) : \forall x^{\mathbf{D}}.\mathbf{P}(x)}\ \forall\textit{-introduction}$$

Note that the assumption 'x : **D**', literally 'x' inhabits '**D**', is discharged by the *introduction* of the universal quantifier, the free occurrences of 'x' in '$f(x)$' being bound by the Λ-abstraction. Observe also that '$f(x)$' may contain none, one, many, etc., occurrences of 'x' and so, according to recipe as to how assumptions are to be discharged, the abstraction might be binding none, exactly one, many, etc., free occurrences of the variable 'x'. This element of vagueness, as we shall see below, will serve to help us determine what kind of universal quantifier we want to have. The characterisation of the connective in question as a universal quantifier will come as its functionality is explained, which is not yet made clear by the mere explanation of its assertability conditions formalised by the *introduction* rule.

The key semantical notion in the functional interpretation is that of *normalisation* of proofs, which is a proof transformation step demonstrating how the *elimination* rules operate on the result of *introduction* inferences. The characterisation of the universal quantifier is then established by the *normalisation* of ∀ proofs: given an arbitrary element 'a' from the domain '**D**', and given a proof-construction of 'for all x in **D**, $\mathbf{P}(x)$ is verified', then the EXTRaction of the proof-construction for $\mathbf{P}(a)$ is made by applying the Λ-term to the argument 'a' eliminating the universal quantifier.

Thus, by fixing the rule of β-normalisation on proofs,[1] namely:

∀-*reduction*

$$\frac{a : \mathbf{D} \qquad \dfrac{\begin{array}{c}[x : \mathbf{D}]\\ f(x) : \mathbf{P}(x)\end{array}}{\Lambda x.f(x) : \forall x^{\mathbf{D}}.\mathbf{P}(x)}\ \forall\textit{-intr}}{\mathtt{EXTR}(\Lambda x.f(x), a) : \mathbf{P}(a)}\ \forall\textit{-elim} \quad \twoheadrightarrow_\beta \quad \begin{array}{c}[a : \mathbf{D}]\\ f(a/x) : \mathbf{P}(a)\end{array}$$

which means that β-equality on terms is fixed by the rule of computation (cf. [Tait, 1965], where the important rôle of convertibility is demonstrated):

$$\mathtt{EXTR}(\Lambda x.f(x), a) =_\beta f(a/x) : \mathbf{P}(a).$$

[1] See [Prawitz, 1965, Chapter II, §2] for an elegant exposition of the notion of proof reduction.

The distinct kinds of universal quantifiers (linear, relevant, etc.) can be obtained by imposing a corresponding Λ-abstraction discipline on the terms allowing (or disallowing) some free occurrences of the variable to be bound (therefore allowing the corresponding assumption to be discharged). It seems natural to take account of this extra-condition imposed on the Λ-abstraction discipline, given that the natural deduction *introduction* rule for conditionals ($\rightarrow$, $\forall$, $\Box$) does leave room for manoeuvre: in the rule of $\forall$-*introduction* above, which is to be read as 'given the assumption that 'x' is an arbitrary element from the domain '**D**' (i.e. a new variable x labelling the formula **D**), and *from this assumption one obtains* an expression '$f(x)$' which depends on x and is a label of the formula '$\mathbf{P}(x)$', one can conclude that '$\Lambda x.f(x)$' is a label for the formula '$\forall x^{\mathbf{D}}.\mathbf{P}(x)$'.' Note that the *emphasized* part does not contain any specific recipe as to if, how or when the assumption '$x : \mathbf{D}$' was actually used to obtain '$f(x) : \mathbf{P}(x)$'. That is to say, the expression '$f(x)$' may have one, many, and perhaps *no* free occurrences of x. In some natural deduction presentations this looseness is made explicit by means of vertical dots that are meant to be understood as 'after a certain number of proof steps', as in:

$$\frac{\begin{array}{c}[x : \mathbf{D}] \\ \vdots \\ f(x) : \mathbf{P}(x)\end{array}}{\Lambda x.f(x) : \forall x^{\mathbf{D}}.\mathbf{P}(x)}$$

One is then supposed to read the rule as saying that, assuming x is an arbitrary individual of domain **D** and, after a certain number of proof steps, we obtain that **P** holds of x, we can conclude that for all individuals in **D**, $\mathbf{P}(x)$ is verified. And the room for manoeuvre, so to speak, as to how to obtain different universal quantifiers through making the three dots a more specific recipe of proof steps, is clearly seen to be given by the (somewhat vague) character of the rule of $\forall$-*introduction*.

2.1 Handling Assumptions with 'World' Variables

Essentially the same procedure of capitalising on this loose aspect of the natural deduction *introduction* rules for conditionals was used in [Gabbay and de Queiroz, 1992] to obtain the functional interpretation of various notions of implication below and above intuitionistic implication. By fixing the normalisation rule on labels for the natural deduction rules for '$\rightarrow$', and parameterising the rule which naturally leaves room for manoeuvre, namely that which involves a λ-abstraction procedure, one can use λ-calculi to obtain the correspondence between normalisation of proofs in implicational calculi and the convertibility of terms in the corresponding term-calculus.

Now, we can look at the modality of necessity, namely '$\Box$', as a sort of second-order universal quantification, in a way such that:

$$\Box\mathbf{A} \equiv \forall \mathbb{W}^{\mathsf{U}}.\mathbf{A}(\mathbb{W}),$$

reading 'for all worlds $\mathbb{W}$ from the (structured) collection of worlds U, **A** is true at (demonstrable at, forced by, etc.) $\mathbb{W}$', with corresponding *introduction* rule:

DEFINITION 2.1 (Necessity introduction)
$\Box$-*introduction*:

$$\frac{\begin{matrix}[\mathbb{W} : \mathsf{U}] \\ F(\mathbb{W}) : \mathbf{A}(\mathbb{W})\end{matrix}}{\Lambda \mathbb{W}.F(\mathbb{W}) : \underbrace{\forall \mathbb{W}^{\mathsf{U}}.\mathbf{A}(\mathbb{W})}_{\Box \mathbf{A}}} \qquad \text{or} \qquad \frac{\begin{matrix}[\mathbb{W} : \mathsf{U}] \\ F(\mathbb{W}) : \mathbf{A}(\mathbb{W})\end{matrix}}{\Lambda \mathbb{W}.F(\mathbb{W}) : \Box \mathbf{A}}$$

where 'U' would be a collection of 'worlds' (where a world can be taken to be, e.g., structured collections (lists, bags, trees, etc.) of labelled formulas) and '$F(\mathbb{W})$' is an expression which may depend on the world-variable '$\mathbb{W}$'.

REMARK 2.2 Besides the need for making formal distinctions between various notions of necessity, a general framework for proofs in modal logics ought to be able to deal with forms of reasoning where such notions as *ordering*, *preference*, *priority* play a definite role in the reasoning process. If one takes the spirit of labelled deduction seriously, one sees those aspects as being part of the so-called 'meta-level' reasoning, and as such must be left to the non-declarative part of the deductive system, which in *LDS* is the calculus on the labels. This is where the notion of *structured* collection of labelled formulas is useful. The framework allows for the incorporation of structural operations to be handled by the functional calculus on the objects denoting collections of formulas. In this paper we will not use the fact that the collection of formulas is structured.

The *reduction* rule is framed in a similar way to the rule of *reduction* for implication and for universal quantification:
DEFINITION 2.3 (Normalisation for necessity)
$\Box$-*reduction*

$$\frac{\mathbb{S} : \mathsf{U} \qquad \dfrac{\begin{matrix}[\mathbb{W} : \mathsf{U}] \\ F(\mathbb{W}) : \mathbf{A}(\mathbb{W})\end{matrix}}{\Lambda \mathbb{W}.F(\mathbb{W}) : \Box \mathbf{A}}\,\Box\textit{-intr}}{\mathcal{EX}(\Lambda \mathbb{W}.F(\mathbb{W}), \mathbb{S}) : \mathbf{A}(\mathbb{S})}\,\Box\textit{-elim} \qquad \twoheadrightarrow_{\beta} \qquad \begin{matrix}[\mathbb{S} : \mathsf{U}] \\ F(\mathbb{S}/\mathbb{W}) : \mathbf{A}(\mathbb{S})\end{matrix}$$

inducing the associated equation on terms:

$$\frac{\mathbb{S} : \mathsf{U} \qquad \begin{matrix}[\mathbb{W} : \mathsf{U}] \\ F(\mathbb{W}) : \mathbf{A}(\mathbb{W})\end{matrix}}{\mathcal{EX}(\Lambda \mathbb{W}.F(\mathbb{W}), \mathbb{S}) =_{\beta} F(\mathbb{S}/\mathbb{W}) : \mathbf{A}(\mathbb{S})}$$

REMARK 2.4 Notice that also here — as in the propositional and the first-order cases [de Queiroz and Gabbay, forthcoming] — it makes good sense to follow the '(operational) semantics of convertibility' mentioned above: the rule of *normalisation*, which shows the effect of the *elimination* operator ('$\mathcal{EX}$', short for 'extract') on terms resulting from *introduction* inferences ('$\Lambda\mathbb{W}.F(\mathbb{W})$') is the key semantical device behind the functional interpretation of '$\Box$'.[2]

So, our $\Box$-*elimination* also follows the pattern established for the other conditionals ($\rightarrow$, $\forall$), and is presented similarly to *modus ponens* and *universal extraction*:

$\Box$-*elimination*

$$\frac{\mathbb{S} : \mathsf{U} \qquad l : \Box\mathbf{A}}{\mathcal{EX}(l, \mathbb{S}) : \mathbf{A}(\mathbb{S})}$$

which can be read as 'given that for all possible worlds **A** is true (where 'l' is the corresponding proof-construction), and $\mathbb{S}$ is a possible world, then **A** is true at $\mathbb{S}$'.[3]

[2]As mentioned in [de Queiroz and Gabbay, forthcoming], there is a connection between the semantics of convertibility and the so-called 'dialogue-game' semantics of [Lorenzen, 1961, 1969] and [Hintikka, 1983]. It suffices to look at the canonical operator (coming from *introduction* steps) as the marks for the 'assertion' ('myself') moves, and the non-canonical operator(s) (coming from *elimination* inferences) as the representatives of the 'attack' ('nature') moves [Lorenzen 1969, pages 25ff; Hintikka 1983, page 3]. The rules of *reduction*, i.e. the explanation of how the *elimination* rule(s) operate(s) on the result of *introduction* rules, then become explanations of how the 'attack' (resp. 'nature') moves operate on 'assertion' (resp. 'myself') moves (cf. [de Queiroz, 1994]).

In a section dedicated to modal logic [Ibid., pages 61ff], Lorenzen explains the dialogical game for the modal necessity by saying how connective is to be *eliminated*. The main limitation of Lorenzen's method for modal logic is similar to many other methods, namely the lack of devices to 'name' worlds: the 'assertion' being 'necessarily A' (notated by Lorenzen as 'ΔA'), the 'attack' operates by attempting to eliminate the 'necessary' connective, but it does not give a name to the world, as it happens with the dialogue rule for the first order universal quantifier where an arbitrary individual is presented by the 'attack': "For the all-quantifier we may now propose the following attack-defence-rule:

Assertion	Attack	Defence
$\bigwedge_x A(x)$	$S?$	$A(S)$

[Ibid., page 25], whereas for the universal quantification over 'worlds' (Lorenzen prefers to think in terms of 'systems of sentences') the rule is given as:

Assertion	Attack	Defence
ΔA	?	A

[Ibid., page 64]. The point here is that the game for 'Δ' is designed to resemble that of the universal quantifier, only the names of 'worlds' do not appear explicitly.

[3]Cf. [Prawitz, 1965] natural deduction rules for Necessity:

$$\text{NI)}\ \frac{A}{NA} \qquad\qquad \text{NE)}\ \frac{NA}{A}$$

where the difference is only that we have in our functional interpretation a procedure to introduce names of worlds, and a procedure to get rid of them by Λ-abstraction on world-variables. Without having some special provisos, such as e.g. the classification of the rule of NI as *improper inference rule* whose assumptions can only be modalised (and negation of modalised) formulas in the case of S4 (resp. S5), Prawitz' NE-rule looks like a trivialisation of the modal Necessity operator.

Observe that, similarly to the $\forall$ case aforementioned, even if we fix the $\Box$-*reduction* rule, the *introduction* rule for the modal conditional '$\Box$' still leaves a 'natural' gap to be explored, namely the three dots below:

$$\frac{\begin{array}{c}[\mathbb{W} : \mathsf{U}]\\ \vdots \\ F(\mathbb{W}) : \mathbf{A}(\mathbb{W})\end{array}}{\Lambda\mathbb{W}.F(\mathbb{W}) : \Box\mathbf{A}}$$

And, indeed, this loose aspect of the *introduction* rule makes room for the characterisation of distinct kinds of 'necessity' (K, T, D, S4, S5, etc.) simply by making more explicit the nature of the three dots (e.g., by saying that the world variable '$\mathbb{W}$' must occur exactly once in the expression '$F(\mathbb{W})$', at least once, etc.), and yet retaining the same general pattern of the *introduction* rule.[4]

Thus, it should be easy to see that the same reasoning made for implication and first-order quantification can be carried through for the case of modal logics: we can fix the β-convertibility on terms (labels) of $\Box$-formulas, and allow for variations on the Λ-abstraction discipline over terms containing variables naming worlds. So, in the case of this higher-order quantification connective we would be distinguishing the kinds of '$\Box$' by the distinct ways of handling assumptions of the

---

Perhaps our formulation of Necessity-*introduction* is better at least in the sense that it shows rather explicitly the fact that the rule must indeed be seen as an *improper* rule (following the terminology of [Prawitz, 1965]), similarly to the *introduction* rule for the other conditionals, namely propositional implication and first-order universal quantifier.

[4]Note that here, where we are trying to explore the 'natural' gaps which exist in the general framework of natural deduction, there is no place for *ad hoc* treatments of modal logics with natural deduction where the distinction from one kind of necessity to another is the presence or absence of certain rules of inference, such as, e.g. in [Fitch, 1966a] and [Fitting, 1993]'s expositions of how one can handle modal logics with natural deduction by means of the so-called 'strict (re)iteration rule' [Fitch, 1966a, page 33]:

"The main difference of the systems S4 and S5 from the systems M and B is that $\ulcorner\Box\Box\phi\urcorner$ is derivable from $\ulcorner\Box\phi\urcorner$ in the former two systems but not in the latter two systems, while the main difference of the systems B and S5 from M and S4 is that $\ulcorner\Box\sim\Box\phi\urcorner$ is derivable from $\ulcorner\sim\phi\urcorner$ in B and S5, but not in M or S4. Also, of these four systems, S5 is the only one in which $\ulcorner\Box\sim\Box\phi\urcorner$ is derivable from $\ulcorner\sim\Box\phi\urcorner$."

"In addition [to the 'strict iteration rule'] there are special rules that vary from logic to logic. These are given below, with the understanding that any logic not mentioned does not need any special rules [Fitting, 1993, page 395]:

B, T, S4 and S5 $\dfrac{\Box X}{X}$

D, DB and D4 $\dfrac{\Box X}{\neg\Box\neg X}$."

Here we are attempting to explore the 'natural' gap: the rule of $\Box$-*introduction* allows room for manoeuvre in the same way that the rules of $\rightarrow$-*introduction* and $\forall$-*introduction* do. So, $\Box$-*introduction* and $\Box$-*elimination* are the only inference rules for all notions of necessity here, with the single exception of a kind of *modal reductio ad absurdum* needed for the modal counterpart of classical propositional logic shown below.

form '$\mathbb{W}$: U', in a similar way that we distinguished the various notions of first-order quantification by varying the handling of assumptions of the form 'x : **D**' ('x is an element from the domain **D**').[5]

Being aware of the 'vagueness' of the three dots in the $\Box$-*introduction* rule, we hope that by making it more specific we can establish:

1. conditions for world-quantification
2. ways of distinguishing one kind of $\Box$ from another

Now, recall:

1. using these guidelines we were able to handle the '$\rightarrow$' case (relevant, linear, intuitionistic, etc.) in [Gabbay and de Queiroz, 1992].
2. the origin of '$\Box$' is usually attributed to Lewis' alternative to classical (material) implication [Lewis and Langford, 1932].

Thus, to summarise, by using labels/terms alongside the formulas, we can:

1. keep track of proof steps (giving local control)
2. handle 'arbitrary' names (via variable abstraction operators)

and our labelled natural deduction system gives us at least two advantages over the usual plain natural deduction systems:

1. it matches

   the functional calculus on terms with

   the logical calculus on formulas

2. it takes care of 'contexts' and 'scopes' in a more explicit fashion.

A question that may naturally arise is 'what are the labels?', and we recall that in the framework of the functional interpretation labels/terms 'name':

[5]Note, however, that we are not trying to find correspondences between first-order formulas and modal formulas, such as in the well-known Tarski's translation of S4 into intuitionistic logic, in [Sahlqvist, 1975]'s correspondence and completeness theorems, and various other well-established 'translation'-based approaches. The objects we are dealing with in our functional interpretation of modal logics are meant to be of a higher-order nature, and we do not intend to treat them as (first-order) individuals: our framework takes the view that by investing in the *orthogonality* between propositional and contextual information we may be in a better position to find useful parallels between implication and necessity. Furthermore, we do not wish to represent the 'accessibility relation' explicitly as most translation methods do. Rather than relying on the device of introducing accessibility relations explicitly, we want to find disciplines of Λ-abstraction over 'world variables' which would allow for the distinction of the various kinds of *necessity* on the basis of how one handles (withdraws) assumptions of the sort '$\mathbb{S}$ is an arbitrary world'.

1. instances of formulas (propositional case)

$$x : \mathbf{A}$$

2. individuals (first-order case)

$$t : \mathbf{D}$$

   and also instances of formulas as in the case above, possibly with open first-order formulas such as

$$f(a) : \mathbf{P}(a)$$

3. structured collections of labelled formulas (higher-order, modal)

$$\mathbb{W} : \mathsf{U}$$

   and, of course, also individuals and instances of formulas, as in the previous cases.

2.2 *Recalling the Propositional Case*

In [Gabbay and de Queiroz, 1992] we gave an answer to the question whether one could refine the functional interpretation of intuitionistic implication in order to obtain an almost as good account of other neighbouring implications, including the so-called 'resource' implications (e.g. linear, relevant, etc.). The observation was that in the definition of λ-terms, apart from the clear distinction between *argument* and *function*, namely in $\mathtt{APP}(M, N)$:

1. 'M' as the *function* is higher than
2. 'N' as the *argument*

there is an element of vagueness in the case of the definition of abstraction for λ-terms. When *abstracting* a variable from a term, one may encounter different situations, namely:

(a) M may have *no* free occurrence of x:

   (a.1) M is an open term, but contains no free occurrence of x;

   (a.2) M is a closed term, thus contains no free variable at all;

(b) M may have *one* free occurrence of x:

   (b.1) M may be of the form $\mathtt{APP}(T, x)$', i.e. x is occurring as an argument;

   (b.2) M may be of the form '$\mathtt{APP}(x, T)$', i.e. x is occurring as a function;

(c) M may have *more than one* free occurrence of x:

(c.1) the λ-*abstraction* may bind *exactly one* of the free occurrences of x;

(c.2) the λ-*abstraction* may bind *all* free occurrences of x;

(c.2.1) all free occurrences in the same *order* (*argument/function*, that is);

(c.2.2) all free occurrences regardless of *order*.

Again, in the rule of application, e.g. '$\texttt{APP}(M, N)$', 'M' is assumed to be of 'higher' order than 'N', i.e. 'M' is supposed to be the 'value-range' of a function, while 'N' is assumed to be the argument. While the rule of *abstraction* leaves some room for varying the λ-abstraction discipline, the rule of application leaves no room for manoeuvre as to how one can have (stronger or weaker) λ-calculi. This seems to match with the fact that all implications have *modus ponens*, each implication differing only in the way assumptions are discharged (one by one, in strict order, relevantly, etc.).

By working with appropriate refinements of the aforementioned vague elements of the rule of *abstraction* we demonstrated how one could use typed λ-calculus together with the Curry-Howard interpretation to formalise a number of systems of implication. The notion of normal form draws on ideas from [Martin-Löf, 1972], where a proof is said to be closed if all assumptions have been discharged. A measure of how far a proof is from a closed proof is what we called 'saturation gap' [Gabbay and de Queiroz, 1992].

DEFINITION 2.5 (Saturation gap) Let t be a term built up by variables, λ-abstraction and application. The *saturation gap* (abbreviation: '$satgap$') of t is defined as the number of occurrences of free variables it contains. It is defined inductively as:

$$\begin{aligned}
&satgap(\lambda\bot.) = 0 \\
&satgap(x) = 1 \qquad \text{where } x \text{ is a variable} \\
&satgap(\texttt{APP}(m, n)) = satgap(m) + satgap(n) \\
&satgap(\lambda x.f(x)) = satgap(f) - nfree(x)
\end{aligned}$$

where '$nfree(x)$' is the number of free occurrences of x in f which are accessible by the appropriate discipline of assumption discharge (which varies according to the assumption-handling procedure of each logic).

DEFINITION 2.6 A proof t of a proposition $\mathbf{P}$ (notation: '$t : \mathbf{P}$') is said to be *categorical* if and only if its proof-construction term m has *saturation gap* equal to 0.

LEMMA 2.7 ([Gabbay and de Queiroz, 1992]) *Proofs containing an* introduction *followed by an* elimination *such as in:*

$$\dfrac{t : \mathbf{A} \qquad \dfrac{\begin{matrix}[x : \mathbf{A}] \\ b(x) : \mathbf{B}\end{matrix}}{\lambda x.b(x) : \mathbf{A} \to \mathbf{B}} \to\textit{-intr}}{\texttt{APP}(\lambda x.b(x), t) : \mathbf{B}} \to\textit{-elim}$$

can be normalised to:

$$\begin{array}{c}[t : \mathbf{A}]\\ b(t/x) : \mathbf{B}\end{array}$$

via β-reduction, and the saturation gap of the label of the end formula (i.e., label '$b(t)$' of formula '$\mathbf{B}$') does not decrease.

2.3 Natural Deduction with an Extra Parameter

There are different ways of dealing with the aforementioned orthogonality between variables denoting individuals in first order logic and those denoting 'worlds' in modal logic. One way is to create another dimension within the framework of the sequent calculus: either (i) by using special structural connectives for modal and Boolean connectives in some kind of 'display logic' [Belnap, 1982; Wansing, 1994]; or (ii) by attaching a calculus of indices to the turnstile symbol in order to keep track of the current 'world' [Benevides and Maibaum, 1992]; or even (iii) by using an extra dimension (a vertical one) within the usual sequent calculus [Masini, 1992]. Another way is by having the meta-language entities which denote 'worlds' manipulated directly into a natural deduction proof theory via a *separate*, yet *harmonious* calculus. As part of the *label-and-formula* perspective, we take the latter as the basis for our interpretation of what a proof theory of modal logics should look like.

Thus, we shall read '$F(\mathbb{W}) : \mathbf{A}(\mathbb{W})$' as '$\mathbf{A}$ is true at $\mathbb{W}$ because $F(\mathbb{W})$ shows how $\mathbf{A}(\mathbb{W})$ was obtained starting from the assumption that $\mathbb{W}$ was an arbitrary world from the universe of worlds (i.e. '$\mathbb{W} : \mathsf{U}$')'. Now, in this perspective the usual rules of deduction (i.e. *introduction* and *elimination*) for the propositional and first-order connectives still apply, though they carry an extra parameter which is handled by the rules of proof for modal connectives. For example:

∧-*introduction*

$$\frac{a_1 : \mathbf{A}_1(\mathbb{W}) \qquad a_2 : \mathbf{A}_2(\mathbb{W})}{\langle a_1, a_2\rangle : (\mathbf{A}_1 \wedge \mathbf{A}_2)(\mathbb{W})}$$

∧-*elimination*

$$\frac{c : (\mathbf{A}_1 \wedge \mathbf{A}_2)(\mathbb{W})}{\mathtt{FST}(c) : \mathbf{A}_1(\mathbb{W})} \qquad \frac{c : (\mathbf{A}_1 \wedge \mathbf{A}_2)(\mathbb{W})}{\mathtt{SND}(c) : \mathbf{A}_2(\mathbb{W})}$$

→-*introduction*

$$\frac{\begin{array}{c}[x : \mathbf{A}(\mathbb{W})]\\ b(x) : \mathbf{B}(\mathbb{W})\end{array}}{\lambda x.b(x) : (\mathbf{A} \to \mathbf{B})(\mathbb{W})}$$

→-*elimination*

$$\frac{a : \mathbf{A}(\mathbb{W}) \qquad y : (\mathbf{A} \to \mathbf{B})(\mathbb{W})}{\mathtt{APP}(y, a) : \mathbf{B}(\mathbb{W})}$$

∀-*introduction*

$$\frac{\begin{array}{c}[x : \mathbf{D}(\mathbb{W})]\\ f(x) : \mathbf{P}(x)(\mathbb{W})\end{array}}{\Lambda x.f(x) : (\forall x^{\mathbf{D}}.\mathbf{P}(x))(\mathbb{W})}$$

∀-*elimination*

$$\frac{a : \mathbf{D}(\mathbb{W}) \qquad y : (\forall x^{\mathbf{D}}.\mathbf{P}(x))(\mathbb{W})}{\mathtt{EXTR}(y, a) : \mathbf{P}(a)(\mathbb{W})}$$

By using the inference rules above we can show that a (logic-dependent) tautology may be proved as 'necessarily' true, or true in 'all possible worlds'. For example, the intuitionistic tautology '$\mathbf{A} \rightarrow (\mathbf{B} \rightarrow \mathbf{A})$' is proved to be '*necessarily* true' in the following way:

$$\dfrac{\dfrac{\dfrac{\dfrac{\dfrac{[\mathbb{W} : \mathsf{U}]}{[y : \mathbf{B}(\mathbb{W})]}}{[x : \mathbf{A}(\mathbb{W})]}}{\lambda y.x : (\mathbf{B} \rightarrow \mathbf{A})(\mathbb{W})}(*)}{\lambda x.\lambda y.x : (\mathbf{A} \rightarrow (\mathbf{B} \rightarrow \mathbf{A}))(\mathbb{W})}}{\Lambda\mathbb{W}.\lambda x.\lambda y.x : \Box(\mathbf{A} \rightarrow (\mathbf{B} \rightarrow \mathbf{A}))}$$

Notice that this is 'necessarily' true if we take the implication to be non-relevant: in the step marked '$(*)$' a non-relevant discharge of assumptions at the propositional level was made.

Strict implication. Using the device of world variables for possible worlds the definition of strict implication in terms of necessity, namely:

$$\mathbf{A} \Rightarrow \mathbf{B} \equiv \Box(\mathbf{A} \rightarrow \mathbf{B})$$

('$\mathbf{A} \Rightarrow \mathbf{B}$' to be read as '**A** strictly implies **B**') comes out naturally from the rule of $\rightarrow$-*introduction* augmented with the extra parameter indicating the use of world variables:

$$\dfrac{\dfrac{\begin{array}{c}[\mathbb{W} : \mathsf{U}] \\ [x : \mathbf{A}(\mathbb{W})] \\ b(x) : \mathbf{B}(\mathbb{W})\end{array}}{\lambda x.b(x) : (\mathbf{A} \rightarrow \mathbf{B})(\mathbb{W})}}{\Lambda\mathbb{W}.\lambda x.b(x) : \underbrace{\Box(\mathbf{A} \rightarrow \mathbf{B})}_{\mathbf{A} \Rightarrow \mathbf{B}}}$$

The $\rightarrow$-*introduction* now says that '$\mathbf{A} \Rightarrow \mathbf{B}$' if for any arbitrary world '$\mathbb{W}$', if **A** is true at (demonstrable at, forced by, etc.) $\mathbb{W}$ then **B** is also true at (resp. demonstrable at, forced by, etc.) $\mathbb{W}$.

Montague's Types, and World-Variables as Senses

Let us for a moment think of a connection to another logical system which would seem to have some similarity in the underlying ontological assumption, at least with respect to the idea of dividing the logical calculus into two dimensions, i.e. the functional and the logical. The semantical framework constructed in [Montague, 1970]'s intensional logic defines the semantic *types* of objects dealt with are given as e, t and s, which in words could be described as: *entities*, *truth-values*, and *senses*. The idea was that logic (language) was supposed to deal with objects of three kinds: names of entities, formulas denoting truth-values, and possible-worlds/contexts of use. Now, here when we say that we wish to have the bi-dimensional calculus, we are saying that the entities which are *namable* (i.e. individuals, possible-worlds, etc.) ought to be dealt with separately from (yet harmoniously with) the logical calculus on the formulas, by a calculus of functional expressions. Whereas the variables for individuals are handled 'naturally' in the interpretation of first-order logic with our labelled natural deduction, the introduction of variables to denote *contexts*, or *possible-worlds* (structured collection of labelled formulas), as in our labelled natural deduction with an extra parameter, is supposed to account for Montague's *senses*.

Open Assumptions and the Extra Parameter

The axiom of implication which corresponds to *permutation* (taking us from concatenation implication to linear implication), namely:

$$(\mathbf{A} \to (\mathbf{B} \to \mathbf{C})) \to (\mathbf{B} \to (\mathbf{A} \to \mathbf{C}))$$

would seem to find a counterpart in an inference rule of the following form:

$$\frac{\vdash \Box(\mathbf{B} \to \mathbf{C})}{\mathbf{B} \to \Box\mathbf{C}}$$

whose proof-tree would look like:

$$\cfrac{\cfrac{\cfrac{[x : \mathbf{B}(\mathbb{S})] \qquad \cfrac{[\mathbb{S} : \mathsf{U}] \qquad l : \Box(\mathbf{B} \to \mathbf{C})}{\mathcal{EX}(l, \mathbb{S}) : (\mathbf{B} \to \mathbf{C})(\mathbb{S})}}{\text{APP}(\mathcal{EX}(l, \mathbb{S}), x) : \mathbf{C}(\mathbb{S})}}{\Lambda\mathbb{S}.\text{APP}(\mathcal{EX}(l, \mathbb{S}), x) : \Box\mathbf{C}}\,(*)}{\lambda x.\Lambda\mathbb{S}.\text{APP}(\mathcal{EX}(l, \mathbb{S}), x) : \mathbf{B} \to \Box\mathbf{C}}$$

Notice, however, that the discharge of the assumption '$[\mathbb{S} : \mathsf{U}]$' was made when there was an open assumption '$x : \mathbf{B}(\mathbb{S})$' with '$\mathbb{S}$' as an extra parameter, namely step '$(*)$'. At that step '$\mathbb{S}$' could not be taken as arbitrary, therefore be universally quantified over, because it had '$x : \mathbf{B}$' in it.

This is an example of how the scope of the connective of 'necessity' might be confused with the scope of 'implication', as is rightly remarked in [Hughes and Cresswell, 1968].[6]

[6]When explaining the equivalence between a definition of strict implication ('$\mathbf{A} \Rightarrow \mathbf{B}$') using 'possibility' ('$\Diamond$') and another using 'necessity' ('$\Box$'), they say [Hughes and Cresswell, 1968, footnote 17 of p. 27]: "It is important not to confuse $L(p \supset q)$, which means that the whole hypothetical 'if p then q' is a necessary truth, or that q follows logically from p, with $p \supset Lq$, which means that if p is true then q is a necessary truth. Unhappily, these are often confused in ordinary discourse, sometimes with disastrous results; and neglect of the distinction is made all the easier by the ambiguity of such common idioms as 'If ... then it must be (*or* is bound to be) the case that—'. To make things worse, the structure of such sentences is more closely analogous to that of $p \supset Lq$, but one suspects that most frequently what the speaker intends to assert (or at least all he is entitled to assert) is something of the form $L(p \supset q)$."

2.4 Working with Subdeductions

One of the basic principles of Gentzen's proof theory is that of the *subformula* principle, which says that a cut-free proof of a sequent $\Gamma \vdash \Delta$ will contain only subformulas of the formulas in Γ and Δ. In the framework of labelled proof theory, however, the emphasis is on the subdeduction, rather than on the subformula property. That is to say, we shall follow a 'backward' procedure similar to the one used in the sequent calculus, but the idea is that when breaking down formulas, we introduce new variables as labels of the subformulas according to the proof rules for the particular connective. The construction of the proof tree then involves 'solving' the unknowns which are introduced in the course of the breaking down the formula(s) into the subdeductions, as soon as we reach atomic formulas, or the unknowns can be solved at a certain point without having to go further in breaking down the formulas we want to prove.

As for normalisation properties, Tait's method [Tait, 1965, 1967] will apply rather straightforwardly in most cases. For the method to work for a wide spectrum of logics, though, one needs to refine the normalisation theorem by working with various abstraction disciplines as we have shown for the case of 'implication' in [Gabbay and de Queiroz, 1992].[7]

Let us give the reader an idea of how the method of subdeduction works. As we have seen from the previous subsection, we work in the framework of *labelled* natural deduction. That is to say, we have an *introduction* and an *elimination* rule for each logical connective, plus a *labelling* calculus alongside the formulas, playing the rôle of the meta-level device. (One can think of the turnstile of the sequent calculus as such a device in Gentzen's logistic calculi.) Thus, we have:

$$\frac{\begin{array}{c}[\mathbb{W} : \mathsf{U}] \\ F(\mathbb{W}) : \mathbf{A}(\mathbb{W})\end{array}}{\Lambda\mathbb{W}.F(\mathbb{W}) : \Box\mathbf{A}}\ \Box\textit{-introduction} \qquad \frac{\mathbb{S} : \mathsf{U} \qquad l : \Box\mathbf{A}}{\mathcal{EX}(l, \mathbb{S}) : \mathbf{A}(\mathbb{S})}\ \Box\textit{-elimination}$$

Let us see how the rules work. In order to check whether the formula

$$\Box(\mathbf{A} \to \mathbf{B}) \to (\Box\mathbf{A} \to \Box\mathbf{B})$$

is valid (i.e. it can be proved without leaving any assumptions undischarged) we analyse the formula into its subparts, and start building the subdeductions. For the formula to be valid we must start from:

(1) $x : \Box(\mathbf{A} \to \mathbf{B})$

as an assumption, and arrive at

[7] The conceptual framework underlying our approach to 'subdeduction' is part of ongoing research on seeking an intuitive account of the *label-and-formula* perspective in terms of Frege's later work [de Queiroz, 1992].

(2) $f(x) : \Box\mathbf{A} \to \Box\mathbf{B}$

(for some $f(x)$) as the premise of an $\to$-*introduction* rule.

Now, analysing (2) we know that the '$f(x)$' must be a λ-term, given that the main connective is an implication. So, we have

(3) $\lambda y.g(x,y) : \Box\mathbf{A} \to \Box\mathbf{B}$

(for some expression g which depends on both x and y) which must have come from an $\to$-*introduction*, with

(4) $y : \Box\mathbf{A}$

as assumption, and

(5) $g(x,y) : \Box\mathbf{B}$

as premise. Now, (5) has 'necessity' as the main connective, so, as we shall see in more detail in the sequel, (5) must be of the form:

(6) $\Lambda\mathbb{W}.h(x,y,\mathbb{W}) : \Box\mathbf{B}$

and it must have come from

(7) $\mathbb{W} : \mathsf{U}$

(where $\mathbb{W}$ is an arbitrary 'world' from the universe U) as an assumption, and

(8) $h(x,y,\mathbb{W}) : \mathbf{B}(\mathbb{W})$

as premise of $\Box$-*introduction*. Now we have:

(1) $x : \Box(\mathbf{A} \to \mathbf{B})$

(4) $y : \Box\mathbf{A}$

(7) $\mathbb{W} : \mathsf{U}$

as assumptions, and we need to obtain $\mathbf{B}(\mathbb{W})$ from those, which will 'solve' our unknown '$h(x,y,\mathbb{W})$' (h is an expression which depends on x, y and $\mathbb{W}$). Now, from (1) and (7) by applying $\Box$-*elimination* we get

(9) $\mathcal{EX}(x,\mathbb{W}) : (\mathbf{A} \to \mathbf{B})(\mathbb{W})$

From (4) and (7) by $\Box$-*elimination* we get

(10) $\mathcal{EX}(y,\mathbb{W}) : \mathbf{A}(\mathbb{W})$

Now, from (9) and (10) by $\to$-*elimination* we get

(11) $\text{APP}(\mathcal{EX}(x,\mathbb{W}), \mathcal{EX}(y,\mathbb{W})) : \mathbf{B}(\mathbb{W})$

and we now have the label of $\mathbf{B}(\mathbb{W})$ as an expression depending on x, y and $\mathbb{W}$ (i.e. our unknown h).

Proceeding towards our deduction of the whole formula out of the subdeductions, by $\Box$-*introduction* we get

(12) $\Lambda\mathbb{W}.\mathtt{APP}(\mathcal{EX}(x,\mathbb{W}),\mathcal{EX}(y,\mathbb{W})) : \Box\mathbf{B}$

By $\rightarrow$-*introduction* we get

(13) $\lambda y.\Lambda\mathbb{W}.\mathtt{APP}(\mathcal{EX}(x,\mathbb{W}),\mathcal{EX}(y,\mathbb{W})) : \Box\mathbf{A} \rightarrow \Box\mathbf{B}$

Again, by $\rightarrow$-*introduction* we obtain

(14) $\lambda x.\lambda y.\Lambda\mathbb{W}.\mathtt{APP}(\mathcal{EX}(x,\mathbb{W}),\mathcal{EX}(y,\mathbb{W})) : \Box(\mathbf{A} \rightarrow \mathbf{B}) \rightarrow (\Box\mathbf{A} \rightarrow \Box\mathbf{B})$

and we are done: all the assumptions introduced in the process of working with subdeductions have been discharged, as the corresponding label-variables reveal (i.e. they are all bound by some abstraction). The actual proof tree can now be reconstructed (see the section on the modal logic K).

2.5 *Connections with Kripke's Truth Definition*

In order to generalise the notion of 'possible world' as defined in an early account of frame-based semantics [Kripke, 1959] where there could be no two worlds in which the same truth-value would be assigned to each atomic formula, [Kripke, 1963] introduced the notion of *accessibility relation*: in addition to a set of possible worlds $\mathcal{K}$ and one distinguished element $\mathbb{G}$ (the 'real world'), the generalised definition of a model would also involve the notion of 'accessibility relation' between worlds. Given two worlds $\mathbb{W}_1$ and $\mathbb{W}_2$, $\mathbb{W}_2$ is *accessible* (relative to $\mathbb{W}_1$) if every proposition true in $\mathbb{W}_2$ is to be possible in $\mathbb{W}_1$. Thus, the 'absolute' notion of possible world (as in [Kripke, 1959]) gives way to a relative notion, of one world being possible relative to another.

Truth Definition. Using the notation $\mathbb{W}_1 \boldsymbol{R} \mathbb{W}_2$ to denote '$\mathbb{W}_2$ is accessible from $\mathbb{W}_1$', the definition of truth in the general model is given in a way such that:

> $\Box\mathbf{A}$ is true at $\mathbb{W}_1$ iff $\mathbf{A}$ is true at $\mathbb{W}_2$ for each $\mathbb{W}_2$ such that $\mathbb{W}_1 \boldsymbol{R} \mathbb{W}_2$.
>
> $\Diamond\mathbf{A}$ is true at $\mathbb{W}_1$ iff there exists $\mathbb{W}_2$, such that $\mathbb{W}_1 \boldsymbol{R} \mathbb{W}_2$, and $\mathbf{A}$ is true at $\mathbb{W}_2$.

As pointed out in [Kripke, 1963], the 'absolute' notion of possible world would be convenient for S5, but would hardly allow room for the treatment of normal propositional calculi in general. For the semantics to be general, there had to be ways of distinguishing various kinds of 'necessity' within the same framework. The introduction of the device of *accessibility relation* was instrumental in that it allowed for the distinction among various notions of necessity within the framework of possible-worlds semantics on the basis of the properties of the relation (reflexive, transitive, Euclidean, etc.).

In setting out a unified framework for the treatment of necessity via the functional interpretation where the distinction of various kinds of necessities is made on the basis of the discipline of handling 'world' variables, we are trying to provide a proof-theoretic counterpart to the notion of *accessibility relation.* The idea is to provide appropriate means to distinguish □'s with the help of the notion of 'assumption handling discipline' in much the same way we have done for the propositional implication [Gabbay and de Queiroz, 1992].

2.6 *Connections with Gentzen's Sequent Calculus*

As the more perspicuous reader will have noticed, our notation for '**A** is true at (demonstrable at, forced by, etc.) $\mathbb{W}$', namely:

$$\mathbf{A}(\mathbb{W})$$

where '$\mathbb{W}$' is a structured collection (list, bag, tree, etc.) of formulas (together with their labels), is another way of writing (using [Gentzen, 1935]'s style sequent calculus):

$$\Gamma \vdash \mathbf{A}$$

where Gentzen's 'Γ' is our '$\mathbb{W}$'. There is an important difference, though: while the operations on the sets (or sequences) of formulas (such as 'Γ') are restricted to the so-called *structural* rules (namely, *weakening*, *contraction* and *exchange*), our structured collections of formulas will be handled according to their own structural properties. They could be sets, multisets, bags, lists, queues, trees, etc., and in each case we shall be able to prescribe particular ways of handling the Γ's (or $\mathbb{W}$'s in our notation). We shall not be restricted to any specific structure (sequence of formulas in Gentzen's sequent calculus, commutative sequence of formulas in [Girard, 1987]'s linear logic, non-commutative string of formulas in [Lambek, 1958]'s calculus, etc.), but we shall simply have a general framework where those collections will have names (identifiers) within the calculus, so that we can refer to them when performing the appropriate operations on them (e.g. removing or inserting an element in a certain Γ which is a tree of formulas, etc.).

3 MODAL LOGICS AND THE FUNCTIONAL INTERPRETATION

Now, within this framework the distinct kinds of '□' will be characterised by the discipline of Λ-abstraction over world-variables, in a similar way to the propositional and first-order conditionals. So, we identify which disciplines of Λ-abstraction on world-variables such that we can prove the axioms of well-known modal logics, namely:

K.	$\Box(\mathbf{A} \to \mathbf{B}) \to (\Box\mathbf{A} \to \Box\mathbf{B})$
D.	$\Box\mathbf{A} \to \Diamond\mathbf{A}$
T.	$\Box\mathbf{A} \to \mathbf{A}$
B.	$\mathbf{A} \to \Box\Diamond\mathbf{A}$
4.	$\Box\mathbf{A} \to \Box\Box\mathbf{A}$
5.	$\Diamond\mathbf{A} \to \Box\Diamond\mathbf{A}$

using the specific discipline of assumption discharge.

3.1 Chellas' RM rule for Standard Normative Logics

Weak Notions of Necessity

As an example of how appropriate the treatment of 'necessity' via an extension of the treatment of propositional implication may prove to be, we can show that the weakest rule characterising the so-called 'standard normative logics' according to [Chellas, 1980], namely:

$$\text{RM} \quad \frac{\vdash \mathbf{A} \to \mathbf{B}}{\Box\mathbf{A} \to \Box\mathbf{B}}$$

(sometimes called 'regularity' principle) mirrors the axiom schema characterising the weakest implication, viz. 'concatenation' implication [Gabbay, forthcoming], and which characterises the weak left-implication of Lambek's [1958] calculus. Similarly to our previous characterisation of the propositional implication via the framework of the functional interpretation, in our calculus here the proof of derivability of Chellas' RM uses the most 'well-behaved' discipline of Λ-abstraction over world-variables. We can see this from the following proof tree:

$$\cfrac{\cfrac{\cfrac{\cfrac{[\mathbb{S} : \mathsf{U}] \quad [w : \Box\mathbf{A}]}{\mathcal{EX}(w, \mathbb{S}) : \mathbf{A}(\mathbb{S})} \qquad p : (\mathbf{A} \to \mathbf{B})(\mathbb{S})}{\texttt{APP}(p, \mathcal{EX}(w, \mathbb{S})) : \mathbf{B}(\mathbb{S})}}{\Lambda\mathbb{S}.\texttt{APP}(p, \mathcal{EX}(w, \mathbb{S})) : \Box\mathbf{B}}}{\lambda w.\Lambda\mathbb{S}.\texttt{APP}(p, \mathcal{EX}(w, \mathbb{S})) : \Box\mathbf{A} \to \Box\mathbf{B}}$$

where '$p : (\mathbf{A} \to \mathbf{B})(\mathbb{S})$' indicates that '$\mathbf{A} \to \mathbf{B}$' is true in the 'actual' world we here name '$\mathbb{S}$', either because is an 'axiom' labelled 'p', or because it is a tautology (according to the notion of implication one is using) and so 'p' represents a closed λ-term, or even because the formula is derivable (with label 'p') by means of the rules of inference of the particular logic.

In the proof-tree mirroring 'RM' the order of discharge of assumptions (and therefore lambda-abstractions) is not violated: the discharge of '$[\mathbb{S} : \mathsf{U}]$' and corresponding Λ-abstraction on '$\mathbb{S}$' is being made before the discharge of '$[w : \Box\mathbf{A}]$' and corresponding λ-abstraction on 'w'. The order of abstraction over free-variables in the 'ordered' term '$\texttt{APP}(p, \mathcal{EX}(w, \mathbb{S}))$' is not violated. (We say that the term is 'ordered' because we take 'APP' and '$\mathcal{EX}$' as functional application operators which take a value-range term and apply it to an argument, the value-range term understood to be of a 'higher' type than the argument.)

Non-Normal Modal Logics

A *normal* modal logic is usually defined as the modal logic which validates:

1. every tautology;
2. all formulas of the form $\Box(\mathbf{A} \to \mathbf{B}) \to (\Box\mathbf{A} \to \Box\mathbf{B})$;
3. $\Box\mathbf{A}$ whenever it validates $\mathbf{A}$.

Observe that the rule of *necessitation*, namely:

$$\frac{\vdash \mathbf{A}}{\Box\mathbf{A}}$$

is left vague as to what '$\vdash \mathbf{A}$' is to mean. The formula may be 'provable' simply because it is a theorem of the logic, regardless of the contextual hypotheses involved in the reasoning. It can also be a 'legitimately context-dependent' theorem, in the sense that some of the assumptions used in the deduction are to remain undischarged.

Then, the definition of *non-normal* modal logics is given by replacing the necessitation rule by a weaker rule, usually the rule of 'regularity', which is defined in [Chellas, 1980] as:

$$\text{RM} \qquad \frac{\vdash \mathbf{A} \to \mathbf{B}}{\Box\mathbf{A} \to \Box\mathbf{B}}$$

As pointed out by various people, when studying the semantics of these logics, it is difficult to avoid the rule of necessitation because its soundness is 'virtually built into the definition of possible world models' [Fitting, 1993]. [Kripke, 1965]'s solution to such semantic difficulty was to account for exceptional ('queer') worlds which would not play by the usual rules.

What would be a reasonable approach to non-normality in our labelled proof theory? To answer this question we need, first of all, to consider what is at stake here, namely the weakening of the standard notion of necessity, in order to make it, in some sense, *relevant*, and perhaps even *linear*. The intended application of a formal characterisation of the non-normal modal logics is the formulation of a logic for notions such as knowledge and belief. In dealing with concepts like these one often needs to make use of the not-so-declarative notion of 'explicitness'. That is to mean one would like to be able to understand how structural aspects (weakening, contraction, permutation, etc.) come into play when modal connectives are concerned.

As we have pointed out earlier on in this paper, when formulated in the pattern of our labelled natural deduction, where names of possible worlds are introduced explicitly in the labels, the rule of necessitation:

$$\dfrac{\begin{array}{c}[\mathbb{W} : \mathsf{U}] \\ \vdots \\ F(\mathbb{W}) : \mathbf{A}(\mathbb{W})\end{array}}{\Lambda \mathbb{W}.F(\mathbb{W}) : \Box\mathbf{A}}$$

will give us a more refined mechanism with which to 'parameterise' our notion of necessity. In a similar way to the characterisation of relevant implication, we will be able to impose a 'relevant' discipline of Λ-abstraction over the label-expression alongside the 'provable **A**' formula, such it requires that the arbitrary world must have been used in order for the formula to be considered 'necessary'. The parallel with the case of propositional implication is then quite immediate, e.g., whereas in the proof displayed in Figure 1 there is an immediate correspondence in terms of abstraction disciplines, even if on variables standing for different kinds of entities, in:

$$\dfrac{\dfrac{\dfrac{[\mathbb{S} : \mathsf{U}]}{[l : \Box\mathbf{A}]}}{\boxed{\Lambda\mathbb{S}}.l : \Box\Box\mathbf{A}}(*)}{\lambda l.\Lambda\mathbb{S}.l : \Box\mathbf{A} \to \Box\Box\mathbf{A}}$$

$$\dfrac{\dfrac{\dfrac{[s : \mathbf{U}]}{[l : \mathbf{A}]}}{\boxed{\lambda s}.l : \mathbf{U} \to \mathbf{A}}(*)}{\lambda l.\lambda s.l : \mathbf{A} \to (\mathbf{U} \to \mathbf{A})}$$

the parallel is seen through a vacuous abstraction on variables. So, neither relevant nor intuitionistic implication are 'resource-aware': while the former, which validates

$$(\mathbf{C} \to (\mathbf{A} \to \mathbf{B})) \to ((\mathbf{C} \to \mathbf{A}) \to (\mathbf{C} \to \mathbf{B})),$$

does not take account of resource counting, the latter, by validating $\mathbf{A} \to (\mathbf{U} \to \mathbf{A})$, leaves the aspect of relevance behind.

In the same vein, neither K nor S4 will be 'normal' modal logics, if we have a $\Box$-*introduction* rule where labels alongside formulas help to control its application by varying the abstraction discipline on the appropriate variables.

3.2 *The Modal Logic K*

The axiom for the modal logic K finds a parallel in the axioms for implication and first-order universal quantification which correspond to the axiom introduced to the

$$\dfrac{\dfrac{\dfrac{\dfrac{\dfrac{[\mathbb{S}:\mathsf{U}] \quad [z:\Box\mathbf{A}]}{\mathcal{EX}(z,\mathbb{S}):\mathbf{A}(\mathbb{S})} \qquad \dfrac{[\mathbb{S}:\mathsf{U}] \quad [w:\Box(\mathbf{A}\to\mathbf{B})]}{\mathcal{EX}(w,\mathbb{S}):(\mathbf{A}\to\mathbf{B})(\mathbb{S})}}{\mathtt{APP}(\mathcal{EX}(w,\mathbb{S}),\mathcal{EX}(z,\mathbb{S})):\mathbf{B}(\mathbb{S})}}{\boxed{\Lambda\mathbb{S}}.\mathtt{APP}(\mathcal{EX}(w,\boxed{\mathbb{S}}),\mathcal{EX}(z,\boxed{\mathbb{S}})):\Box\mathbf{B}}(*)}{\lambda z.\Lambda\mathbb{S}.\mathtt{APP}(\mathcal{EX}(w,\mathbb{S}),\mathcal{EX}(z,\mathbb{S})):\Box\mathbf{A}\to\Box\mathbf{B}}}{\lambda w.\lambda z.\Lambda\mathbb{S}.\mathtt{APP}(\mathcal{EX}(w,\mathbb{S}),\mathcal{EX}(z,\mathbb{S})):\Box(\mathbf{A}\to\mathbf{B})\to(\Box\mathbf{A}\to\Box\mathbf{B})}$$

$$\dfrac{\dfrac{\dfrac{\dfrac{\dfrac{[s:\mathbf{C}] \quad [z:\mathbf{C}\to\mathbf{A}]}{\mathtt{APP}(z,s):\mathbf{A}} \qquad \dfrac{[s:\mathbf{C}] \quad [w:\mathbf{C}\to(\mathbf{A}\to\mathbf{B})]}{\mathtt{APP}(w,s):\mathbf{A}\to\mathbf{B}}}{\mathtt{APP}(\mathtt{APP}(w,s),\mathtt{APP}(z,s)):\mathbf{B}}}{\boxed{\lambda s}.\mathtt{APP}(\mathtt{APP}(w,\boxed{s}),\mathtt{APP}(z,\boxed{s})):\mathbf{C}\to\mathbf{B}}(*)}{\lambda z.\lambda s.\mathtt{APP}(\mathtt{APP}(w,s),\mathtt{APP}(z,s)):(\mathbf{C}\to\mathbf{A})\to(\mathbf{C}\to\mathbf{B})}}{\lambda w.\lambda z.\lambda s.\mathtt{APP}(\mathtt{APP}(w,s),\mathtt{APP}(z,s)):(\mathbf{C}\to(\mathbf{A}\to\mathbf{B}))\to((\mathbf{C}\to\mathbf{A})\to(\mathbf{C}\to\mathbf{B}))}$$

Figure 1.

axiomatics of linear implication to obtain relevant implication, needs a discipline of Λ-abstraction over world-variables such that it binds more than one occurrence of the variable, similarly to the λ-abstraction discipline for the relevant implication:

$$\text{K.} \quad \Box(\mathbf{A}\to\mathbf{B})\to(\Box\mathbf{A}\to\Box\mathbf{B})$$

which could be rewritten as:

$$\forall\mathbb{W}^{\mathsf{U}}.((\mathbf{A}\to\mathbf{B})(\mathbb{W}))\to(\forall\mathbb{W}^{\mathsf{U}}.\mathbf{A}(\mathbb{W})\to\forall\mathbb{S}^{\mathsf{U}}.\mathbf{B}(\mathbb{S}))$$

where the proof-tree would look like:

$$\dfrac{\dfrac{\dfrac{\dfrac{\dfrac{[\mathbb{S}:\mathsf{U}] \quad [z:\Box\mathbf{A}]}{\mathcal{EX}(z,\mathbb{S}):\mathbf{A}(\mathbb{S})} \qquad \dfrac{[\mathbb{S}:\mathsf{U}] \quad [w:\Box(\mathbf{A}\to\mathbf{B})]}{\mathcal{EX}(w,\mathbb{S}):(\mathbf{A}\to\mathbf{B})(\mathbb{S})}}{\mathtt{APP}(\mathcal{EX}(w,\mathbb{S}),\mathcal{EX}(z,\mathbb{S})):\mathbf{B}(\mathbb{S})}}{\Lambda\mathbb{S}.\mathtt{APP}(\mathcal{EX}(w,\mathbb{S}),\mathcal{EX}(z,\mathbb{S})):\Box\mathbf{B}}(*)}{\lambda z.\Lambda\mathbb{S}.\mathtt{APP}(\mathcal{EX}(w,\mathbb{S}),\mathcal{EX}(z,\mathbb{S})):\Box\mathbf{A}\to\Box\mathbf{B}}}{\lambda w.\lambda z.\Lambda\mathbb{S}.\mathtt{APP}(\mathcal{EX}(w,\mathbb{S}),\mathcal{EX}(z,\mathbb{S})):\Box(\mathbf{A}\to\mathbf{B})\to(\Box\mathbf{A}\to\Box\mathbf{B})}$$

and the $\Lambda\mathbb{S}$-abstraction at step marked '$(*)$' binds two free-occurrences of the world-variable '$\mathbb{S}$' in the term '$\mathtt{APP}(\mathcal{EX}(w,\mathbb{S}),\mathcal{EX}(z,\mathbb{S}))$', in the increasing order: first the 'lower' variable '$\mathbb{S}$', then the one a step higher ('z'), and finally the higher 'w'.

3.3 The Modal Logic D

The modal logic D is characterised by the existence of an accessible possible world, wherever world one is supposed to be in. In axiomatic presentation it becomes:

$$\text{D.} \quad \Box\mathbf{A} \to \Diamond\mathbf{A}$$

which reads 'if for all possible worlds we have **A** then there exists a possible world where we have **A**'. Now, we draw your attention to the structural similarity of such an axiom to the axiom which distinguishes first-order logic from *inclusive* logics (i.e. those logics which deal with non-empty domains)[8]: $\forall x.\mathbf{P}(x) \to \exists x.\mathbf{P}(x)$

$$\dfrac{\dfrac{\dfrac{[\forall x.\mathbf{P}(x)]}{\mathbf{P}(t)}(*)}{\exists x.\mathbf{P}(x)}}{\forall x.\mathbf{P}(x) \to \exists x.\mathbf{P}(x)}$$

Notice that at the step marked '$(*)$' an assumption was made, namely that the domain over which the '$\forall$' is quantifying is a non-empty domain containing at least an element which is being named 't'. This assumption does not get recorded in the system of plain natural deduction.

Using the functional interpretation, where the presence of terms, and the expliciting of domains of quantification make the framework a much richer instrument for deduction calculi, we have:

$$\dfrac{\dfrac{\dfrac{[t:\mathbf{D}] \qquad [z:\forall x^{\mathbf{D}}.\mathbf{F}(x)]}{\mathtt{EXTR}(z,t):\mathbf{P}(t)}}{\varepsilon x.(\mathtt{EXTR}(z,x),t):\exists x^{\mathbf{D}}.\mathbf{P}(x)}}{\lambda z.\varepsilon x.(\mathtt{EXTR}(z,x),\boxed{t}):\forall x^{\mathbf{D}}.\mathbf{P}(x) \to \exists x^{\mathbf{D}}.\mathbf{P}(x)}$$

Here the existence of a free variable (namely 't') indicates that the assumption '$[t : \mathbf{D}]$' remains to be discharged. By making the domain of quantification explicit one does not have the antecedent (vacuously) true and the consequent trivially false in the case of empty domain: the proof of the proposition is still depending on the assumption '$[t : \mathbf{D}]$', i.e. that the type '**D**' is non-empty. To be categorical the

[8]Cf. [Fine, 1985, Chapter 21, page 205]: "An *inclusive* logic is one that is meant to be correct for both empty and non-empty domains. There are certain standard difficulties in formulating a system of inclusive logic. If, for example, we have the usual rules of UI, EG and conditional proof, then the following derivation of the theorem $\forall xFx \supset \exists xFx$ goes through (...) But the formula $\forall xFx \supset \exists xFx$ is not valid in the empty domain; the antecedent is true, while the consequent is false."

above proof would still have to proceed one step, as in:

$$\cfrac{\cfrac{\cfrac{\cfrac{[t : \mathbf{D}] \qquad [z : \forall x^{\mathbf{D}}.\mathbf{P}(x)]}{\mathtt{EXTR}(z,t) : \mathbf{P}(t)}}{\varepsilon x.(\mathtt{EXTR}(z,x),t) : \exists x^{\mathbf{D}}.\mathbf{P}(x)}}{\lambda z.\varepsilon x.(\mathtt{EXTR}(z,x),\boxed{t}) : \forall x^{\mathbf{D}}.\mathbf{P}(x) \rightarrow \exists x^{\mathbf{D}}.\mathbf{P}(x)}}{\underbrace{\lambda t.\lambda z.\varepsilon x.(\mathtt{EXTR}(z,x),t)}_{\text{no free variable}} : \mathbf{D} \rightarrow (\forall x^{\mathbf{D}}.\mathbf{P}(x) \rightarrow \exists x^{\mathbf{D}}.\mathbf{P}(x))}$$

At this point we look at the proof-construction ('$\lambda t.\lambda z.\varepsilon x.(\mathtt{EXTR}(z,x),t)$') and we can see no free variables, thus the corresponding proof is categorical, i.e. does not depend on any assumption. And the result is that for the proof to be categorical, the domain of individuals must be non-empty.

Now, by considering variables (such as 'x') as ranging over possible worlds, and the domain (such as '**D**') as being the universe of possible worlds, we immediately see the parallel: in order for the axiom to be true the domain '**D**', now seen as the universe of possible worlds, must be non-empty. In other words, there is at least one possible world accessible from wherever one is, which in standard modal Kripke-style semantics represents the seriality of the accessibility relation.

3.4 The Modal Logic T

The modal logic T is characterised axiomatically by the schema:

$$\text{T.} \quad \Box\mathbf{P} \rightarrow \mathbf{P}$$

but because we are working with 'closed' formulas (sentences)[9] and the above axiom implicitly leaves the 'actual' ('current') world free (non-quan-tified), let us look at the proof construction for the axiom schema '$\Box(\Box\mathbf{P} \rightarrow \mathbf{P})$' which is the universal closure of the T-axiom schema. The proof tree is constructed in the

[9] Note that the axiom schema '$\Box\mathbf{P} \rightarrow \mathbf{P}$', when translated to a formula using universal quantification over world-variables becomes: $\forall \mathtt{W}^{\mathsf{U}}.\mathbf{P}(\mathtt{W}) \rightarrow \mathbf{P}(\mathtt{S})$, where '$\mathtt{S}$' is supposed to name the 'current' world. To obtain a closed formula, which would not name the 'current' world, but make use of a bound variable, we would be able to quantify over '$\mathtt{S}$':

$$\forall \mathtt{S}^{\mathsf{U}}.(\forall \mathtt{W}^{\mathsf{U}}.\mathbf{P}(\mathtt{W}) \rightarrow \mathbf{P}(\mathtt{S}))$$

which when translated back gives '$\Box(\Box\mathbf{P} \rightarrow \mathbf{P})$'. The reading would then be 'for any possible world (including the 'current' one), if **P** is true in all possible worlds, then **P** is true in 'this' world.' Obviously, if '$\Box\mathbf{P}$' is interpreted as '**P** is provable', then we may start to encounter incongruencies if we are not careful with the notion of provability we actually mean. [Gödel, 1933] remarked that $\Box(\Box\mathbf{P} \rightarrow \mathbf{P})$, where '$\Box$' is taken to be 'provable in an axiomatic theory', making it possible to read the axiom as 'it is provable that (if **P** is provable then **P** is true)', could not be part of the axiomatics of the notion of provability itself because it would be contradicting his Second Incompleteness Theorem.

following way:

$$\dfrac{\dfrac{[\mathbb{S}:\mathsf{U}] \qquad \dfrac{[\mathbb{S}:\mathsf{U}]}{[l:\Box\mathbf{P}(\mathbb{S})]}}{\mathcal{EX}(l,\mathbb{S}):\mathbf{P}(\mathbb{S})}}{\dfrac{\lambda l.\mathcal{EX}(l,\mathbb{S}):(\Box\mathbf{P}\to\mathbf{P})(\mathbb{S})}{\Lambda\mathbb{S}.\lambda l.\mathcal{EX}(l,\mathbb{S}):\Box(\Box\mathbf{P}\to\mathbf{P})}}$$

and the distinguishing characteristic of the Λ-abstraction over the world-variable '$\mathbb{S}$' is that it breaks the order of Λ-abstraction in the sense that it is first abstracting over the variable at the 'function' place, namely 'l', when the variable at the 'argument' place, namely '$\mathbb{S}$' is still free. Recall that the weak implication called 'concatenation' [Gabbay, forthcoming] admits the withdrawing of assumptions only in the increasing order, which is being obeyed by the Λ-abstraction discipline for K.

3.5 The Modal Logic B

The 'Brouwerian' modal logic is characterised as the modal logic of reflexive and symmetric accessibility relation. It is often characterised by the axiom schema:

$$\text{B.}\quad \mathbf{A}\to\Box\Diamond\mathbf{A}$$

for any formula $\mathbf{A}$. The proof of this axiom can be constructed in the following way:

$$\dfrac{\dfrac{\dfrac{[\mathbb{T}:\mathsf{U}] \qquad \dfrac{[\mathbb{S}:\mathsf{U}]}{[x:\mathbf{A}(\mathbb{T})(\mathbb{S})]}}{\varepsilon\mathbb{W}.(x,\mathbb{T}):\Diamond\mathbf{A}(\mathbb{S})}(*)}{\Lambda\mathbb{S}.\varepsilon\mathbb{W}.(x,\mathbb{T}):\Box\Diamond\mathbf{A}}}{\lambda x.\Lambda\mathbb{S}.\varepsilon\mathbb{W}.(x,\mathbb{T}):\mathbf{A}\to\Box\Diamond\mathbf{A}}$$

where the world variable '$\mathbb{T}$' is still free. In order to close the proof, we need to bind that free occurrence of '$\mathbb{T}$' and this means adding an antecedent saying that the universe 'U' is non-empty:

$$\mathsf{U}\to(\mathbf{A}\to\Box\Diamond\mathbf{A})$$

If the universe 'U' of worlds is not empty, then the B-axiom schema holds.

We notice that, similarly to the non-intuitionistic existential quantifier (cf. [de Queiroz and Gabbay, 1995]) the ε-abstraction made at step marked '$(*)$' is a vacuous abstraction — the term 'x' did not have any free occurrence of the world variable '$\mathbb{W}$'. Thus, as we would expect, this axiom schema is not valid in S4 (the 'modal' counterpart to intuitionistic logic).[10]

[10] In [de Queiroz and Gabbay, 1995], we show how to deal with (weakened versions of) classical first-order theorems which are not provable in first-order intuitionistic logic. This is done by allowing

3.6 The Modal Logic S4 and a Parallel with Intuitionistic Logic

In [Gabbay and de Queiroz, 1992], in order to obtain intuitionistic implication from relevant implication we had to introduce the device of 'vacuous' abstraction. By doing this we were then able to construct a derivation of the axiom '$\mathbf{A} \to (\mathbf{B} \to \mathbf{A})$' which is provable in intuitionistic logic but it is not provable in relevant logic.

Now, the parallel between intuitionistic logic and modal logic S4 appears rather naturally in the framework of the functional interpretation: in order to obtain S4 from K (which is characterised by an axiom schema similar to the one which characterises relevant implication), we shall allow vacuous Λ-abstractions on world-variables, i.e. $\Lambda\mathbb{S}$-like abstractions over terms which may not contain '$\mathbb{S}$' free. By extending the Λ-abstraction discipline on world-variables in this way we can construct a derivation of the axiom scheme which is valid in S4 but is not valid in K, namely:

$$4. \quad \Box\mathbf{A} \to \Box\Box\mathbf{A}$$

and the proof-tree will look like:

$$\dfrac{\dfrac{\dfrac{[\mathbb{S} : \mathsf{U}]}{[l : \Box\mathbf{A}]}}{\Lambda\mathbb{S}.l : \Box\Box\mathbf{A}}(*)}{\lambda l.\Lambda\mathbb{S}.l : \Box\mathbf{A} \to \Box\Box\mathbf{A}}$$

which involves a proof step where a vacuous $\Lambda\mathbb{S}$-abstraction is being made, namely the step marked with '$(*)$' where $\mathbb{S}$ does not occur free in the term 'l' over which the $\Lambda\mathbb{S}$-abstraction is being performed.

---

vacuous ε-abstraction in the rule of $\exists$-*introduction*. E.g.:

$$(\forall x^{\mathbf{D}}.\mathbf{P}(x) \to \mathbf{C}) \to \exists y^{\mathbf{D}}.(\mathbf{P}(y) \to \mathbf{C}) \text{ (classical)}$$

is proved as:

$$\dfrac{\dfrac{\dfrac{[t : \mathbf{D}] \qquad \dfrac{\dfrac{\dfrac{\dfrac{[t : \mathbf{D}]}{[f(t) : \mathbf{P}(t)]}}{\Lambda t.f(t) : \forall t^{\mathbf{D}}.\mathbf{P}(t)} \qquad [z : \forall x^{\mathbf{D}}.\mathbf{P}(x) \to \mathbf{C}]}{\mathtt{APP}(z, \Lambda t.f(t)) : \mathbf{C}}}{\lambda f.\mathtt{APP}(z, \Lambda t.f(t)) : \mathbf{P}(t) \to \mathbf{C}}}{\varepsilon y.(\lambda f.\mathtt{APP}(z, \Lambda y.f(y)), t) : \exists y^{\mathbf{D}}.(\mathbf{P}(y) \to \mathbf{C})}(*)}{\lambda z.\varepsilon y.(\lambda f.\mathtt{APP}(z, \Lambda y.f(y)), t) : (\forall x^{\mathbf{D}}.\mathbf{P}(x) \to \mathbf{C}) \to \exists y^{\mathbf{D}}.(\mathbf{P}(y) \to \mathbf{C})}}{\lambda t.\lambda z.\varepsilon y.(\lambda f.\mathtt{APP}(z, \Lambda y.f(y)), t) : \mathbf{D} \to ((\forall x^{\mathbf{D}}.\mathbf{P}(x) \to \mathbf{C}) \to \exists y^{\mathbf{D}}.(\mathbf{P}(y) \to \mathbf{C}))}$$

and the step '$(*)$' is made by means of an $\varepsilon y.$-abstraction over a term which does not contain any free occurrence of the variable 'y'. Note that here we have a weakened version of the classical case where the full classical version is conditional to the domain '$\mathbf{D}$' being non-empty.

3.7 Grzegorczyk's Extension of S4

In [Grzegorczyk, 1967], an extension of S4 which has as theorems precisely the sentences valid on all finite partially ordered frames, is characterised by the introduction of the following schema to the axioms of S4:

$$\text{Grz.}\quad \Box(\Box(\mathbf{A}\to\Box\mathbf{A})\to\mathbf{A})\to\mathbf{A}$$

whose proof construction is the following:

$$\dfrac{\dfrac{\dfrac{\dfrac{\begin{matrix}[\mathbb{S}:\mathsf{U}]\\ [u(\mathbb{S}):(\Box(\mathbf{A}\to\Box\mathbf{A}))(\mathbb{S})]\end{matrix}\quad \dfrac{[\mathbb{S}:\mathsf{U}]\quad [l:\Box(\Box(\mathbf{A}\to\Box\mathbf{A})\to\mathbf{A})]}{\mathcal{EX}(l,\mathbb{S}):(\Box(\mathbf{A}\to\Box\mathbf{A})\to\mathbf{A})(\mathbb{S})}}{\mathtt{APP}(\mathcal{EX}(l,\mathbb{S}),u(\mathbb{S})):\mathbf{A}(\mathbb{S})}\quad \dfrac{\begin{matrix} & [\mathbb{S}:\mathsf{U}]\\ [\mathbb{S}:\mathsf{U}] & [u(\mathbb{S}):(\Box(\mathbf{A}\to\Box\mathbf{A}))(\mathbb{S})]\end{matrix}}{\mathcal{EX}(u(\mathbb{S}),\mathbb{S}):(\mathbf{A}\to\Box\mathbf{A})(\mathbb{S})}}{\mathtt{APP}(\mathcal{EX}(u(\mathbb{S}),\mathbb{S}),\mathtt{APP}(\mathcal{EX}(l,\mathbb{S}),u(\mathbb{S}))):\Box\mathbf{A}(\mathbb{S})}\,(*)}{\Lambda\mathbb{S}.\lambda u.\mathtt{APP}(\mathcal{EX}(u(\mathbb{S}),\mathbb{S}),\mathtt{APP}(\mathcal{EX}(l,\mathbb{S}),u(\mathbb{S}))):\mathbf{A}}}{\lambda l.\Lambda\mathbb{S}.\lambda u.\mathtt{APP}(\mathcal{EX}(u(\mathbb{S}),\mathbb{S}),\mathtt{APP}(\mathcal{EX}(l,\mathbb{S}),u(\mathbb{S}))):\Box(\Box(\mathbf{A}\to\Box\mathbf{A})\to\mathbf{A})\to\mathbf{A}}$$

Now, the interesting characteristic of this extension of S4 is that it differs from all the previous notions of necessity (even S5) in the sense that it admits a 'modal *reductio ad absurdum*', namely the inference step marked with '$(*)$' where both the world variable '$\mathbb{S}$' and the label 'u' were bound because they were used (in conjunction) as major premise and as minor premise: they appear in the term both in the function place and in the argument place, characterising a restricted self-application (a higher-order parallel to our generalised propositional *reductio ad absurdum* defined in [Gabbay and de Queiroz, 1992]). The formula '$(\Box(\mathbf{A}\to\Box\mathbf{A}))(\mathbb{S})$' was used both as minor (in the left hand corner of the proof tree) in an $\to$-*elimination* (*modus ponens*), and as major (in the right hand corner) in an $\Box$-*elimination* (universal *extraction*). And the modal counterpart to our extended propositional *reductio ad absurdum*:

$$\dfrac{\begin{matrix}[x:\mathbf{A}\to\mathbf{B}]\\ b(x,\ldots,x):\mathbf{B}\end{matrix}}{\lambda x.b(x,\ldots,x):\mathbf{A}}\quad \boxed{\begin{matrix}\mathbf{A}\to\mathbf{B}\\ \text{minor + major}\end{matrix}}$$

introduced in [Gabbay and de Queiroz, 1992], appears to reveal the parallel between the propositional and the modal versions of the device of *reductio ad absurdum*. The idea that the rule gives us classical positive implication finds support in the fact that there is a kind of restricted self-application of 'x' to itself, given that it labels a formula that was used both as minor premise and as major premise of an application of *modus ponens*.[11]

The modal counterpart would then be introduced as:

$$\dfrac{\begin{array}{c}[\mathbb{S} : \mathsf{U}] \\ [u : (\Box(\mathbf{A} \to \Box\mathbf{A}))(\mathbb{S})] \\ F(G(\mathbb{S}, u), \ldots, H(\mathbb{S}, u)) : \Box\mathbf{A}(\mathbb{S})\end{array}}{\Lambda\mathbb{S}.\lambda u.F(G(\mathbb{S}, u), \ldots, H(\mathbb{S}, u)) : \mathbf{A}} \quad \boxed{\begin{array}{c}(\Box(\mathbf{A} \to \Box\mathbf{A}))(\mathbb{S}) \\ \text{minor + major}\end{array}}$$

Note that the term '$\texttt{APP}(\mathcal{EX}(u, \mathbb{S}), \texttt{APP}(\mathcal{EX}(l, \mathbb{S}), u))$' in the step '$(*)$', like

$$F(G(\mathbb{S}, u), \ldots, H(\mathbb{S}, u))$$

has both '$\mathbb{S}$' and 'u' involved in a sort of restricted self-application: the functional expression '$\mathcal{EX}(u, \mathbb{S})$' is being applied to another functional expression where '$\mathbb{S}$' and 'u' are themselves the arguments.

We recall that our propositional *RAA* (*reductio ad absurdum*) makes sure that for any propositional formulas '**A**' and '**B**', either '$(\mathbf{A} \to \mathbf{B}) \to \mathbf{A}$' or '$\mathbf{A} \to \mathbf{B}$' is provable. Now, the modal logic S4Grz characterises the set of modal formulas which are valid on all finite partially ordered frames, meaning that for any frame, and for any formula '**A**' in that frame it must be the case that either '**A**' is provable, or that '$\mathbf{A} \to \mathbf{B}$', for some '**B**', is provable.

3.8 The Modal Logic S5 and its Parallel with Classical Logic

Usually, the modal logic S5 is said to differ from the modal logic S4 at least in the sense that the connective of *possibility* is defined as:

$$\Diamond\mathbf{A} \equiv \neg\Box\neg\mathbf{A}$$

for any formula **A**.

Sometimes S5 is also characterised as an extension of S4 where negations of modal formulas can be introduced and later discharged as assumptions in a proof of a modal theorem.[12] In terms of axiom schema this is reflected in that the axiom

[11] With that special proviso we can prove *Peirce's axiom* in the following way:

$$\dfrac{\dfrac{\dfrac{\dfrac{[y : \mathbf{A} \to \mathbf{B}] \quad [x : (\mathbf{A} \to \mathbf{B}) \to \mathbf{A}]}{\texttt{APP}(x, y) : \mathbf{A}} \quad [y : \mathbf{A} \to \mathbf{B}]}{\texttt{APP}(y, \texttt{APP}(x, y)) : \mathbf{B}}}{\lambda y.\texttt{APP}(y, \texttt{APP}(x, y)) : \mathbf{A}}(*)}{\lambda x.\lambda y.\texttt{APP}(y, \texttt{APP}(x, y)) : ((\mathbf{A} \to \mathbf{B}) \to \mathbf{A}) \to \mathbf{A}}$$

where in the step marked with '$(*)$' we have applied our *generalised reductio ad absurdum*. In the resulting term '$\lambda x.\lambda y.\texttt{APP}(y, \texttt{APP}(x, y))$' notice the (restricted) self-application which is being uncovered: 'y' is being applied to the result of an application of 'x' to 'y' itself.

[12] Here we are referring to Prawitz' proviso that 'an instance of the deduction rule for NI [Necessity Introduction] in S4 (resp. S5) is to have the form $\langle\langle\Gamma, A\rangle, \langle\Gamma, NA\rangle\rangle$, where every formula of Γ is modal (resp. modal or *he negation of a modal formula*)' [Prawitz, 1965, page 74].

characterising S4, namely:

$$\Box \mathbf{A} \to \Box\Box \mathbf{A}$$

becomes:

$$\Box\neg \mathbf{B} \to \Box\Box\neg \mathbf{B}$$

for any formula **B**, which can be rewritten to the more commonly used axiom schema:

$$\Diamond \mathbf{C} \to \Box\Diamond \mathbf{C}$$

by putting $\Box \equiv \neg\Diamond\neg$ and $\neg\Diamond\mathbf{B} \equiv \Diamond\mathbf{C}$. Here we want to see S5 through its parallel with first-order classical logic, and for that we intend to work with the following equivalent axiom schema:

$$5. \quad \Diamond \mathbf{P} \to \Box\neg\Box\neg \mathbf{P} \qquad (*)$$

whose translation into a formula with quantification over world variables gives us:

$$\exists \mathbb{S}^{\mathsf{U}}.\mathbf{P}(\mathbb{S}) \to \forall \mathbb{T}^{\mathsf{U}}.(\forall \mathbb{W}^{\mathsf{U}}.(\mathbf{P}(\mathbb{W}) \to \mathcal{F}) \to \mathcal{F})$$

where $\neg\mathbf{P}(\mathbb{S})$ is written as $\mathbf{P}(\mathbb{S}) \to \mathcal{F}$, '$\mathcal{F}$' being the *falsum*.

The proof-tree can be constructed in the following way:

$$\cfrac{\cfrac{\cfrac{[\mathbb{W}:\mathsf{U}]}{[u:\Diamond\mathbf{P}]} \qquad \cfrac{\cfrac{\cfrac{\cfrac{[\mathbb{T}:\mathsf{U}]}{[G(\mathbb{T}):\mathbf{P}(\mathbb{T})]}}{\Lambda\mathbb{T}.G(\mathbb{T}):\Box\mathbf{P}} \qquad [z:\Box\mathbf{P}\to\mathcal{F}]}{\mathtt{APP}(z,\Lambda\mathbb{T}.G(\mathbb{T})):\mathcal{F}}}{\lambda z.\mathtt{APP}(z,\Lambda\mathbb{T}.G(\mathbb{T})):(\Box\mathbf{P}\to\mathcal{F})\to\mathcal{F}}}{\cfrac{\mathtt{INST}(u,\sigma G.\sigma\mathbb{T}.\lambda z.\mathtt{APP}(z,\Lambda\mathbb{T}.G(\mathbb{T}))):(\Box\mathbf{P}\to\mathcal{F})\to\mathcal{F}}{\Lambda\mathbb{W}.\mathtt{INST}(u,\sigma G.\sigma\mathbb{T}.\lambda z.\mathtt{APP}(z,\Lambda\mathbb{T}.G(\mathbb{T}))):\Box((\Box\mathbf{P}\to\mathcal{F})\to\mathcal{F})}}}{\lambda u.\Lambda\mathbb{W}.\mathtt{INST}(u,\sigma G.\sigma\mathbb{T}.\lambda z.\mathtt{APP}(z,\Lambda\mathbb{T}.G(\mathbb{T}))):\Diamond\mathbf{P}\to\Box((\Box\mathbf{P}\to\mathcal{F})\to\mathcal{F})}$$

and here we note that the assumption '$z : \Box\mathbf{P} \to \mathcal{F}$' is not a (fully) modalised formula but the negation of a fully modalised formula.

3.9 *Problematic Cases in Predicate Modal Logic: the Scope of '□'*

Here we shall attempt to demonstrate that the aspect of the methodology enforced by the functional interpretation, namely that which involves introducing names for 'worlds' (world variables), might help throwing some light on issues which have been traditionally looked at mostly from the denotational semantics point of view.

The Barcan Formula. The Barcan formula is said to be true in the denotational semantics approach when the domain of individuals is either constant or decreasing. We can perhaps understand this interpretation by constructing a proof-tree as in the previous case.

The Barcan formula is the following:

$$\forall x_1^{\mathbf{D}}.\Box\mathbf{P}(x) \to \Box\forall x_2^{\mathbf{D}}.\mathbf{P}(x)$$

which can be proved using our methodology as follows:

$$\dfrac{\dfrac{\dfrac{\dfrac{[\mathbb{S}:\mathsf{U}] \quad \dfrac{[t:\mathbf{D}_1] \quad [u:\forall x^{\mathbf{D}_1}.\Box\mathbf{P}(x)]}{\mathtt{EXTR}(u,t):\Box\mathbf{P}(t)}}{\mathcal{EX}(\mathtt{EXTR}(u,t),\mathbb{S}):\mathbf{P}(t)(\mathbb{S})}}{\Lambda t.\mathcal{EX}(\mathtt{EXTR}(u,t),\mathbb{S}):(\forall t^{\mathbf{D}_2}.\mathbf{P}(t))(\mathbb{S})}(*)}{\Lambda\mathbb{S}.\Lambda t.\mathcal{EX}(\mathtt{EXTR}(u,t),\mathbb{S}):\Box\forall t^{\mathbf{D}_2}.\mathbf{P}(t)}}{\lambda u.\Lambda\mathbb{S}.\Lambda t.\mathcal{EX}(\mathtt{EXTR}(u,t),\mathbb{S}):\forall x^{\mathbf{D}_1}.\Box\mathbf{P}(x) \to \Box\forall t^{\mathbf{D}_2}.\Box\mathbf{P}(t)}$$

and, similarly to the case below, the universal quantification (step '$(*)$') over the domain of individuals '$\mathbf{D}_2$' can only be safely made if this domain is contained in the domain '$\mathbf{D}_1$'. Unlike the previous case, the domain of individuals must either remain constant or else decrease as one moves from the 'current' world to any next possible world.

The Converse of the Barcan Formula. Another example is the so-called Converse of the Barcan formula, namely:

$$\Box\forall x^{\mathbf{D}_1}.\mathbf{P}(x) \to \forall x^{\mathbf{D}_2}.\Box\mathbf{P}(x),$$

where the debate concerns the relation between $\mathbf{D}_1$ and $\mathbf{D}_2$, the domains of individuals. Let us construct a proof tree for the Converse of the Barcan formula using the proof methodology advocated in this paper:

$$\dfrac{\dfrac{\dfrac{\dfrac{[t:\mathbf{D}_1(\mathbb{S})] \quad \dfrac{[\mathbb{S}:\mathsf{U}] \quad [l:\Box\forall x^{\mathbf{D}_1}.\mathbf{P}(x)]}{\mathcal{EX}(l,\mathbb{S}):(\forall x^{\mathbf{D}_1}.\mathbf{P}(x))(\mathbb{S})}}{\mathtt{EXTR}(\mathcal{EX}(l,\mathbb{S}),t):\mathbf{P}(t)(\mathbb{S})}}{\Lambda\mathbb{S}.\mathtt{EXTR}(\mathcal{EX}(l,\mathbb{S}),t):\Box\mathbf{P}(t)}}{\Lambda t.\Lambda\mathbb{S}.\mathtt{EXTR}(\mathcal{EX}(l,\mathbb{S}),t):\forall t^{\mathbf{D}_2}.\Box\mathbf{P}(t)}(*)}{\lambda l.\Lambda t.\Lambda\mathbb{S}.\mathtt{EXTR}(\mathcal{EX}(l,\mathbb{S}),t):\Box\forall x^{\mathbf{D}_1}.\mathbf{P}(x) \to \forall t^{\mathbf{D}_2}.\Box\mathbf{P}(t)}$$

where in the step marked with '$(*)$' it is clear that the universal quantification over $\mathbf{D}_2$' can only be made if all elements arbitrarily chosen from '$\mathbf{D}_1$' also inhabit '$\mathbf{D}_2$'. In other words, for the Converse of the Barcan formula to be true, the domain of individuals must be either constant or increase as one moves from the 'current' world to any next possible world.

4 FINALE

Parallel to the functional interpretation of the (first-order) existential quantifier (which is developed in [de Queiroz and Gabbay, 1995]), we are also investigating how to characterise the functional interpretation of the modal connective of *possibility*.[13] (In fact, we have already given the reader some idea of how the parallel between the $\exists$ and the $\Diamond$ is dealt with when we discussed the modal logic S5 above.) By introducing names for worlds in conjunction with procedures for Skolemising on world names, we shall also be able to demonstrate how to obtain natural extensions of fundamental theorems, such as Skolem's and Herbrand's theorems, to modal logics.

Ruy J.G.B. de Queiroz
Departamento de Informática
Universidade Federal de Pernambuco (UFPE)

Dov M. Gabbay
Department of Computing
Imperial College

Acknowledgements. Many thanks are due to the anonymous referees for their excellent suggestions for improvement of the content and the presentation of the paper.

Part of the research was carried out in the context of the ESPRIT BRA MEDLAR (Mechanised Deduction in the Logics of Practical Reasoning). Ruy de Queiroz wishes to thank the Brazilian Research Council CNPq for his 'Pesquisador

[13] In [Prawitz, 1965] the parallel between the *introduction* and *elimination* rules for the *possibility* connective and those for the first-order existential quantifier is already suggested [Prawitz, 1965, page 75]: "One may add an additional modal operator, $\Diamond$ expressing *possibility*, and inference rules indicated by the figures:

$$\Diamond\text{I})\ \frac{A}{\Diamond A} \qquad \Diamond\text{E})\ \frac{\Diamond A \qquad \begin{matrix}(A)\\ B\end{matrix}}{B}.$$

By using a similar pattern to the presentation of our $\exists$-type of ε-terms developed in [de Queiroz and Gabbay, 1995], and considering that now we are dealing with names of 'worlds' instead of names of (first-order) individuals, we put: $\Diamond\mathbf{A} \equiv \exists \mathtt{W}^{\mathsf{U}}.\mathbf{A}(\mathtt{W})$ and we have the following counterparts to Prawitz' rules for *possibility*:

$\Diamond$-*introduction* $\qquad$ $\Diamond$-*elimination*

$$\frac{\mathtt{S} : \mathsf{U} \qquad F(\mathtt{S}) : \mathbf{A}(\mathtt{S})}{\varepsilon\mathtt{W}.(F(\mathtt{W}), \mathtt{S}) : \Diamond\mathbf{A}} \qquad \frac{e : \Diamond\mathbf{A} \qquad \begin{matrix}[\mathtt{T} : \mathsf{U}, G(\mathtt{T}) : \mathbf{A}(\mathtt{T})]\\ D(G(\mathtt{T}), \mathtt{T}) : \mathbf{C}\end{matrix}}{\mathcal{INST}(e, \sigma G.\sigma\mathtt{T}.D(G(\mathtt{T}), \mathtt{T})) : \mathbf{C}}$$

1C' (Senior Research Fellow) grant. Dov Gabbay acknowledges a Senior Research Fellowship from the British 'Engineering and Physical Sciences Research Council' (EPSRC).

REFERENCES

[Belnap, 1982] N. Belnap. Display logic. *Journal of Philosophical Logic*, 11:14–25, 1982.

[Benevides and Maibaum, 1992] M. Benevides and T. Maibaum. A constructive presentation for the modal connective of necessity (□). *Journal of Logic and Computation*, 2:31–50, 1992.

[Chellas, 1980] B.F. Chellas. *Modal Logic*. Cambridge University Press, 1980.

[Curry, 1934] H.B. Curry. Functionality in combinatory logic. *Proceedings of the National Academy of Sciences of USA*, 20:584–590, 1934.

[de Queiroz, 1992] R.J.G.B. de Queiroz. *Grundgesetze* alongside *Begriffsschrift* (abstract). In *Abstracts of Fifteenth International Wittgenstein Symposium*, pages 15–16, 1992.

[de Queiroz, 1994] R.J.G.B. de Queiroz. Normalisation and language-games. *Dialectica*, 48:83–123, 1994.

[de Queiroz and Gabbay, 1995] R.J.G.B. de Queiroz and D.M. Gabbay. The functional interpretation of the existential quantifier. *Bulleting of the IGPL*, 3(3):243–290, 1995.

[de Queiroz and Gabbay, forthcoming] R.J.G.B. de Queiroz and D.M. Gabbay. An introduction to labelled natural deduction. To appear in *Festschrift for Dov Gabbay*, Kluwer Academic Publishers.

[Fariñas del Cerro and Herzig, 1990] K. Fariñas del Cerro and A. Herzig. Modal deduction with applications in epistemic and temporal logics. Research report, LSI-IRIT, Toulouse, 1990.

[Fine, 1985] K. Fine. *Reasoning with Arbitrary Objects*, volume 3 of *Aristotelian Society series*. Basil Blackwell, Oxford, 1985.

[Fitch, 1966a] F.B. Fitch. Natural deduction rules for obligation. *American Philosophical Quarterly*, 3:27–38, 1996.

[Fitch, 1966b] F.B. Fitch. Tree proofs in modal logic. *Journal of Symbolic Logic*, 31:152, 1966.

[Fitting, 1972] M. Fitting. An epsilon-calculus system for first-order S4. *Lecture Notes in Mathematics*, pages 103–110. Springer-Verlag, 1972.

[Fitting, 1975] M. Fitting. A modal logic epsilon-calculus. *Notre Dame Journal of Formal Logic*, 16:1–16, 1975.

[Fitting, 1981] M. Fitting. *Proof Methods for Modal and Intuitionistic Logics*, volume 169 of *Synthese Library*. D. Reidel, Dordrecht, 1981.

[Fitting, 1989] M. Fitting. Modal logic should say more than it does. In J. Lassez and G. Plotkin, editors, *Computational Logic. Essays in Honor of A. Robinson*. MIT Press, 1989.

[Fitting, 1993] M. Fitting. Modal logic. In D. Gabbay, C. Hogger, and J. Robinson, editors, *Handbook of Logic in Artificial Intelligence and Logic Programming. Vol. I: Logical Foundations*, pages 65–448. Oxford University Press, 1993.

[Frege, 1893] G. Frege. *Grundgesetze der Arithmetik. Begriffsschriftlich abgeleitet. I.* Verlag von Hermann Pohle, Jena, 1893.

[Frege, 1903] G. Frege. *Grundgesetze der Arithmetik. Begriffsschriftlich abgeleitet. II.* Verlag von Hermann Pohle, Jena, 1903.

[Gabbay, forthcoming] D.M. Gabbay. *Labelled Deductive Systems. Part I: Foundations*. Oxford University Press, forthcoming.

[Gabbay and de Queiroz, 1992] D.M. Gabbay and R.J.G.B. de Queiroz. Extending the Curry-Howard intepretation to linear, relevant and other resource logics. *Journal of Symbolic Logic*, 57:1319–1365, 1992.

[Gabbay and Hodkison, 1990] D.M. Gabbay and I. Hodkinson. An axiomatization of the temporal logic with Until and Since over the real numbers. *Journal of Logic and Computation*, 1:229–259, 1990.

[Gentzen, 1935] G. Gentzen. Untersuchungen über das logische Schliessen. *Mathematische Zeitschrift*, 39:176–210 and 405–431, 1935.

[Girard, 1987] J.-Y. Girard. Linear logic. *Theoretical Computer Science*, 50:1–102, 1987.

[Gödel, 1933] K. Gödel. Eine intepretation des intuitionistischen aussagenkalküls. *Ergebnisse eines mathematischen Kolloquiums*, 4:39–40, 1933.

[Grzegorczyk, 1967] A. Grzegorczyk. Some relational systems and the associated topological spaces. *Fundamenta Mathematicae*, 60:223–231, 1967.

[Hintikka, 1983] K.J. Hintikka. Game-theoretical semantics: Insights and prospects. In J. Hintikka and J. Kulas, editors, *The Game of Language*, volume 22 of *Synthese Language Library*, pages 1–31. D. Reidel, Dordrecht, 1983.

[Howard, 1980] W.A. Howard. The formulae-as-types notion of construction. In J.P. Seldin and J.R. Hindley, editors, *To H. B. Curry: Essays on Combinatory Logic, Lambda Calculus and Formalism*, pages 79–490. Academic Press, London, 1980.

[Hughes and Cresswell, 1968] G. Hughes and M. Cresswell. *Introduction to Modal Logic*. Methuen, London, 1968.

[Kripke, 1959] S. Kripke. A completeness theorem in modal logic. *Journal of Symbolic Logic*, 24:1–14, 1959.

[Kripke, 1963] S. Kripke. Semantic analysis of modal logic. I: normal propositional calculi. *Zeitschrift für mathematische Logik und Grundlagen der Mathematik*, 9:67–96, 1963.

[Kripke, 1965] S. Kripke. Semantical analysis of modal logic II: non-normal modal pro positional calculi. In J.W. Addison, L. Henkin, and A. Tarski, editors, *The Theory of Models*, pages 206–220. North-Holland, Amsterdam, 1965.

[Lambek, 1958] J. Lambek. The mathematics of sentence structure. *American Mathematical Monthly*, 65:154–170, 1958.

[Lewis and Langford, 1932] C.I. Lewis and C.H. Langford. *Symbolic Logic*. The Century Co., New York, 1932.

[Lorenzen, 1961] P. Lorenzen. Ein dialogisches Konstruktivitätskriterium. In *Infinitistic Methods*, Pergamon Press, Oxford, 1961.

[Lorenzen, 1969] P. Lorenzen. *Normative Logic and Ethics*. B.I. Hochschultaschenbücher. Systematische Philosophie, vol. 236*. Bibliographisches Institut, Mannheim/Zürich, 1969.

[Martin-Löf, 1972] P. Martin-Löf. Infinite terms and a system of natural deduction. *Compositio Mathematica*, 24:93–103, 1972.

[Masini, 1992] A. Masini. 2-Sequent calculus: a proof theory of modalities. *Annals of Pure and Applied Logic*, 58:229–246, 1992.

[Montague, 1970] R. Montague. Universal grammar. *Theoria*, 36:373–398, 1970.

[Ohlbach, 1991] H.-J. Ohlbach. Semantics based translation methods for modal logics. *Journal of Logic and Computation*, 1:691–746, 1991.

[Prawitz, 1965] D. Prawitz. *Natural Deduction. A Proof-Theoretical Study*, volume 3 of *Acta Universitatis Stockholmiensis*. Almqvist & Wiksell, Stockholm, 1965.

[Sahlqvist, 1975] H. Sahlqvist. Completeness and correspondence in the first and second order semantics for modal logic. In S. Kanger, editor, *Proceedings of the Third Scandinavian Logic Symposium. Uppsala 1973*, pages 110–143, North-Holland, Amsterdam, 1975.

[Tait, 1965] W.W. Tait. Infinitely long terms of transfinite type. In J.N. Crossley and M.A.E. Dummett, editors, *Formal Systems and Recursive Functions*, Studies in Logic and The Foundations of Mathematics, pages 176–185, North-Holland, Amsterdam, 1965.

[Tait, 1967] W.W. Tait. Intensional interpretations of functionals of finite type I. *Journal of Symbolic Logic*, 32:198–212, 1967.

[Thomason and Stalnaker, 1968] R. Thomason and R. Stalnaker. Modality and reference. *Noûs*, 2:359–372, 1968.

[Wansing, 1994] H. Wansing. Sequent calculi for normal propositional logics. *Journal of Logic and Computation*, 4:125–142, 1994.

VLADIMIR V. RYBAKOV

LOGICS OF SCHEMES FOR FIRST-ORDER THEORIES AND POLY-MODAL PROPOSITIONAL LOGIC

INTRODUCTION

Schemes of theorems are a well-known object of study in logic; general properties of first-order theories are expressible exactly by formula schemes. The study of such schemes goes back to the famous monograph [Henkin *et al.*, 1971] which contains a thorough investigation of this field from an algebraic point of view. Monk's [1965] study of substitutionless predicate logic goes in the same direction, as does the study of algebraic counterparts of logical systems (cf. [Blok and Pigozzi, 1989]). A significant part of the (algebraic) results about formula schemes known at present is contained in [Németi, 1987].

The content of this paper is closely connected to the investigations mentioned above, but it aims at the study of schemes of theorems in a more logical manner. The algebraic approach considers varieties of cylindric and diagonal-free algebras; varieties are generated by free cylindric (diagonal-free) algebras corresponding to the predicate calculus. Our approach is based on a simple observation that the set of all formula schemes for a given first-order theory forms a poly-modal propositional logic. It is natural to clarify the recursive complexity of these logics, and to study which semantics is adequate for them etc. Another interest in logics of schemes emerges from the possibility to describe the notion of admissible inference rules within a poly-modal language. The study of inference rules can be motivated by various arguments (cf. [Fagin *et al.*, 1992], for instance). And a number of results about admissible inference rules for propositional logics have so far been obtained. For instance, Harvey Friedman's problem [Friedman, 1975] (whether there exists an algorithm which recognizes admissible inference rules of intuitionistic propositional logic) has been affirmatively solved in [Rybakov, 1989]. Some related problems for modal logics are studied in [Rybakov, 1990b, 1987].

New inference rules enable us to build new, more expressive formal systems. For instance, [Venema, 1991] presents a detailed study of cylindric modal logics; completeness results obtained by Venema give rise to axiomatizations for some logics of schemes. The question of axiomatizing frame classes in (multi-) modal logic by means of derivation rules has been considered in [Venema, 1993], where a metatheorem on completeness is given. Thus, inference rules for propositional logics form an active domain of research. And it would be interesting to study

M. de Rijke (ed.), Advances in Intensional Logic, 93–106.

questions about admissible inference rules for first-order theories having a much more expressible language.

In this short paper we study each of the questions mentioned above. We find a complete description for first-order theories which have an algorithm for recognizing admissibility. Furthermore, it is shown that almost all first-order theories have no algorithm recognizing admissible rules. As a corollary, a description for first-order theories with a decidable logic of schemes is obtained. We also consider the connection between the complexity of a first-order theory and the complexity of its logic of schemes. We give examples of decidable, finitely axiomatizable first-order theories without recursively enumerable logic of schemes. In addition, we define a Kripke style semantics for the logic of schemes of pure predicate calculus (predicate calculus without equality). Some results of this paper can be extracted from [Henkin *et al.*, 1971] and [Németi, 1987]. However, our proofs are based on other ideas and, in a sense, they are shorter and more straightforward. Some results of this paper were announced in [Rybakov 1990a, 1991].

1 BACKGROUND

We assume the reader to be familiar with the basic concepts, definitions and notations of first-order logic and propositional modal logic (see, for instance [Chang and Keisler, 1973; Rautenberg, 1979; Goldblatt, 1976; Fine, 1974]).

What is a scheme of theorems for any first-order theory T? As is well known, a scheme is an expression (or formula) of the form $A(z_i)$ which is built up from metavariables z_i by means of logical connectives, quantifiers and equalities between variables with the following property. For any tuple of formulas F_i in the first-order language of the theory T, $A(F_i) \in T$ holds. Therefore, we introduce the language LS of schemes of theorems in the following way. It has a countable set of metavariables z_i for formulas, the usual logical connectives, a countable set of unary logical connectives $\forall x_i$ (which simulate the universal quantifier) and a countable set of 0-place logical connectives (or logical constants) $(x_i = x_j)$ (which simulate equality). The formulas of LS are built up from the metavariables by using the connectives mentioned above in the usual way. So

$$\forall x_1(z_1 \,\&\, z_2) \equiv (\forall x_1 z_1) \,\&\, (\forall x_1 z_2), \quad \forall x_1 \exists x_1 z_1 \to \exists x_1 z_1$$

are some formulas of LS.

DEFINITION 1.1 The *logic of schemes* $LS(T)$ of a first-order theory T is the set

$$\{A(z_i) \mid A(F_i) \in T \text{ for arbitrary formulas } F_i \text{ in the language of } T\}.$$

We can view each logic $LS(T)$ as a propositional logic with propositional variables z_i and the logical connectives mentioned above (including the logical constants). The logical connectives $\forall x_i$ can be viewed as modal *necessity* connectives

$\Box_i$. It is clear that each $\Box_i$ satisfies all laws of the modal system $S4$. Moreover, for each theory T we evidently have

$$\begin{array}{rcl}(\exists x_i z_1 \to \forall x_i \exists x_i z_1) & \in & LS(T) \\ (\forall x_i \forall x_j z_1 \equiv \forall x_j \forall x_i z_1) & \in & LS(T).\end{array}$$

This immediately yields

PROPOSITION 1.2 *For any first-order theory T, LS(T) forms a polymodal propositional logic with constants which extends the poly-modal analog $S5_\infty$ (with constants) of Lewis' modal logic $S5$ with the law of commutating modalities $\Box_i\Box_j p \equiv \Box_j\Box_i p$. In other words:*

$$S5_\infty + (\Box_i\Box_j p \equiv \Box_j\Box_i p) \subseteq LS(T).$$

An *inference rule* for a logic (or theory) T is an expression of the form

$$\frac{A_1(z_i), \ldots, A_n(z_i)}{B(z_i)}$$

(also written as $A_1(z_i), \ldots, A_n(z_i)/B(z_i)$), where $A_j(z_i)$, $B(z_i)$ are formulas of LS.

DEFINITION 1.3 An inference rule $A_1(z_i), \ldots, A_n(z_i)/B(z_i)$ is *admissible* for the logic (or theory) T if for each tuple of formulas B_i in the language of T we have

$$A_1(B_i) \in T, \ldots, A_n(B_i) \in T \Rightarrow B(B_i) \in T.$$

It is clear that admissible inference rules are exactly the ones which we can add to the set of postulated inference rules of T while preserving the set of its theorems (but increasing the power of deductive tools).

Of course, the rule $z \equiv z/A(z_i)$ is admissible in T iff $A(z_i) \in LS(T)$. Therefore, if $LS(T)$ is undecidable then there is no algorithm recognizing admissibility in T. This is in fact our second motivation to study $LS(T)$.

2 LOGICS OF SCHEMES

DEFINITION 2.1 Let M_1, M_2 be models of signature Σ, and let g be a homomorphism from M_1 onto M_2. We call g *strong* if for all predicates $P_i(x_1, \ldots, x_m)$ of Σ we have

$$M_1 \models P_i(a_1, \ldots, a_m) \text{ if and only if } M_2 \models P_i(g(a_1), \ldots, g(a_m)).$$

The next lemma is well-known (and can easily be verified).

LEMMA 2.2 *Let g be a strong homomorphism of M_1 onto M_2, and let A be a formula that has no occurrences of equality, then $M_1 \models A(a_1, \ldots, a_m)$ iff $M_2 \models A(g(a_1), \ldots, g(a_m))$.*

Let K_1, K_2 be classes of models of signatures Σ_1 and Σ_2, respectively, where Σ_1 has no constants or functional symbols.

DEFINITION 2.3 We say that K_1 is *weakly reflected* in K_2 if for each model $M_1 \in K_1$ there exists a model $M_2 \in K_2$ and there is a tuple $a_1, \ldots, a_m$ of elements of M_2 such that the following holds.

> For all predicates $P_i(v_1, \ldots, v_n)$ from Σ_1, there are some first-order formulas $f_i(v_1, \ldots, v_n, y_1, \ldots, y_m)$ in the signature Σ_2 with the following property. There is a strong homomorphism of the model
>
> $$\langle M_2, f_i(v_1, \ldots, v_n, a_1, \ldots, a_m)\rangle$$
>
> (where the f_i interpret the predicates P_i) onto the model M_1. (Of course, not all of the variables or constants displayed in the above formulas need to occur.)

Readers experienced in model theory will easily recognize a special case of the notion of relatively elementary definability in the above definition.

We recall that sets X, Y, where $X \subseteq Y$, are called *recursively undetached* if there is no decidable set Z such that $X \subseteq Z \subseteq Y$. Let K be a class of algebraic systems (or models). Then $Th(K)$ denotes the elementary theory of K. $Th_p(K)$ denotes the pure elementary theory of K (the set of all first-order formulas without equality, in the signature of K, which are true on K); PPC_Σ will denote the pure predicate calculus (without equality) of the signature Σ (without functional symbols).

Let K_1 be a class of models in the signature Σ_1 which has no functional and constant symbols.

THEOREM 2.4 *If K_1 is weakly reflected in a class K_2, and $Th_p(K_1)$ and PPC_{Σ_1} are recursively undetached, then $LS(Th(K_2))$ is undecidable.*

Proof. Let A be a formula without equality in the signature Σ_1, let $Var(A)$ be the variables having occurrences in A, and let $P_i(x_{j1}, \ldots, x_{jn})$ be all predicate letters with given occurrences of variables from A. We define a translation of A into a formula A^t of the logic of schemes, where

$$A^t := (P(A) \to V(A))$$

and $V(A)$ is obtained from A by replacing all formulas $P_i(x_{j1}, \ldots, x_{jn})$ by the corresponding meta-variables $z_{P_i(x_{j1},\ldots,x_{jn})}$, and $P(A)$ is the conjunction over all $z_{P_i(x_{j1},\ldots,x_{jn})}, z_{P_i(x_{k1},\ldots,x_{kn})}$ of all formulas

$$(1)\quad \forall x_{j1}\ldots\forall x_{jn}\,[\forall x_1\ldots\forall x_d z_{P_i(x_{j1},\ldots,x_{jn})} \Longleftrightarrow \exists x_1\ldots\exists x_d z_{P_i(x_{j1},\ldots,x_{jn})}],$$

where $\{x_1,\ldots,x_d\} := Var(A) - \{x_{j1},\ldots,x_{jn}\}$, and

$$(2)\quad \begin{aligned}&\forall x_{j1}\ldots\forall x_{jn}\forall x_{k1}\ldots\forall x_{kn}\forall y_1\ldots\forall y_p\,[\bigwedge(x_{jr}=x_{kr})\rightarrow\\ &\qquad(z_{P_i(x_{j1},\ldots,x_{jn})} \Longleftrightarrow z_{P_i(x_{k1},\ldots,x_{kn})})],\end{aligned}$$

where $\{y_1,\ldots,y_p\} := Var(A) - \{x_{j1},\ldots,x_{jn},x_{k1},\ldots,x_{kn}\}$.
We need the following

LEMMA 2.5 *If A is a theorem of* PPC$_{\Sigma_1}$ *then for each first-order theory T, A^t belongs to $LS(T)$.*

Proof. Suppose that $A^t \notin LS(T)$. This means that there is a tuple

$$B_{P_i(x_{j1},\ldots,x_{jn})}$$

of first-order formulas in the signature of the theory T with the following property. Consider the formula

$$(A^t)(B_{P_i(x_{j1},\ldots,x_{jn})} \hookleftarrow z_{P_i(x_{j1},\ldots,x_{jn})})$$

which is obtained from A^t by substituting the formulas $B_{P_i(x_{j1},\ldots,x_{jn})}$ for the meta-variables $z_{P_i(x_{j1},\ldots,x_{jn})}$; it is not a theorem of T. Therefore, there is a model M of T, and there is an interpretation $y_l \mapsto b_l$ of the free variables of

$$(A^t)(B_{P_i(x_{j1},\ldots,x_{jn})} \hookleftarrow z_{P_i(x_{j1},\ldots,x_{jn})})$$

into M such that

$$M \not\models (A^t)(B_{P_i(x_{j1},\ldots,x_{jn})} \hookleftarrow z_{P_i(x_{j1},\ldots,x_{jn})})(b_l).$$

Then the premise of the formula on the right-hand side of the relation above is true on M. We introduce new predicate relations $Q_i(x_1,\ldots,x_n)$ on M by

$$\begin{aligned}&\forall c_{j1}\in M,\ldots,\forall c_{jn}\in M(M \models Q_i(c_{j1},\ldots,c_{jn})) \Longleftrightarrow\\ &\qquad M \models B_{P_i(x_{j1},\ldots,x_{jn})}\binom{x_{j1},\ldots,x_{jn}}{c_{j1},\ldots,c_{jn}}\binom{y}{c}\binom{y_l}{b_l},\end{aligned}$$

where

- y are any variables from $(Var(A) - \{x_{j1},\ldots,x_{jn}\})$,
- y_l are all free variables from the formula $B_{P_i(x_{j1},\ldots,x_{jn})}$ which differ with variables from $Var(A)$,
- c is a fixed element of M,
- b_l are given above interpretation of y_l,

Taking into account both the form of (1) and (2) and the fact that the premise is true, it can easily be verified that this definition is correct. (In particular, it does not depend on the choice of the variables $x_{j1}, \ldots, x_{jn}$ in $B_{P_i(x_{j1},\ldots,x_{jn})}$.) So the truth of Q_i in M is expressible by means of formulas $B_{P_i(x_{j1},\ldots,x_{jn})}(y_l)$. The formula obtained from $V(A)$ by substituting $B_{P_i(x_{j1},\ldots,x_{jn})}(y_l)$ for the metavariables $z_{P_i(x_{j1},\ldots,x_{jn})}$ is not true in M under the given interpretation $y_l \mapsto b_l$. This implies that the formula A is not true on the model of the signature Σ_1 which is obtained from M by introducing predicates $Q_i(x_1, \ldots, x_n)$. But this contradicts the fact that the formula A is a theorem of PPC_{Σ_1}. ⊣

To continue the proof of the theorem, we suppose that $LS(Th(K_2))$ is decidable. The translation t is a one to one map that is effectively defined. Therefore the set $t^{-1}(LS(Th(K_2)))$ is decidable too. We proceed to prove that

(3) $t^{-1}(LS(Th(K_2))) \subseteq Th_p(K_1)$

Indeed, let A be a member of the set on left-hand side. It follows that $A^t \in LS(Th(K_2))$. Suppose that A does not belong to $Th_p(K_1)$. Then there exists a model M_1 from K_1 on which this formula is not true: $M_1 \not\models A(b_r)$. According to the supposition, K_1 is weakly reflected in K_2. Now, M_1 is given. Therefore there is a model M_2 from K_2 (related to M_1) with properties as required by the definition of weak reflection. In particular, there is a strong homomorphism g of the model $\langle M_2, f_i(v_1, \ldots, v_n, a_1, \ldots, a_m)\rangle$ onto the model M_1. According to Lemma 2.2 we have that

$$\langle M_2, f_i(v_1, \ldots, v_n, a_1, \ldots, a_m)\rangle \not\models A(g(b_r)).$$

The latter immediately yields

$$\langle M_2, f_i(v_1, \ldots, v_n, a_1, \ldots, a_m)\rangle \not\models A^{P_i}_{(f_i(v_1,\ldots,v_n,a_1,\ldots,a_m))}(g(b_r)).$$

We take the interpretation of the meta-variables from A^t to be the following:

$$z_{P_i(x_{j1},\ldots,x_{jn})} \mapsto f_i(x_{j1}, \ldots, x_{jn}, a_1, \ldots, a_m).$$

It can easily be verified directly that the premise of A^t will be true on the model $\langle M_2, f_i(v_1, \ldots, v_n, a_1, \ldots, a_m)\rangle$ under this interpretation of the meta-variables and the above interpretation of the free variables. However, the conclusion of A^t is not true as we have seen above. Therefore, the formula A^t does not belong to the set $LS(Th(K_2))$, which gives us a contradiction. So the inclusion (3) holds. From this and Lemma 2.5 we obtain that

(4) $\mathrm{PPC}_{\Sigma_1} \subseteq t^{-1}(LS(Th(K_2))) \subseteq Th_p(K_1)$,

which contradicts to the recursive undetachment of the first and last set in (4).

This completes the proof of the theorem. ⊣

We will now deduce some corollaries from this theorem. First, it is interesting to compare the complexity of a first-order theory T and its logics of schemes $LS(T)$. At first glance, $LS(T)$ is much simpler than T because $LS(T)$ is just a poly-modal propositional logic which extends the poly-modal analog of the logic $S5$ (which, in a sense, is the simplest modal logic). Moreover, the logic of schemes does not have the ability to express concrete properties of the models, because all predicates letters are removed from the language. Surprisingly enough, all that does not really matter. To be more precise, let us consider the following theorems.

THEOREM 2.6 *Let K be a class of models and assume that for all $m \in \mathbb{N}$ there is a model M in K such that K has at least m elements. Then $LS(Th(K))$ is undecidable.*

Proof. Consider the class K_0 of all finite models of a binary irreflexive symmetric predicate. It is well-known that the pure first-order elementary theory $Th_p(K_0)$ of this class and the pure predicate calculus PPC of a binary predicate letter are recursively undetached (cf., for instance [Ershov *et al.*, 1965]). Now we will use Theorem 2.4. To this end it suffices to show that the class K_0 is weakly reflected in the class K.

Let $M_0 = \langle\{1, \ldots, m\}, P(x, y)\rangle$ be a model of K_0 . By supposition, there is a model M from K which has m different elements, say $a_1, \ldots, a_m$. We introduce the set R and the formula F as follows

$$\begin{aligned} R &:= \{j \mid P(1, j) \wedge (1 \leq i \leq m)\} \\ F(x, y, a_1, \ldots, a_m) &:= \bigvee\{(x = a_i) \wedge (y = a_j) \mid P(i, j)\} \vee \\ &\quad (\bigwedge_{a_i} \neg(x = a_i) \wedge \bigvee_{j \in R} (y = a_j)) \vee \\ &\quad (\bigwedge_{a_i} \neg(y = a_i) \wedge \bigvee_{j \in R} (x = a_j)). \end{aligned}$$

Furthermore, we define the map g of M onto the given model M_0 from K_0 as follows: $g(a_i) := i$, and for all $b \neq a_i$, $1 \leq i \leq m$, we put $g(b) := 1$. It is easy to see that for all $c, d \in M$

$$M \models F(c, d) \Longleftrightarrow M_0 \models P(g(c), g(d)).$$

The latter means that g is a strong onto homomorphism. Thus by Theorem 2.4 we have that the logic of schemes $LS(Th(T))$ is undecidable. ⊣

COROLLARY 2.7 *If the first-order theory T is undecidable, then $LS(T)$ is undecidable too.*

The proof is evident because every theory of a finite number of finite models is decidable; hence every undecidable theory satisfies the condition of Theorem 2.6.

THEOREM 2.8 *Let T be a decidable first-order theory, and assume that for all $m \in \mathbb{N}$ there is a model of the theory T which has at least m elements (or equivalently T has infinite models). Then $LS(T)$ is not recursively enumerable.*

Proof. That T is decidable implies that the complement of the logic of its schemes $LS(T)$ is recursively enumerable. By Theorem 2.6, $LS(T)$ is not decidable. Therefore $LS(T)$ can not be recursively enumerable. ⊣

Note that the same result for theories in monadic signature only follows from Theorem 2.1(iv)(b) of [Németi, 1987].

EXAMPLE 2.9 It is well-known that the following first-order theories are decidable [Chang and Keisler, 1973; Ehrenfeucht, 1959; Ershov *et al.*, 1965; Feferman and Vaught, 1959] and have infinite models:

- the theory BA of boolean algebras,
- the theory DLO of dense unbounded linear order,
- the theory AG of abelean groups.

Hence by Theorem 2.8 all these theories do not have recursively enumerable logics of schemes. The same is true for theories of all finite boolean algebras or all finite cyclic groups, for instance. Thus, contrary to our initial intuitions, logics of schemes for first-order theories are much more complicated than the theories themselves.

THEOREM 2.10 *If M is a finite model, then $LS(Th(M))$ is decidable. In addition, there is an algorithm for recognizing admissibility of inference rules in $Th(M)$.*

Proof. Assume that the base set of M has k elements $\{1, \ldots, k\}$. Let $P(x_1, \ldots, x_n)$ be any predicate relation on $\{1, \ldots, k\}$ (which need not be a predicate relation of the model M). Any such relation P can be expressed on M by first-order formulas with only equalities. More precisely, we put

$$\begin{aligned} &F_P(x_1, \ldots, x_n, y_1, \ldots, y_k) := \\ &\quad \bigwedge\{\neg(y_i = y_j) \mid 1 \leq i < j \leq k\} \wedge \\ &\quad \bigvee\{(x_1 = y_{i_1}) \wedge \ldots \wedge (x_n = y_{i_n}) \mid \\ &\qquad\qquad i_1, \ldots, i_n \in \{1, \ldots, k\}, M \models P(i_1, \ldots, i_n)\}. \end{aligned}$$

Under the interpretation $y_i \mapsto i$ the truth of F_P in M coincides with the truth of P in M :

$$(5) \quad \forall a_i \in M (M \models F_P(a_1, \ldots, a_n, 1, \ldots, k) \Leftrightarrow M \models P(a_1, \ldots, a_n))$$

Taking (5) into account, we infer the following. For any formula $A(z_i)$ (in the language of the logic of schemes)

(6) $A(z_i) \in LS(Th(M)) \Longleftrightarrow \forall Q_i(A(Q_i) \in Th(M))$,

where Q_i are arbitrary atomic formulas of form $P_i(x_{i1}, \ldots, x_{in})$, where P_i are any predicate letters (not only from the signature of M). Let given formula $A(z_i)$ has quantifiers (connectives) $\forall x_1, \ldots \forall x_m$. It can be easily seen that it is sufficient to consider in (6) only atomic formulas Q_i which depends on some variables from $x_1, \ldots, x_n$ only. This gives an algorithm for recognizing whether $A(z_i) \in LS(Th(M))$ is true. Hence $LS(Th(M))$ is decidable.

A given inference rule A/B is admissible in $Th(T)$ iff the following holds: if the formula A is true on the model M under some interpretation of the metavariables by means of predicates that are expressible by formulas, then B is also true with respect to this interpretation. As we have seen above, all predicates on M can be expressed by first-order formulas with only equalities. Therefore we can simply replace 'predicates expressible by formulas' in the above condition describing admissibility by 'predicates.' Reasoning as before we conclude that it is suffices to consider only predicates with bounded dimension. $\dashv$

Thus we have a complete description for first-order theories which are decidable with respect to admissibility of inference rules.

THEOREM 2.11 *A first-order theory T has an algorithm recognizing the admissibility of inference rules if and only if*

$$T = \bigcap_{1 \leq i \leq n} Th(M_i)$$

where all M_i are finite.

Proof. This follows immediately from Theorems 2.10 and 2.6. $\dashv$

In particular, as a corollary, we have a description of first-order theories with decidable logics of schemes.

COROLLARY 2.12 *A first-order theory T has a decidable logic of schemes if and only if*

$$T = \bigcap_{1 \leq i \leq n} Th(M_i)$$

where all M_i are finite.

Note that this corollary can be also derived from a result in algebraic logic announced in [Németi, 1987, page 248]. (The result concerns the question of decidability for logics of schemes of first-order theories that can be expressed in the language of algebraic logic. It is the question of solvability of the equational theory of varieties generated by cylindric-like algebras.)

3 PURE LOGICS OF SCHEMES

The results of the previous section show us that logics of schemes are very complicated from an algorithmic point of view. We can make the language of the logic of schemes weaker by removing logical constants simulating equalities $(x_i = x_j)$ from the language. It is natural to consider such a language because equality can simply be understood as a binary predicate relation. So we can not distinguish between the predicate of equality and other predicates. A second reason is the existence a number of well-known and well studied first-order theories in the language of pure predicate calculus. A third motivation is to study to which extent the expressive power of the language of the logics of schemes is influenced by the presence of logical constants simulating equalities.

Let PLS be the language of the pure logic of schemes, i.e., obtained from LS by removing all logical constants $x_i = x_j$. The pure logic of schemes PLS(T) of any first order theory T (with arbitrary first-order language, — with equality or without it, and in arbitrary signatures) is defined just like $LS(T)$, but in its own language.

The question that arises is which results of Section 2 remain true for PLS. A part of the answers is evident. For instance,

PROPOSITION 3.1 *If* M_1, M_2 *are models in pure predicate signature and there is a strong homomorphism of* M_1 *onto* M_2 *then* PLS$(Th_p(M_1))$ = PLS$(Th_p(M_2))$.

Proof. By Lemma 2.2 M_1 and M_2 agree on all formulas without equality. Hence, the pure logics of schemes on M_1, M_2 coincide too. ⊣

Because of the fact that there are strong homomorphisms from some infinite models onto finite models, the last proposition and Theorem 2.10 tell us that there exist infinite models with decidable pure logics of schemes. Thus there are counterexamples to the Theorem 2.6, Corollary 2.12 in the case of PLS. So the language of PLS is really weaker than LS.

Can we obtain analogs for PLS of the results we obtained in the case of LS? We proceed to answer this question for the pure predicate calculus PPC itself. (From the point of view of algebraic logic we will now investigate equational theories of diagonal-free cylindric algebras (cf. [Henkin *et al.*, 1971])).

Let $A(z_i)$ be a metaformula in the language PLS which has only connectives $\forall x_1, \ldots, \forall x_n$ among its connectives simulating quantifiers. Then we call it an n-*metaformula*. Let $t(A)$ be a first-order formula obtained from A by replacing all metavariables z_i by corresponding predicate letters $P_i(x_1, \ldots, x_n)$. We call $t(A)$ the *first-order track* of A.

LEMMA 3.2 $A(z_i) \in$ PLS(PPC) *if and only if* $t(A) \in$ PPC.

Proof. The implication from left to right follows immediately. Suppose that the left-hand side of the proposition is not true. Then there is a tuple of formulas $g_i(x_1, \ldots, x_n, y_i)$ in the language of PPC such that

$$A(g_i(x_1, \ldots, x_n, y_i)) \notin \text{PPC}.$$

Then there is a model M of PPC and some valuation of the free variables of this formula such that

$$M \not\models A(g_i(a_1, \ldots, a_n, b_i)).$$

On the set M we introduce a new model M_1 by taking new predicates P_i, where $M_1 \models P_i(c_1, \ldots, c_n)$ iff $M \models g_i(c_1, \ldots, c_n, b_i))$. One can easily show by induction on the length of subformulas $B(z_i)$ of the formula $A(z_i)$, that

$$M_1 \models B(P_i(c_1, \ldots, c_n)) \text{ iff } M \models B(g_i(c_1, \ldots, c_n, b_i))).$$

Thus $M_1 \not\models A(P_i(c_1, \ldots, c_n))$ and $t(A) \notin$ PLS(PPC). ⊣

Thus, PLS(PPC) can be embedded into PPC itself. This allows us to use known results about PPC for describing its pure logic of schemes.

THEOREM 3.3 *The subclass* PLS_3(PPC) *of* PLS(PPC) *which consists of all formula schemes having occurrences of connectives* $\forall x_i$ *from the set* $\{\forall x_1, \forall x_2, \forall x_3\}$ *only is undecidable, but* PLS(PPC) *itself is recursively enumerable and undecidable.*

Proof. By Lemma 3.2 the question about decidability of PLS(PPC) can be reduced to the question about decidability a subclass of PPC which consists of first-order tracks of metaformulas. More precisely, they are first-order formulas all predicates of which have 'frozen' occurrences of all variables. These formulas are called homogeneous in [Pieczkowski, 1968]. In that paper it is proved to be an undecidable problem whether a given homogeneous formula of degree 3 (i.e., having only ternary predicates) is valid. This together with Lemma 3.2 gives us the undecidability of PLS_3(PPC). ⊣

Note that by Lemma 3.2 the pure logic of schemes PLS(PPC) is recursively enumerable. Dr. Max Urchs has drawn my attention to the possibility of using this result of [Pieczkowski, 1968] together with Lemma 3.2 in a private conversation. Note that Theorem 3.3 can be extracted also from Theorem 5.1.66 (about equational theories of diagonal-free cylindric algebras) of [Henkin *et al.*, 1971, Part 2, page 202]. We refer here to [Pieczkowski, 1968] as earlier publication.

Theorem 3.3 has some consequences for poly-modal propositional logic. As we have seen above, PLS(PPC) forms a poly-modal propositional logic without any logical constants and with commutating modalities. Therefore, Theorem 3.3 gives

COROLLARY 3.4 *For all $n > 2$, there are undecidable n-poly-modal propositional logics extending $S5_n + (\Box_i\Box_j p \equiv \Box_j\Box_i p)$.*

It is interesting to look at Kripke semantics for PLS(PPC). Let M be a model of PPC with n-placed predicate letters $P_1, \ldots, P_m$. We call such a model an *nm-model.* The associated *n-poly-modal Kripke model* $K(M)$ for M (or n-cube, cf. [Venema, 1991]) is the set $\{(a_1, \ldots, a_n) \mid a_i \in M\}$ with n accessibility relations $R_1, \ldots, R_n$, where

$$(a_1, \ldots, a_n) R_i (b_1, \ldots, b_n) \Longleftrightarrow \forall j((i \neq j) \Rightarrow (a_j = b_j)),$$

and with the valuation V of variables $z_1, \ldots, z_m$ on this model given by

$$(a_1, \ldots, a_n) \Vdash_V z_i \Longleftrightarrow M \models P_i(a_1, \ldots, a_n).$$

We are now in a position to define truth on $K(M)$ of metaformulas $A(z_i)$ of the pure logic of schemes in variables $z_1, \ldots, z_m$ where the connectives $\forall x_i$ are viewed as modal logical connectives $\Box_i$.

Let M be an nm-model, let $K(M)$ be its associated n-poly-modal Kripke model, and let $A(z_i)$ be an n-metaformula in the variables displayed.

LEMMA 3.5 $K(M) \Vdash A(z_i) \Longleftrightarrow M \models t(A)$.

Proof. Let $B(z_i)$ be a subformula of $A(z_i)$. Then $(a_1, \ldots, a_n) \Vdash_V B(z_i) \Longleftrightarrow M \models t(B)^{a_1,\ldots,a_n}_{x_1,\ldots,x_n}$. The proof can be easily given by induction on the length of B. Indeed, the base of induction and the cases of the nonmodal logical connectives are evident. Suppose that $(a_1, \ldots, a_n) \Vdash_V \Box_r B(z_i)$. Then

$$\forall b_r (a_1, \ldots, a_{r-1}, b_r, a_{r+1}, \ldots, a_n) \Vdash_V B(z_i).$$

By induction hypotheses this gives us

$$\forall b_r M \models t(B)^{x_1,\ldots,x_r,\ldots,x_n}_{a_1,\ldots,b_r,\ldots,a_n}$$

This relation yields $M \models t(\forall x_r B)^{x_1,\ldots,x_n}_{a_1,\ldots,a_n}$

The converse implication can be shown by a similar argument, but in the inverse order. ⊣

From Lemmas 3.2, 3.5 we obtain a semantic description of PLS(PPC):

THEOREM 3.6 *Let $A(z_i)$ be an arbitrary n-metaformula of* PLS *with m metavariables. Then $A(z_i) \in$* PLS(PCC) *if and only if $K(M) \Vdash A(z_i)$ holds for every nm-model M.*

So, in particular, the class of $S5$-n-poly-modal Kripke frames mentioned in Theorem 3.6 has an undecidable modal theory for every $n > 2$ (by Theorem 3.3). Note that the same description is possible for logics of schemes of PC.

Here, we have only briefly touched on work on poly-modal Kripke frames (only in order to give some applications of results about logics of schemes). More detailed investigations of such frames (n-cubes) can be found in [Venema, 1991]).

To conclude, as we have seen above, some of the results about logics of schemes are not preserved for pure logic of schemes, but some results have analogs (see Theorem 3.3, for instance).

Mathematics Department
Krasnoyarsk University

Acknowledgments. This research was supported by the Alexander von Humboldt Foundation during a visit of the author at the Free University of Berlin.

REFERENCES

[Blok and Pigozzi, 1989] W.J. Blok and D. Pigozzi. Algebraizable logic. *Memoirs of the Amer. Math. Soc.*, 396, 1989.

[Chang and Keisler, 1973] C.C. Chang and H.J. Keisler. *Model Theory*. North-Holland, Amsterdam, 1973.

[Ehrenfeucht, 1959] A. Ehrenfeucht. Decidability of the theory of linear ordering relation. *Notices Amer. Math. Soc.*, 6:268–269, 1959.

[Ershov *et al.*, 1965] Yu.L. Ershov, I.A. Lavrov, A.D. Taimanov, and M.A. Taislin. Elementary theories. *Uspechi Matematicheskich Nauk*, 20:37–108, 1965. In Russian.

[Fagin *et al.*, 1992] R. Fagin, J.Y. Halpern, and M.Y. Vardi. What is an inference rule? *Journal of Symbolic Logic*, 57:1018–1045, 1992.

[Feferman and Vaught, 1959] S. Feferman and R. Vaught. The first-order properties of algebraic systems. *Fundamenta Mathematicae*, 47:57–103, 1959.

[Fine, 1974] K. Fine. An ascending chain of S4 logics. *Theoria*, 40:110–116, 1974.

[Friedman, 1975] H. Friedman. One hundred and two problems in mathematical logic. *Journal of Symbolic Logic*, 40:113–130, 1975.

[Goldblatt, 1976] R.I. Goldblatt. Metamathematics of modal logic I. *Reports on Mathematical Logic*, 6:41–78, 1976.

[Henkin *et al.*, 1971] L. Henkin, J.D. Monk, and A¿ Tarski. *Cylindric Algebras. Part 1. Part 2.* North-Holland, Amsterdam, 1971.

[Monk, 1965] J.D. Monk. Substitutionless predicate logic with identity. *Arch. Math. Logic Grundl.*, 7:353–358, 1965.

[Németi, 1987] I. Németi. On varieties of cylindric algebras with applications to logic. *Annals of Pure and Applied Logic*, 36:235–277, 1987.

[Pieczkowski, 1968] A. Pieczkowski. Undecidability of the homogeneous formulas of degree 3 of the predicate calculus. *Studia Logica*, 22:7–14, 1968.

[Rautenberg, 1979] W. Rautenberg. *Klassische und nichtklassische Aussagenlogik*. Braunschweig, Wiesbaden, Deutschland, 1979.

[Rybakov, 1987] V.V. Rybakov. Bases of admissible rules of the modal system *grz* and of intuitionistic logic. *Math. USSR Sbornic*, 56:311–331, 1987.

[Rybakov, 1989] V.V. Rybakov. Problems of admissibility and substitution, logical equations and restricted theories of free algebras. In *Proc. of the 8-th Inter. Congress of Logic, Method. and Phil. of Science*, pages 121–139, 1989.

[Rybakov, 1990a] V.V. Rybakov. Metatheories of first-order theories. In *Proc. of the Fourth Asian Logic Conference*, pages 16–17, CSK Educational Center, Tokyo, 1990.

[Rybakov, V.V., 1990b] V.V. Rybakov. Problems of substitution and admissibility in the modal system *grz* and intuitionistic calculus. *Annals of Pure and Applied Logic*, 50:71–106, 1990.

[Rybakov, 1991] V.V. Rybakov. Poly-modal logic as metatheory of pure predicate calculus. In *Abstracts of 9-th Intern. Congress of Logic, Meth. and Phil. of Science*, page 158, 1991.

[Venema, 1991] Y. Venema. *Many-Dimensional Modal Logic*. PhD thesis, Department of Mathematics and Computer Science, University of Amsterdam, 1991.

[Venema, 1993] Y. Venema. Derivation rules as anti-axioms in modal logic. *Journal of Symbolic Logic*, 59:1003–1034, 1993.

JERRY SELIGMAN

THE LOGIC OF CORRECT DESCRIPTION

1 TRUTH AND CORRECT DESCRIPTION

A basic tenet of situation semantics [Barwise and Perry, 1983; Barwise and Etchemendy, 1987; Barwise, 1989], deriving from Austin's theory of truth [Austin, 1950], is that every statement is about a situation. To make a statement by uttering the sentence 'Michel tossed the salad,' I must also be referring to a specific situation — in this case, an event — which, if my statement is true, is one in which Michel tossed the salad.

There are problems with Austin's account of truth. To say exactly how a particular situation is picked out as the reference of a given statement is notoriously hard. Austin suggested that this is achieved by what he called the *demonstrative conventions* of language-use, but he said little about what such conventions are, or how they succeed in identifying the right situation. The theory is most convincing for true statements about concrete events, such as the salad-tossing mentioned above. It fairs less well when applied to false statements ('Albert tossed the salad'), negative statements ('Jon doesn't own an electric toothbrush'), and has serious problems with universal statements ('Dictionaries are heavier than poetry books') and non-empirical statements ('$2 + 2 = 4$').

For these and similar reasons, Austin's theory has been heavily criticised, and a response to the critics is clearly needed. However, the purpose of this paper is not to give a defense of Austin. Instead, we wish to examine the consequences of the theory for logic, asking what, if anything, would change in our analysis of logical consequence if Austin's theory were true.

One obvious difficulty arises from the fact that statements, not sentences, are regarded as the primary bearers of truth. For logic, this would matter little if the truth-conditions of a statement depended only on which sentence was used in making the statement, but on Austin's account, at least in principle, the situation to which the statement refers must also be considered.

For this reason, we supplement Austin's theory, by introducing the more overtly relational semantic predicate of *correct description*:

> A sentence φ is a correct description of a situation s iff any statement about s using sentence φ would be true.

For example, the sentence 'Michel tossed a salad' correctly describes those salad-tossing events of which Michel is the agent, because any statement made about one of those situations using the sentence would be true.

M. de Rijke (ed.), Advances in Intensional Logic, 107–135.

Turning this definition on its head, we will take correct description to be the primary semantic relation, and define the truth of a statement as follows:

> A statement about s using sentence φ is true iff φ is a correct description of s.

For example, a statement about a specific event s made by uttering 'Michel tossed a salad' is true just in case s is one of those events which the sentence correctly describes.

It might well be objected that in focusing on correct description we have departed from Austin's account in an important respect. Austin maintains that a statement is related to a type of situation by *descriptive conventions* of language-use, and that the statement is true if the situation to which it refers is of the specified type. For our purposes, this formulation has the disadvantage that one has to refer explicitly both to statements and to types of situations. By rephrasing the account in terms of correct description, one can study the relational character of the semantic theory without becoming entangled in questions about the ontology of statements and situation-types.

A potential disadvantage of this move is that the two accounts are equivalent only on the assumption that the type of situation specified by the descriptive conventions depends only on the sentence used in making the statement. In that case, we may suppose that the type of situation specified by the descriptive conventions is the type of situation correctly described by the sentence used in making the statement. However, the assumption fails for sentences containing indexicals, demonstratives, or any other elements whose meaning depends on extra-linguistic factors of the context in which they are used.

Once this problem is recognized, it poses no great threat. Our concern is to study the logic of a theoretical language for reasoning about situations, based on semantic principles deriving from Austin's analysis. But that language need not have all the features of natural language. Like Frege, we may ban the use of indexicals and other context-dependent expressions, at least for scientific purposes. This is not to say that we abandon the hope of providing an analysis of such expressions *within* the theory. One of the main purposes of a language for reasoning about situations is the development of a theory of meaning [Situation Semantics; Barwise and Perry, 1983], but that is a secondary project.

2 A DEFINITION OF CORRECT DESCRIPTION

In the previous section we defined truth in terms of the relation of correct description between sentences and situations. In this section, we will provide a recursive definition of correct description, and so complete the definition of truth. The definition is in the manner of Tarski's definition of truth for first-order languages, but is motivated by Austin's analysis instead than Tarski's. For languages without semantic vocabulary, this involves only minor changes. They become significant when the language is extended to enable one to talk about the relation of correct description itself.

We start with an ordinary first-order language, containing function-symbols and relation-symbols of arbitrary arity, with a distinguished binary relation-symbol for identity, and a countably infinite number of variables. We suppose that each situation is associated with a determinate set of objects consisting of the objects occurring in it. One object may occur in many different situations, and have different properties in each; thus the reference of predicates may vary from situation to situation. The reference of terms is invariant from one situation to another: each constant-symbol names a unique object, which may occur in different situations, and each function-symbol refers to a unique function mapping objects to objects.

We are already in a position to define correct description for atomic sentences:

> s is correctly described by $\ulcorner R(t_1, \ldots t_n)\urcorner$ iff the objects $\underline{t_1}$, ..., $\underline{t_n}$ stand in the relation $\underline{R}$ in s.[1]

But this is not enough. We also need a way of dealing with formulae containing free variables. Given a sequence σ of objects and a term t, let $t[\sigma]$ be the object which t would denote if each variable x_i occurring in t were taken to be a name for the ith object in σ. More precisely, $x_i[\sigma]$ is the ith object in the sequence σ, and $\ulcorner f(t_1, \ldots, t_n)\urcorner[\sigma]$ is a the object to which the function $\underline{f}$ maps the objects $t_1[\sigma]$, ..., $t_n[\sigma]$. Now we may define correct description of atomic formulae:

> $\ulcorner R(t_1, \ldots t_n)\urcorner$ is a correct description of σ in s iff the objects $t_1[\sigma]$, ..., $t_n[\sigma]$ stand in the relation $\underline{R}$ in s.

A fundamental problem concerns the interpretation of negation. It is unclear how the conditions for a sentence $\ulcorner\neg\varphi\urcorner$ to describe a situation correctly depend on the conditions for φ. About all one can say in general is that it is inconsistent to say that both φ and $\ulcorner\neg\varphi\urcorner$ correctly describe the same situation. Beyond that different stories may be told to justify a range of logical principles. The classical principle,

> $\ulcorner\neg\varphi\urcorner$ correctly describes s iff φ does not correctly describe s

is the strongest, and may be thought too strong. Consider the event s of Michel tossing a salad, while preparing a light lunch in his kitchen. If he performs this culinary feat unassisted, then it fair to say that the sentence 'Albert chopped mushrooms' does not correctly describe s. But what are we to make of the sentence 'Albert did not chop mushrooms'? On the classical view, this sentence describes s correctly, even if Albert was busy chopping mushrooms in another kitchen at exactly the same time.

[1] We shall follow the convention of using corner-quotes in the text, but ignoring them when formulae are displayed in proofs. The reference of an expression is indicated using underlining.

It is worth noting that the classical view is not entirely unwelcome here. There *is* a sense in which the situation is correctly described by the sentence 'Albert did not chop mushrooms,' if the situation is taken to be the ultimate arbitrator of mushroom-chopping in the vicinity. This observation suggests that our language might usefully be strengthened with a classical negation. Alternatively, it might be part of the meaning of particular predicates that they obey the classical principles, even if classical reasoning is inappropriate in the general case. We will see an example of this in a little while.

We opt for a minimal solution, supposing that for each n-ary predicate R, there is another n-ary predicate $\sim R$, which one may read as 'not-R.' We will not legislate on the matter of which objects lie in the relation denoted by $\sim R$ in a given situation, except in so far as to claim that it is inconceivable for a sequence of objects to stand in both the relation $\underline{R}$ and $\underline{\sim R}$.

We may now give a recursive definition of what in general it is for a formula φ to be a correct description of a sequence σ of objects in a situation s. Then, we say that a sentence φ *correctly describes* s iff φ is a correct description of every sequence σ of objects in s.

For each s and sequence σ of objects, we have the following:

1. For each n-ary predicate R, and terms $t_1, \ldots, t_n$, $\ulcorner R(t_1, \ldots, t_n) \urcorner$ is a *correct description* of σ in s iff the objects $t_1[\sigma], \ldots, t_n[\sigma]$ stand in the relation denoted by R in s.

2. For each n-ary predicate R, and terms $t_1, \ldots, t_n$, $\ulcorner \neg R(t_1, \ldots, t_n) \urcorner$ is a *correct description* of σ in s iff the objects $t_1[\sigma], \ldots, t_n[\sigma]$ stand in the relation denoted by $\sim R$ in s.

3. $\ulcorner \varphi \wedge \psi \urcorner$ is a *correct description* of σ in s iff both φ and ψ are correct descriptions of σ in s.

4. $\ulcorner \neg(\varphi \wedge \psi) \urcorner$ is a *correct description* of σ in s iff either $\ulcorner \neg\varphi \urcorner$ or $\ulcorner \neg\psi \urcorner$ is a correct description of σ in s.

5. $\ulcorner \varphi \vee \psi \urcorner$ is a *correct description* of σ in s iff either φ or ψ is a correct description of σ in s.

6. $\ulcorner \neg(\varphi \vee \psi) \urcorner$ is a *correct description* of σ in s iff both $\ulcorner \neg\varphi \urcorner$ and $\ulcorner \neg\psi \urcorner$ are correct descriptions of σ in s.

7. $\ulcorner \forall x_i \varphi \urcorner$ is a *correct description* of σ in s iff φ is a correct description in s of each sequence σ' of objects differing from σ in at most the ith place.

8. $\ulcorner \neg\forall x_i \varphi \urcorner$ is a *correct description* of σ in s iff $\ulcorner \neg\varphi \urcorner$ is a correct description in s of some sequence σ' of objects differing from σ in at most the ith place.

9. $\ulcorner \exists x_i \varphi \urcorner$ is a *correct description* of σ in s iff φ is a correct description in s of some sequence σ' of objects differing from σ in at most the ith place.

10. $\ulcorner \neg \exists x_i \varphi \urcorner$ is a *correct description* of σ in s iff $\ulcorner \neg \varphi \urcorner$ is a correct description in s of each sequence σ' of objects differing from σ in at most the ith place.

There is not much to surprise and excite in this definition. In fact, for each situation s, if φ is a formula all of whose closed terms refer to objects in s, then a systematic replacement of the phrase 'is a *correct description* of σ in s' by 'is satisfied by σ' would yield the definition of satisfaction for partial first-order logic, as suggested by [Kleene, 1952].

To the basic language, we add a binary operator δ which takes a sentence as its first argument and a term as its second. If φ is a formula and t is a term, then for each s and each sequence σ,

11. $\ulcorner \delta(\varphi, t) \urcorner$ is a *correct description* of σ in s iff $t[\sigma]$ occurs in s and φ is a correct description of σ in $t[\sigma]$

12. $\ulcorner \neg \delta(\varphi, t) \urcorner$ is a *correct description* of σ in s iff $t[\sigma]$ occurs in s and φ is not a correct description of σ in $t[\sigma]$.

In particular, if φ is a sentence and t occurs in s then $\ulcorner \delta(\varphi, t) \urcorner$ is a correct description of s iff φ is a correct description of $\underline{t}$.

Implicit in the above is the assumption that the terms of the language range over situations as well as objects. We have not distinguished between situation-terms and object-terms in the syntax of the language, and so the definition of correct description applies to ordinary objects as well as to situations. Whether or not such descriptions make sense depends on the basic vocabulary of the language. One way of incorporating object-descriptions in a sensible way would be to allow property names to occur as zero-arity predicates. With 'red' and 'blue' in the language, the expression 'red $\vee$ blue' describes an object a correctly just in case a is either red or blue.

Nonetheless, there is a real distinction between situations and other objects; it lies in the fact that only situations may have other objects occurring within them. We say that

s is a *situation* iff there is some object which occurs in s.

The definition is purely stipulative, but useful. It enables us to distinguish between situations and non-situations on the basis of the sentences describing them. A predicate of arity zero just expresses a property of the things it describes, and so may apply to situations and non-situations alike. But if φ is a sentence containing a predicate-symbol of positive (i.e., non-zero) arity, and φ correctly describes s then s is a situation. For example, if $\ulcorner R(t_1, t_2) \urcorner$ is a correct description of s then $\underline{t_1}$ and $\underline{t_2}$ occur in s, and so s is a situation.

A consequence of the description-conditions for δ-sentences is that if $\ulcorner\delta(\varphi, t)\urcorner$ correctly describes a situation s, then $\underline{t}$ occurs in s. Situations which contain other situations are called *semantic* situations. If s is a semantic situation, then for each term t referring to a situation $\underline{t}$ occurring in s and each formula φ,

> $\ulcorner\neg\delta(\varphi, t)\urcorner$ is a correct description of s iff $\ulcorner\delta(\varphi, t)\urcorner$ is not a correct description of s

In other words, matters of semantic fact are treated classically.

Finally, we will need some way of naming the current situation or object — the thing that the statement is about. We do this by allowing terms to occur as formulae. A term t may be used as a formula to say that the reference-situation (the situation to which the whole statement refers) is $\underline{t}$.

13. t is a *correct description* of σ in s iff $t[\sigma]$ is identical to s

14. $\ulcorner\neg t\urcorner$ is a *correct description* of σ in s iff $t[\sigma]$ and s are distinct.

Again, we assume that our language behaves classically with respect to terms occurring as formulae. The description-conditions for a complex formula containing terms as subformulae is given by the other clauses of the definition. For example, the formula $\ulcorner t \wedge \varphi\urcorner$ is a correct description of σ in s iff $s = t[\sigma]$ and φ is a correct description of σ in s.[2]

3 SITUATED CONSEQUENCE AND INDEXICALITY

In assessing the validity of arguments, we need a semantic analysis of logical relationships among sentences. In particular, we are interested in characterizing when a move from sentence φ to sentence ψ is *truth-preserving*. But if truth is a property of statements, rather than sentences, how are we to determine if this is the case? For all that has been said, φ and ψ may be used to make statements about quite different situations.

To see the problem more clearly, consider the sentences 'Michel tossed a salad gracefully' and 'Michel tossed a salad.' To justify the inference from the first of these to the second, truth-preservation alone is not enough, because the two sentences may be used to make statements about different occasions, one in which Michel did toss a salad gracefully, and the other in which he did not toss a salad at all.

The problem is easily solved by basing consequence on correct description instead of truth. We say that one sentence is a *(situated) consequence* of another

[2]The use of terms as formulae stems from [Prior, 1967] and has been more recently investigated by [Blackburn, 1993] and [Gargov and Goranko, 1993].

if every conceivable situation correctly described by the one is also correctly described by the other. In our example, 'Michel tossed a salad' is a situated consequence of 'Michel tossed a salad gracefully,' because every conceivable situation in which Michel tossed a salad gracefully, is one in which he tossed a salad.

We use the expression 'situated consequence' rather than 'logical consequence' to preserve the traditional association between the latter and the property of truth, which — on the present account — can only be had by statements. This may seem somewhat pedantic. After all, it is common to use the latter expression for consequence between sentences containing indexical items or demonstratives, which rarely have a truth-value independent of their context of use.

Despite a superficial similarity between the two cases, they are really quite different. Sentences containing context-dependent elements, such as indexicals and demonstratives, are regarded as incomplete only in so far as they lack a reference for those elements. The inadequacy is a local matter, to be settled by specific aspects of the context in which they are used: 'I' and 'you' refer to the speaker and addressee; 'here' refers to the location of the speaker. Moreover, according to the traditional view, sentences without such elements do not depend on the context to provide them with a truth-value.

By contrast, according to Austin, *every* statement refers to a situation, and this reference is required to determine truth-value. The result is a much more radical departure from previous theories of truth. To appreciate the subtleties it is worth considering the statement

(S) The sentence 'Michel tossed a salad' correctly describes the situation s.

On the Austinian theory, (S) can only be true if it refers to a situation s' which is correctly described by the sentence "The sentence 'Michel tossed a salad' correctly describes the situation s." It is difficult to say exactly which situation s' is, but it must be something which involves facts about language and meaning, and so is unlikely to be identical to s, a mere salad-tossing event. Whether or not this results in a problematic regress, a hierarchy of semantic concepts, or genuine semantic relativism is a matter for further consideration, but it is safe to say that nothing like this occurs in the analysis of indexicals.

This is not to say that the situation to which a statement refers is never determined indexically. An example, which will be used extensively in Section 5, is the indexicality of many expressions with regard to spatial location. Statements made using the sentence 'There is a chill in the air' and 'It's night-time' usually refer to the situation in the vicinity of the speaker, and may often be paraphrased by 'there is a chill in the air here' and 'it's night-time here.' In such cases, the reference of the statements may be said to be determined indexically.

Yet even in these cases, one can see a distinction between the indexically determined location and the reference-situation. If, for example, we are watching a TV news-broadcast of a reporter shivering outside the state capitol, then we may make statements using the sentences 'There is a chill in the air' and 'It's night-time,' which are about the reporter's situation, rather than ours, and which may not be paraphrased using 'here.'

The distinction between the reference of indexicals in statements and the reference of the statements themselves is most clearly illustrated by the use of the past tense. Reichenbach famously distinguished between speech-time and reference-time, and used the distinction to provide a semantic classification of the tenses. In the terms of the present account, the reference-time of a statement is just the time of the event to which the statement refers, and so is determined by "demonstrative conventions" which may or may not have an indexical element.[3] By contrast, the determination of speech-time is purely indexical; it is just the time at which the statement is made.

We may also use past tense statements to bring out the difference between situated consequence and logical consequence. The statement that

(1) Albert started making the mayonnaise

about an event e is a logical consequence of the statement that

(2) Albert had almost finished making the mayonnaise.

which is about a different event e' in the later stages of the culinary process initiated by e. The truth of (1) is necessitated by the truth of (2) because there can be no event which is correctly described by the sentence used in (2) without there being another, appropriately related, event which is correctly described by the sentence used in (1). However, e and e' cannot be the same event: even a cook of Albert's great virtuosity cannot start and have almost finish making mayonnaise at once.[4] For this reason, the sentence used in (1) is *not* a situated consequence of that used in (2).

To evaluate whether a sentence ψ is a situated consequence of a sentence φ, one must ask whether there is conceivable situation s which is correctly described by φ but not by ψ; if there is then it isn't, and if there isn't then it is. It is a commonplace that such tasks of imagination are best analyzed by providing a mathematical model of (conceivable) situations and providing a rigorous definition of a relation $\models$ between models of situations and sentences, such that if m models a conceivable situation s then $m \models \varphi$ iff φ is a correct description of s. We hope that it is clear to the reader how this could be done for the language introduced in the previous section. We will not fill in the details, in part because of they are routine, but also because the level of generality involved will not be needed. In the next

[3] In the case of past-tense statements, there is always *some* indexical element, because the event described is required to have occurred before the statement is uttered; but such constraints are very far from sufficient for reference.

[4] The analysis of the meaning of these statements is by no means uncontroversial. See [Seligman and ter Meulen, 1994] for more details.

section we will impose some severe restrictions on the domain of "conceivable situations."

4 OMNISCIENT SITUATIONS

The language, or rather class of languages, introduced in Section 2 was designed to be reasonably close to what would be required for expressing a theory of situations at the elementary level — no higher-order entities are involved. It differs from previous languages for talking about situations (notably [Westerståhl, 1990]), in that we analyze the truth-conditions of statements using Austin's theory of truth, instead of Tarski's. Moreover, because of the foundational nature of the subject matter, the language is designed to reflect this semantic analysis, by including a predicate for correct description, rather than the more usual predicate of 'support' between a situation and some non-linguistic entity, like an "infon" or "state of affairs" (e.g., [Fernando, 1990; Barwise, 1989]).

The language is very flexible and full of logical intricacies: a weak negation, restricted quantifiers, etc. Each of these is motivated by the general analysis of correct description, and the rôle required of such a language in expressing a theory of situations. However, a full treatment of these matters would be both overly complicated and somewhat tedious. Partial logic is now sufficiently well-understood for it to be desirable to side-step its detailed consideration if at all possible, and that is what we shall do.

Subtracting partiality, and with it restricted quantification, we are left with the δ predicate itself, terms which occur as formulae, and classical predicate logic with identity. All predicates will be assumed to behave classically: for each n-ary predicate R, objects $a_1, \ldots, a_n$, and situation s,

> $a_1, \ldots, a_n$ stand either in the relation denoted by R in s or in the relation denoted by $\sim R$ in s.

We may characterize the results of this surgery in semantic terms by restricting our attention to *omniscient situations*: situations in which all other situations occur. Only terms denoting omniscient situations will be used, the quantifiers will be taken to range over omniscient situations, and for a situation to count as "conceivable," it will have to be omniscient. Working with this restriction has two effects. Every object is required to be a situation, and every situation can "see" every other situation. Under these assumptions, if a situation s is correctly described by φ, then every other situation can see that this is so, and is correctly described by $\ulcorner\delta(\varphi, s)\urcorner$.

Implausible as these restrictions are in the general case, they allow us to study the logic of correct description in an environment free from the distractions of partiality. What's more, the logic we will be considering is considerably stronger that its more realistic cousins, and so suggestions of inconsistency rebuffed here have no chance of affecting the weaker logics. Finally, it is not difficult to reintroduce partiality; standard techniques are available in both model-theory and proof-theory [Langholm, 1988, 1989].

In subsequent sections of the paper, we will look more closely at the proof-theory of the restricted language, but for now it it useful to give a straightforward axiomatization. First, we note that all principles of reasoning from classical first-order logic are sound in this setting. More precisely, temporarily regarding the new formulae — those of the form $\ulcorner\delta(\varphi, t)\urcorner$ and t — as atomic formulae, we may say that

> if a formula φ is a classical consequence of a set of formulae Γ, then φ is a situated consequence of Γ.

On top of this classical base, we may add all instances of the following schema:

1. $\forall x(\delta(\varphi \wedge \psi, x) \leftrightarrow \delta(\varphi, x) \wedge \delta(\psi, x))$
2. $\forall x(\delta(\varphi \vee \psi, x) \leftrightarrow \delta(\varphi, x) \vee \delta(\psi, x))$
3. $\forall x(\delta(\neg\varphi, x) \leftrightarrow \neg\delta(\varphi, x))$
4. $\forall x(\delta(\forall y\varphi, x) \leftrightarrow \forall y\delta(\varphi, x))$
5. $\forall x(\delta(\exists y\varphi, x) \leftrightarrow \exists y\delta(\varphi, x))$
6. $\forall x\forall y(\delta(x, y) \leftrightarrow x = y)$
7. $\forall x\forall y(\delta(\delta(\varphi, y), x) \leftrightarrow \delta(\varphi, y))$

(with telescoping abbreviations: $\ulcorner\varphi_1 \rightarrow \varphi_2\urcorner$ for $\ulcorner\neg\varphi_1 \vee \varphi_2\urcorner$ and $\ulcorner\varphi_1 \leftrightarrow \varphi_2\urcorner$ for $\ulcorner\varphi_1 \rightarrow \varphi_2 \wedge \varphi_2 \rightarrow \varphi_1\urcorner$).

That instances of the first seven axiom schema correctly describe every omniscient situation follows directly from the definition of correct description given in Section 2. For the last schema, we also need the requirement that every situation is omniscient. The completeness of the above axiomatization may be proved by a straightforward modification of the proof of the completeness of classical first-order logic. Let Σ be a sets of sentences which are classically consistent w.r.t. the above axioms, and let Σ^* be a Henkin-complete and consistent extension of Σ. We know that such sets exist by the usual argument from Zorn's Lemma. Now for each closed term t, "construct" a situation t^* such that

1. $t_1^* = t_2^*$ iff the sentence $\ulcorner t_1 = t_2\urcorner$ is in Σ^*, and
2. objects $t_1^*, \ldots, t_n^*$ stand in the relation denoted by R in t^* iff the sentence $\ulcorner\delta(R(t_1, \ldots, t_n), t)\urcorner$ is in Σ^*.

If we had given a formal semantics for our language, we could perform this construction using the usual set-theoretic methods, given the closure of Σ^* under the usual laws of identity. In any case, we only require that for each term t, there is a conceivable situation satisfying the above; and that seems unproblematic. It only remains to show that for each sentence φ and each term t,

$$\varphi \text{ is a correct description of } t^* \text{ iff } \ulcorner\delta(\varphi, t)\urcorner \text{ is in } \Sigma^*.$$

The proof of this fact follows the similar proof for classical first-order logic, by induction on the logical complexity of φ, using each of the axioms schematized above. The axioms directly reflect the conditions for correct-description, given the completeness of classical first-order logic, and so it is unsurprising that they are also complete.

As an illustration of how the restricted languages may be of use, despite their limitations, consider a language containing the binary predicate '$\trianglelefteq$,' used to denote the relation of one situation's being part of another. The requirement that $\trianglelefteq$ behave classically does not seem to be too repugnant.

The theory of $\trianglelefteq$ may be developed in various ways, but a central concern is that of the persistence of information from a situation to one containing it. We may express the fact that a formula φ is persistent by

$$\ulcorner\forall x \forall y\ (\delta(\varphi, x) \wedge x \trianglelefteq y) \rightarrow \delta(\varphi, y)\urcorner$$

According to one school of thought, every sentence is persistent. Call this the Strong Theory of persistence. It is axiomatized by all sentences instantiating the above schema.

It should be clear that the Strong Theory is rather too strong in the present setting. For suppose that $s_1 \trianglelefteq s_2$ and that s_1 and s_2 are distinguishable. Then there is a sentence φ which correctly describes s_2 but not s_1, and so $\ulcorner\neg\varphi\urcorner$ describes s_1 correctly but not s_2, contradicting the claimed persistence of $\ulcorner\neg\varphi\urcorner$. Thus the Strong Theory entails that all $\trianglelefteq$-related situations are indistinguishable.

There are independent reasons for rejecting the Strong Theory [Barwise, 1989; Cooper, 1991; Cooper and Kamp, 1991], so we need not be too sad about its failure in the present context. Moreover, something of persistence may be salvaged by restricting it to formulae not containing negation. In fact, we may do a little better. We have already seen how to state the persistence of a given formula; now we define the *anti-persistence* of φ by

$$\ulcorner\forall x \forall y\ (\delta(\varphi, x) \wedge y \trianglelefteq x) \rightarrow \delta(\varphi, y)\urcorner$$

Now observe that

1. $\ulcorner\neg\varphi\urcorner$ is persistent iff φ is anti-persistent
2. $\ulcorner\neg\varphi\urcorner$ is anti-persistent iff φ is persistent

3. $\ulcorner\varphi \wedge \psi\urcorner$ and $\ulcorner\varphi \vee \psi\urcorner$ are persistent if φ and ψ are persistent

4. $\ulcorner\varphi \wedge \psi\urcorner$ and $\ulcorner\varphi \vee \psi\urcorner$ are anti-persistent if φ and ψ are anti-persistent

5. $\ulcorner\exists x\varphi\urcorner$ and $\ulcorner\forall x\varphi\urcorner$ are persistent if φ is persistent

6. $\ulcorner\exists x\varphi\urcorner$ and $\ulcorner\forall x\varphi\urcorner$ are anti-persistent if φ is anti-persistent

We may dub this the Conditional Theory of persistence. It shows how any initial classification of atomic formulae according to their persistence properties may be extended to a wide class of other formulae.

5 SITUATED REASONING: THE SPATIAL ANALOGY

In Section 3 we mentioned that the situation to which a statement refers is sometimes fixed by an indexically determined spatial location. Paradigm examples are statements concerning the weather and other prevailing conditions. In this section we will focus exclusively on such statements, using them to motivate natural rules of deduction.

So, from this point on, until the end of the section, we will assume that the situation described by a statement is the one the stater is *in*. Thus, the statement that it's raining refers to a suitably encompassing situation, determined by my present location, which happens to be Bloomington, Indiana. The situation is correctly described by the sentence 'It's raining' just in case it's raining in Bloomington, Indiana.

This last equivalence permits us to look to our usual patterns of reasoning using sentence of the form $\ulcorner$In l, $\varphi\urcorner$ or $\ulcorner\varphi$ in $l\urcorner$ (where l is a term referring to a spatial location) in order to motivate more general principles of reasoning about correct description. Furthermore, sentences of the form $\ulcorner$This is $l\urcorner$ play an analogous role to our situation-terms: they correctly describe a situation iff it is the situation determined by the location l. These two observations form the basis of what we shall call the Spatial Analogy.

If t_l is a term referring to the situation determined by the location l, we capture the central correspondence of the Spatial Analogy as follows:

1. $\ulcorner$In l, $\varphi\urcorner$ is equivalent to $\ulcorner\eth(\varphi, t_l)\urcorner$,

2. $\ulcorner$This is $l\urcorner$ is equivalent to t_l, and

3. $\ulcorner l_1$ is $l_2\urcorner$ is equivalent to $\ulcorner\eth(t_{l_1}, t_{l_2})\urcorner$.

Our method will be to examine how the expressions on the left hand side are used in constructing valid arguments, and to formulate natural-deduction rules which characterize those arguments. By analogy, the rules will transfer to rules involving the expressions on the right hand side. Particular attention will be given to arguments which either "introduce" a spatial expression in the conclusion which does not occur in the premises or "eliminate" an expression which does occur in the premises by drawing a conclusion which does not contain the expression. For the primary spatial word 'in' the following arguments are examples of introduction and elimination:

(1) The sun is shining; this is Bloomington, so the sun is shining in Bloomington.

(2) In Tokyo, people drive on the left; this is Tokyo, so people drive on the left.

They suggest the following introduction and elimination rules for 'in':

$$\frac{\varphi \qquad \text{This is } l}{\text{In } l, \varphi}\,\textit{In}\text{-I} \qquad\qquad \frac{\text{In } l, \varphi \qquad \text{This is } l}{\varphi}\,\textit{In}\text{-E}$$

These rules should be supplemented with a natural-deduction calculus for classical logic (such as that found in [Prawitz, 1965]) so that the logical connectives behave in the usual way. Are they sufficient? Unfortunately not. The rules for 'in' are sound — no invalid arguments can be made using them — but there are intuitively valid arguments which cannot be represented. The most important arguments are those that depend on the reasoner imagining, as a hypothetical premise, that they are somewhere else. For example:

(3) Alcohol is forbidden in Abu Dabi; Sake contains alcohol; so Sake is forbidden in Abu Dabi.

This argument can be justified using the introduction and elimination rules for 'in' if, in addition, the hypothetical premise 'This is Abu Dabi' is allowed for the duration of the argument:

$$\frac{[\text{This is Abu Dabi}] \qquad \dfrac{\dfrac{\text{In Abu Dabi, alcohol is forbidden} \qquad [\text{This is Abu Dabi}]}{\text{Alcohol is forbidden}}\,\textit{In}\text{-E} \qquad \text{Sake contains alcohol}}{\text{Sake is forbidden}}\,\text{C.S.}}{\text{In Abu Dabi, sake is forbidden}}\,\textit{In}\text{-I}$$

(The rule C.S. is an abbreviation for a sub-proof which could be constructed from a "common sense" understanding of the logical form of the example sentences.) At the last step of the proof, the hypothetical premise is *discharged* — removed from the set of genuine premises — and this is indicated by enclosing any occurrences of the premise in brackets.

The strategy works well for arguments like (3), but it may lead us to make mistakes if applied more widely. For example, the invalid argument

(4) In Islamabad, it only rains during monsoon; it's raining; so it's monsoon in Islamabad

appears to have a similar proof:

$$\dfrac{[\text{This is Islamabad}] \qquad \dfrac{\dfrac{\text{In Islamabad, it only rains during monsoon} \quad [\text{This is Islamabad}]}{\text{It only rains during monsoon}}\,\mathit{In}\text{-E} \qquad \text{It's raining}}{\text{It's monsoon}}\,\text{C.S.}}{\text{In Islamabad, it's monsoon}}\,\mathit{In}\text{-I}$$

It is easy to see what is going wrong. The premise 'It's raining' is spatially-indexical: said here, one may infer that it's raining here but not that it's raining in Islamabad. On hypothetical journeys one should avoid burdening oneself with premises which are true at home but false abroad.

This kind of mistake may be avoided if care is taken in specifying the circumstances under which a hypothetical premise of the form $\ulcorner$This is $l\urcorner$ may be discharged. A sufficient condition is that the "context" of the proof — the set of premises and conclusion after the discharge — does not contain any spatially-indexical sentences. This condition is met in the proof of argument (3) since the context consists of only 'Alcohol is forbidden in Abu Dabi,' 'Sake contains alcohol' and 'Sake is forbidden in Abu Dabi,' none of which are spatially-indexical. The condition is *not* met by the proof of argument (4) since that contains the spatially-indexical sentence 'It's raining.'

We are not done yet. There is an alternative way of using hypothetical premises to obtain valid arguments which cannot be proved, even using the discharge rule described above. Consider the argument:

(5) It's raining; wherever it's raining, it's wet, so it's wet.

Suppose that 'wherever' is a universal quantifier ranging over locations. Then the second premise is equivalent to the following sentence:

$\ulcorner$For all locations l, if it's raining in l then it's wet in $l\urcorner$.

In order to use this premise, a reasoner must find a location at which to instantiate it. Of course, the reasoner would like to instantiate it at her current location, where it's raining. To do so, she must first give a name to her current location — 'X,' say. This may be done by assuming hypothetically 'This is X.' She may then infer 'It's raining in X' (by 'in'-introduction), instantiate the second premise to 'If it's raining in X then it's wet in X,' deduce 'It's wet in X' and conclude 'It's wet' (by 'in'-elimination). Finally, she should discharge the hypothetical premise 'This is X':

$$\dfrac{[\text{This is } X] \qquad \dfrac{\dfrac{\text{It's raining} \quad [\text{This is } X]}{\text{In } X\text{, it's raining}}\,\mathit{In}\text{-I} \qquad \dfrac{\text{Wherever it's raining, it's wet}}{\text{If it's raining in } X \text{ then it's wet in } X}\,\forall\text{-E}}{\text{In } X\text{, it's wet}}\,\to\text{-E}}{\text{It's wet}}\,\mathit{In}\text{-E}$$

Such a discharge is not licensed by the condition discussed above because the context contains the spatially-indexical sentences 'It's raining' and 'It's wet.' The discharge of 'This is X' is only safe because it uses an invented name, 'X.' Considerations of this kind suggest that there is another sufficient condition on discharges of premises of the form $\ulcorner$This is $l\urcorner$: that l is a place-name which does not occur in any sentence of the context.

The discharge in the proof of (5) is licensed because its context consists of the sentences 'It's raining,' 'Whenever it's raining, it's wet' and 'It's wet,' none of which contain an occurrence of 'X.' The discharge of 'This is Islamabad' in the non-proof of (4) is *not* licensed because its context contains the sentences 'In Islamabad, it only rains in the monsoon' and 'It's monsoon in Islamabad,' both of which contain occurrences of 'Islamabad.'

The introduction and elimination rules for 'in' together with rules for the usual classical connectives and the two discharge rules mentioned above are jointly sufficient to capture all spatial consequences involving spatial expressions of the form $\ulcorner$In $l, \varphi\urcorner$ and $\ulcorner$This is $l\urcorner$.

The attribution of properties to places by expressions of the form $\ulcorner P(l)\urcorner$ has not been mentioned explicitly, but many of the logical properties of such expressions are captured by the usual rules of classical logic; in particular, the use of spatial quantifiers (such as 'wherever' and 'somewhere') to express generalizations concerning places can be captured by the same rules as those for individual quantifiers.

Special to the spatial case is the attribution of *relational* properties to places. Being on the other side of the world is a relation property, so expressions like 'Tokyo is on the other side of the world' are spatially-indexical: they relate the subject, Tokyo, to the indexically-determined current location of the stater. The logical properties of expressions using relational predicates is governed by the rules given above. For example, the following examples are proved using the introduction and elimination rules for 'in':

(7) This is Bloomington; Edinburgh is far away, so in Bloomington, Edinburgh is far away.

(8) In Edinburgh, London is to the south; this is Edinburgh, so London is to the South.

(9) SY 026 is across campus; the office of the Philosophy Dept. is SY 026, so the office of the Philosophy Dept. is across campus.

(7) and (8) are both provable using a single application of the *In*-elimination rule, and (9) require the principle of indiscernibility of identicals.

This completes the analysis of natural deduction using the spatial expressions $\ulcorner$ In $l, \varphi\urcorner$, $\ulcorner$This is $l\urcorner$ and $\ulcorner l_1$ is $l_2\urcorner$. We have omitted much, because we are only concerned to motivate those patterns of reasoning which may be lifted to the general case. Disanalogous aspects of the correspondence have been ignored, and there are some further possibilities for the Spatial Analogy which have not been explored, such as the analogy between $\trianglelefteq$ and spatial-part.

6 TWO CALCULI

The patterns of reasoning displayed in the previous section may be transferred to a more general setting, by virtue of the Spatial Analogy. First, we will need a few syntactic details.

In formulating a natural-deduction calculus it is convenient to have a stock of *formal parameters*. These are terms which do not occur in any formula of the language, but which may be used in hypothetical reasoning. We use 'a' to range over parameters. A formula (term) is *parametric* if it contains a parameter; otherwise, it is *non-parametric*. A term is *closed* if it contains no free variables; it may contain parameters. Given a formula φ, a parameter a and a term t, the formula φ^a_t is the result of replacing every occurrence of a in φ by t. A *δ-formula* is any formula of the form $\ulcorner\delta(\varphi, t)\urcorner$.

In addition to the usual introduction and elimination rules for the classical connectives (as given in [Prawitz, 1965], for example), we formulate rules for the predicate δ. They are directly analogous to those for $\ulcorner$In $l\urcorner$, φ.'

$$\frac{\varphi \quad t}{\delta(\varphi, t)}\,\delta\text{-I} \qquad\qquad \frac{\delta(\varphi, t) \quad t}{\varphi}\,\delta\text{-E}$$

In addition, we have two new ways in which premises may be discharged:

Term rule If t is a term which occurs as a premise of a proof, then all occurrences of t among the premises may be discharged at once, so long as the premises and conclusion of the resulting proof are all δ-formulae.

Parameter rule If a is a parameter which occurs as a premise of a proof, then all occurrences of a among the premises may be discharged at once, so long as the premises and conclusion of the resulting proof do not contain occurrences of a.

The Term rule permits the use of terms as hypothetical premises in a proof by stating that to prove an argument, it is legitimate to make the hypothetical assumption that the situation being described is named by a given term. The rule's side-condition is analogous to the first condition on hypothetical premises of the form $\ulcorner$This is $l\urcorner$ discussed in the previous section. The Parameter Rule enforces a condition analogous to the second condition on hypothetical spatial premises: it allows the reasoner to "name" the situation being described with a new parameter.

Finally, we need a rule capturing the fact that $\ulcorner\delta(t_1, t_2)\urcorner$ means that t_1 and t_2 have the same reference. By the Spatial Analogy, expression of the form

$\ulcorner\delta(t_{l_1}, t_{l_2})\urcorner$ correspond to identity statements of the form $\ulcorner l_1$ is $l_2\urcorner$. To capture the logic of the former expressions, we will need to import rules corresponding to the classical rules for identity. Reflexivity is already derivable:

$$\dfrac{\dfrac{[t]^1 \quad [t]^1}{\delta(t,t)}\delta\text{-I}}{\delta(t,t)}\text{Term}^1$$

But to capture the substitutivity of co-referring terms, we need the following Substitution rules:[5]

$$\dfrac{\delta(t_1, t_2) \quad \varphi^a_{t_1}}{\varphi^a_{t_2}}\text{Sub-1} \qquad\qquad \dfrac{\delta(t_2, t_1) \quad \varphi^a_{t_1}}{\varphi^a_{t_2}}\text{Sub-2}$$

The resulting natural-deduction calculus, which we shall call NK$_\delta$ is sound with respect to the restricted interpretation of our languages, as we hope the reader of the previous section is now convinced.[6] It is complete, because all the axioms listed in Section 4 are provable. For example,

$$\dfrac{\dfrac{\dfrac{\dfrac{\dfrac{[\delta(\delta(\varphi,a),b)]^2 \quad [b]^1}{\delta(\varphi,a)}\delta\text{-E}}{\delta(\varphi,a)}\text{Term}^1}{\delta(\delta(\varphi,a),b) \rightarrow \delta(\varphi,a)}\rightarrow\text{-I}^2 \quad \dfrac{\dfrac{\dfrac{[\delta(\varphi,a)]^4 \quad [b]^3}{\delta(\delta(\varphi,a),b)}\delta\text{-I}}{\delta(\delta(\varphi,a),b)}\text{Term}^3}{\delta(\varphi,a) \rightarrow \delta(\delta(\varphi,a),b)}\rightarrow\text{-I}^4}{\dfrac{\delta(\delta(\varphi,a),b) \leftrightarrow \delta(\varphi,a)}{\forall y(\delta(\delta(\varphi,a),y) \leftrightarrow \delta(\varphi,a))}\forall\text{-I}}\leftrightarrow\text{-I}}{\forall x \forall y(\delta(\delta(\varphi,x),y) \leftrightarrow \delta(\varphi,x))}\forall\text{-I}$$

Our final offering is a Gentzen-style sequent-calculus for the logic. Throughout, as above, 'φ' and 'ψ' range over sentences, 's' and 't' range over closed terms, 'a' ranges over parameters, and 'Γ' and 'Δ' range over sets of sentences. Standard abbreviations using ',' for set union on either side of '$\vdash$' will be used.

Structural Rules

$$\dfrac{}{\varphi \vdash \varphi}\text{I} \qquad\qquad \dfrac{\Gamma \vdash \Delta}{\Gamma', \Gamma \vdash \Delta, \Delta'}\text{W}$$

[5] The parameter a does *not* occur in any formula occurring in an application of one of these rules. The parameter is used merely as a placeholder to indicate the position in which the terms t_1 and t_2 are placed.

[6] If any doubt remains, the rules may be checked against the definition of correct-description given in Section 2, taking into account the additional assumptions of the restricted interpretation.

The familiar rules of Identity and Weakening are both sound, for the usual reasons.

Term Rules

$$\frac{a, \Gamma \vdash \Delta}{\Gamma \vdash \Delta}\text{P } (a \text{ new}) \qquad\qquad \frac{a, \Gamma \vdash \Delta}{\Gamma \vdash \Delta}\text{T } (\Gamma, \Delta \text{ are } \delta)$$

The P-rule is the sequent-calculus analogue of the Parameter rule. The side-condition requires a to be a parameter which does not occur in Γ or in Δ. It is sound because any situation s which is a counterexample to the validity of $\ulcorner\Gamma \vdash \Delta\urcorner$ may be named by a, so long as a is not already used for some other purpose.

The T-rule is the sequent-calculus analogue of the Term rule. The side-condition requires that Γ and Δ consist only of δ-formulae. It is sound, because any counterexample to $\ulcorner\Gamma \vdash \Delta\urcorner$ is a situation correctly described by each formula in Γ and no formula in Δ. But the formulae in Γ and Δ are required to be δ-formulae, so each formula in Γ and no formula in Δ is a correct description of $\underline{t}$, and so $\underline{t}$ is a counterexample to $\ulcorner t, \Gamma \vdash \Delta\urcorner$.

Substitution Rules

$$\frac{\Gamma^a_{t_1} \vdash \Delta^a_{t_1}}{\delta(t_1, t_2), \Gamma^a_{t_2} \vdash \Delta^a_{t_2}}\text{S}_1 \qquad\qquad \frac{\Gamma^a_{t_1} \vdash \Delta^a_{t_1}}{\delta(t_2, t_1), \Gamma^a_{t_2} \vdash \Delta^a_{t_2}}\text{S}_2$$

The S-rules are the sequent-calculus analogues of the Substitution rules. The rule S_1 is sound because if $\ulcorner\delta(t_1, t_2), \Gamma^a_{t_2} \vdash \Delta^a_{t_2}\urcorner$ has a counterexample s then $\ulcorner\delta(t_1, t_2)\urcorner$ correctly describes s and so $\underline{t_1} = \underline{t_2}$. Consequently, each formula $\varphi^a_{t_1}$ in $\Gamma^a_{t_1}, \Delta^a_{t_1}$ correctly describes s iff $\varphi^a_{t_2}$ does, and so s is a counterexample to $\ulcorner\Gamma^a_{t_1} \vdash \Delta^a_{t_1}\urcorner$.

Description Rules

$$\frac{\Gamma \vdash \Delta, t \quad \varphi, \Gamma \vdash \Delta}{\delta(\varphi, t), \Gamma \vdash \Delta}\text{L}\delta\!\downarrow \qquad\qquad \frac{\Gamma \vdash \Delta, t \quad \Gamma \vdash \Delta, \varphi}{\Gamma \vdash \Delta, \delta(\varphi, t)}\text{R}\delta\!\downarrow$$

$$\frac{\delta(\varphi, t), \Gamma \vdash \Delta}{t, \varphi, \Gamma \vdash \Delta}\text{L}\delta\!\uparrow (\varphi \text{ is non-}\delta) \qquad\qquad \frac{\Gamma \vdash \Delta, \delta(\varphi, t)}{t, \Gamma \vdash \Delta, \varphi}\text{R}\delta\!\uparrow (\varphi \text{ is non-}\delta)$$

The four rules for description are related to the natural-deduction rules δ-I and δ-E. We will demonstrate their soundness in turn.

$\text{L}\delta\!\downarrow$ rule. A counterexample to a sequent of the form $\ulcorner\delta(\varphi, t), \Gamma \vdash \Delta\urcorner$ is a situation s which is correctly described by all the formulae in $\ulcorner\delta(\varphi, t)\urcorner$, Γ and none of the formulae in Δ. In particular, s is correctly described by $\ulcorner\delta(\varphi, t)\urcorner$, and so $\underline{t}$ is correctly described by φ. There are two possibilities: either t names s or it doesn't. If $\underline{t} = s$ then s is a counterexample to $\ulcorner\varphi, \Gamma \vdash \Delta\urcorner$. Alternatively, if $\underline{t} \neq s$ then s is a counterexample to $\ulcorner\Gamma \vdash \Delta, t\urcorner$.

L$\delta\uparrow$ rule. A counterexample to a sequent of the form $\ulcorner t, \varphi\Gamma \vdash \Delta\urcorner$ is a situation s which is correctly described by all the formulae in t, φ, Γ and none of the formulae in Δ. Thus $\underline{t}$ (=s) is correctly described by φ, and so every situation, including s, is correctly described by $\ulcorner\delta(\varphi, t)\urcorner$. Hence s is a counterexample to $\ulcorner\delta(\varphi, t), \Gamma \vdash \Delta\urcorner$.

The right rules R$\delta\downarrow$ and R$\delta\uparrow$ by be seen to be sound by very similar arguments. Note that two *upward* rules, L$\delta\uparrow$ and R$\delta\uparrow$, fail to have the subformula property: the formula $\ulcorner\delta(\varphi, t)\urcorner$ need not occur as a subformula of any formula in the conclusion. This is unfortunate, but unavoidable (See [Seligman, 1991], for a discussion of this point). The damage is limited by the side-condition that φ be a non-δ-formula, which is not actually necessary for soundness.

Together with the standard rules for conjunction, disjunction, negation, universal and existential quantification (all of which are repeated in the Appendix) the above rules form a calculus which we call LK$_\delta$.

That each natural deduction in NK$_\delta$ is provable in LK$_\delta$ follows from the derivability of each of the natural-deduction rules, and the following theorem. The completeness of the calculus LK$_\delta$ therefore follows from the completeness of the natural-deduction system.

THEOREM 6.1 *If* $\Gamma \vdash \Delta, \varphi$ *and* $\varphi, \Gamma' \vdash \Delta'$ *then* $\Gamma, \Gamma' \vdash \Delta, \Delta'$.

A sketch of a proof of this theorem — an extension of Gentzen's *Haupstatz* for the system LK of classical first-order logic — is given in the Appendix.

By inspecting the rules of LK$_\delta$, we see that the logic is almost *analytic* in the proof-theoretic sense: whether or not a sequent has a proof is determined by whether or not its parts have proofs. To make this condition more precise, we say that a formula φ is a *quasi-subformula* of a set Σ of formulae if either

1. φ is a subformula of one of the formulae in Σ, or

2. $\varphi = \ulcorner\delta(\psi, t)\urcorner$ and ψ and t are subformulae of formulae in Σ.

THEOREM 6.2 *If* π *is a proof of* $\ulcorner\Gamma \vdash \Delta\urcorner$ *then there is a finite set* P *of parameters such that every formula occurring in* π *is a quasi-subformula of* P, Γ, Δ.

Proof. Let P be the set of parameters occurring in a proof π of $\ulcorner\Gamma \vdash \Delta\urcorner$. All the rules have the subformula property except the the P-rule and the upward δ-rules. The P-rule is covered by the inclusion of P in the set P, Γ, Δ. Hence every non-δ-formula occurring in π is a subformula of some formula in P, Γ, Δ. But now consider applications of the upward δ-rules:

$$\frac{\delta(\varphi, t), \Gamma \vdash \Delta}{t, \varphi, \Gamma \vdash \Delta}\text{L}\delta\uparrow \qquad \frac{\Gamma \vdash \Delta, \delta(\varphi, t)}{t, \Gamma \vdash \Delta, \varphi}\text{R}\delta\uparrow$$

The formula φ is required to be a non-δ-formula, and so is a subformula of some formula in P, Γ, Δ, by our previous remark. Likewise, the term t is subformula of some formula in the same set, and so the formula $\ulcorner\delta(\varphi, t)\urcorner$ is a quasi-subformula of P, Γ, Δ. $\dashv$

The theorem may be used to prove decidability and interpolation properties of various fragments of the logic. A discussion of fragments of the logic is given in [Seligman, 1992].

7 CONCLUDING REMARKS

We have examined the concept of correct description suggested by Austin's 1950 theory of truth. In the first part of the paper we gave an analysis of the concept in the form of a recursive definition, initially for a standard first-order language, and then for a language containing terms referring to situations and a correct-description predicate 'δ'. We made some idealizing assumptions about situations and gave a complete axiomatization of the resulting logic. In the second part we studied natural patterns of reasoning using 'δ' by way of the example of spatially-indexical language. This was developed into two formal calculi: a natural-deduction calculus NK_δ and a sequent-calculus LK_δ. The latter was shown to admit Cut and to be (almost) analytic.

APPENDIX: *HAUPSTATZ* FOR LK_δ

The System LK_δ. (We omit the rules for implication and identity for the sake of brevity. Each may be considered to be defined in terms of the other connectives: $\ulcorner\varphi \to \psi\urcorner$ is just $\ulcorner\neg\varphi \vee \psi\urcorner$ and $\ulcorner t_1 = t_2\urcorner$ is $\ulcorner\delta(t_1, t_2)\urcorner$. In both cases, the equivalence between definiens and definiendum is provable in the system augmented with the appropriate rules.)

$$\frac{}{\varphi \vdash \varphi}\text{I} \qquad \frac{\Gamma \vdash \Delta}{\Gamma', \Gamma \vdash \Delta, \Delta'}\text{W} \qquad \frac{a, \Gamma \vdash \Delta}{\Gamma \vdash \Delta}\text{P}\ (a \text{ new}) \qquad \frac{a, \Gamma \vdash \Delta}{\Gamma \vdash \Delta}\text{T}\ (\Gamma, \Delta \text{ are } \delta)$$

$$\frac{\Gamma^a_{t_1} \vdash \Delta^a_{t_1}}{\delta(t_1, t_2), \Gamma^a_{t_2} \vdash \Delta^a_{t_2}}\text{S}_1 \qquad \frac{\Gamma^a_{t_1} \vdash \Delta^a_{t_1}}{\delta(t_2, t_1), \Gamma^a_{t_2} \vdash \Delta^a_{t_2}}\text{S}_2$$

$$\frac{\Gamma \vdash \Delta, t \qquad \varphi, \Gamma \vdash \Delta}{\delta(\varphi, t), \Gamma \vdash \Delta}\text{L}\delta\downarrow \qquad \frac{\Gamma \vdash \Delta, t \qquad \Gamma \vdash \Delta, \varphi}{\Gamma \vdash \Delta, \delta(\varphi, t)}\text{R}\delta\downarrow$$

$$\frac{\delta(\varphi, t), \Gamma \vdash \Delta}{t, \varphi, \Gamma \vdash \Delta}\text{L}\delta\uparrow\ (\varphi \text{ are non-}\delta) \qquad \frac{\Gamma \vdash \Delta, \delta(\varphi, t)}{t, \Gamma \vdash \Delta, \varphi}\text{R}\delta\uparrow\ (\varphi \text{ non-}\delta)$$

$$\frac{\varphi, \psi, \Gamma \vdash \Delta}{\varphi \wedge \psi, \Gamma \vdash \Delta}\text{L}\wedge \qquad \frac{\varphi, \Gamma \vdash \Delta \qquad \psi, \Gamma \vdash \Delta}{\varphi \vee \psi, \Gamma \vdash \Delta}\text{L}\vee \qquad \frac{\Gamma \vdash \Delta, \varphi}{\neg\varphi, \Gamma \vdash \Delta}\text{L}\neg$$

$$\frac{\Gamma \vdash \Delta, \varphi \qquad \Gamma \vdash \Delta, \psi}{\Gamma \vdash \Delta, \varphi \wedge \psi}\text{R}\wedge \qquad \frac{\Gamma \vdash \Delta, \varphi, \psi}{\Gamma \vdash \Delta, \varphi \vee \psi}\text{R}\vee \qquad \frac{\varphi, \Gamma \vdash \Delta}{\Gamma \vdash \Delta, \neg\varphi}\text{R}\neg$$

$$\frac{\varphi^x_t, \Gamma \vdash \Delta}{\forall x.\varphi, \Gamma \vdash \Delta}\text{L}\forall \qquad \frac{\Gamma \vdash \Delta, \varphi}{\Gamma \vdash \Delta, \forall x.\varphi^a_x}\text{R}\forall\ (a \text{ new})$$

$$\frac{\varphi, \Gamma \vdash \Delta}{\exists x.\varphi^a_x, \Gamma \vdash \Delta}\text{L}\exists\ (a \text{ new}) \qquad \frac{\Gamma \vdash \Delta, \varphi^x_t}{\Gamma \vdash \Delta, \exists x.\varphi}\text{R}\exists$$

In order to demonstrate the admissibility of Cut in the system LK_δ, we extend it to the system LK^*_δ by adding the *S-axioms*

$$\frac{}{\delta(t_1, t_2), \varphi^a_{t_2} \vdash \psi^a_{t_2}}\text{S}_1\text{ax} \qquad \frac{}{\delta(t_2, t_1), \varphi^a_{t_2} \vdash \psi^a_{t_2}}\text{S}_2\text{ax}$$

for formulae φ and ψ such that $\varphi^a_{t_1} = \psi^a_{t_1}$, and three *Cut-rules* of the form

$$\frac{\Gamma \vdash \Delta, \varphi_1 \qquad \varphi_2, \Gamma' \vdash \Delta'}{\varphi_3\Gamma, \Gamma' \vdash \Delta, \Delta'}C(\varphi_1, \varphi_2, \varphi_3)$$

given by

1. C = all instances of $\text{C}(\varphi, \varphi, \emptyset)$
2. $\text{C}\delta\text{L}$ = all instances of $\text{C}(\varphi, \delta(\varphi, t), t)$
3. $\text{C}\delta\text{R}$ = all instances of $\text{C}(\delta(\varphi, t), \varphi, t)$

In each case, we call φ the *cut-formula* of the rule.

We show that any proof in LK^*_δ may be replaced by a proof in LK_δ with the same end-sequent. We order formulae in complexity by the number of connectives they contain. A *cut* in a proof is an application of one of the Cut-rules. A cut c in a proof is *at least as complex as* another cut c' if

1. the cut-formula of c is at least as complex as that of c', and
2. if the cut-formula of c is of the same complexity as that of c', then the sum of the heights of the proofs of the premises c is at least as great as the corresponding sum for c'.

A proof π is *at least as complex as* another proof π' if there is a cut occurring in π which is at least as complex as all the cuts occurring in π'. A proof is *S-reduced* if it contains no application of the S-rules (it may have S-axioms).

CLAIM 1 *For any proof π in LK^*_δ there is an S-reduced proof π' in LK^*_δ with the same end-sequent and of no greater complexity.*

Proof of Claim. It is routine to check that the S-rules commute with every other rule in LK^*_δ. Such transformations do not change the complexity of the proof. The only case worthy of suspicion is that of the T-rule, because of its side-condition; but $\delta(t_1, t_2)$ is a δ-formula, and so the rules commute without difficulty. In this way, applications of the S-rules may be pushed up the proof-tree until they reach the leaves. At the leaves, they may be replaced by S-axioms as follows:

$$\dfrac{\dfrac{}{\varphi^a_{t_1} \vdash \psi^a_{t_1}}\,\mathrm{I}}{\delta(t_1, t_2), \varphi^a_{t_2} \vdash \psi^a_{t_2}}\,\mathrm{S}_1 \quad \rightsquigarrow \quad \dfrac{}{\delta(t_1, t_2), \varphi^a_{t_2} \vdash \psi^a_{t_2}}\,\mathrm{S}_1\mathrm{ax}$$

where $\varphi^a_{t_1} = \psi^a_{t_1}$, as required for the I-rule and for the S-axiom. The transformation for S_2 is similar. Note that this process does not change the logical structure of the cut-formulae of any of the cuts in the proof, and so the complexity is not increased. $\dashv$

CLAIM 2 *Suppose there is an S-reduced proof π in LK^*_δ ending in a cut*

$$\dfrac{\begin{matrix}\pi_1 \\ \Gamma \vdash \Delta, \varphi_1\end{matrix} \qquad \begin{matrix}\pi_2 \\ \varphi_2, \Gamma' \vdash \Delta'\end{matrix}}{\varphi_3\Gamma, \Gamma' \vdash \Delta, \Delta'}$$

of maximal complexity among the cuts in π. Then there is an S-reduced proof in LK^*_δ of the sequent $\varphi_3\Gamma, \Gamma' \vdash \Delta, \Delta'$ which is strictly less complex than π.

Proof of Claim. The structure of the proof is very similar to that of cut-elimination for classical logic, so we will only give a sketch, indicating the points of difference. There are various cases, according to the final steps of π_1 and π_2.

1. π_1 is an S-axiom. W.l.o.g. we may assume that it is an S_1-axiom. Depending on the Cut-rule, we may replace π as follows.

$$
\dfrac{\dfrac{}{\delta(t_1,t_2),\varphi^a_{t_2}\vdash\psi^a_{t_2}}\,\mathrm{S_1ax}\qquad \dfrac{\pi_2}{\psi^a_{t_2},\Gamma'\vdash\Delta'}}{\delta(t_1,t_2),\varphi^a_{t_2},\Gamma'\vdash\Delta'}\,\mathrm{C}
\quad\leadsto\quad
\dfrac{\begin{array}{c}\dfrac{\dfrac{\pi_2}{\psi^a_{t_2},\Gamma'\vdash\Delta'}}{\delta(t_1,t_2),\psi^a_{t_1},\Gamma'\vdash\Delta'}\,\mathrm{S_2}\\ = \\ \delta(t_1,t_2),\varphi^a_{t_1},\Gamma'\vdash\Delta'\end{array}}{\delta(t_1,t_2),\varphi^a_{t_2},\Gamma'\vdash\Delta'}\,\mathrm{S_1}
$$

$$
\dfrac{\dfrac{}{\delta(t_1,t_2),\varphi^a_{t_2}\vdash\psi^a_{t_2}}\,\mathrm{S_1ax}\qquad \dfrac{\pi_2}{\delta(\psi^a_{t_2},t),\Gamma'\vdash\Delta'}}{t,\delta(t_1,t_2),\varphi^a_{t_2},\Gamma'\vdash\Delta'}\,\mathrm{C\delta L}
\quad\leadsto\quad
\dfrac{\begin{array}{c}\dfrac{\dfrac{\dfrac{\pi_2}{\delta(\psi^a_{t_2},t),\Gamma'\vdash\Delta'}}{\psi^a_{t_2},t,\Gamma'\vdash\Delta'}\,\mathrm{L\delta\uparrow}}{\delta(t_1,t_2),\psi^a_{t_1},t,\Gamma'\vdash\Delta'}\,\mathrm{S_2}\\ = \\ \delta(t_1,t_2),\varphi^a_{t_1},t,\Gamma'\vdash\Delta'\end{array}}{\delta(t_1,t_2),\varphi^a_{t_2},t,\Gamma'\vdash\Delta'}\,\mathrm{S_1}
$$

$$
\dfrac{\dfrac{}{\delta(t_1,t_2),\delta(\varphi^a_{t_2},t^a_{t_2})\vdash\delta(\psi^a_{t_2},u^a_{t_2})}\,\mathrm{S_1ax}\qquad \dfrac{\pi_2}{\psi^a_{t_2},\Gamma'\vdash\Delta'}}{u^a_{t_2},\delta(t_1,t_2),\delta(\varphi^a_{t_2},t^a_{t_2}),\Gamma'\vdash\Delta'}\,\mathrm{C\delta L}
\quad\leadsto
$$

$$
\dfrac{\begin{array}{c}\dfrac{\dfrac{\dfrac{\dfrac{}{u^a_{t_2}\vdash u^a_{t_2}}\,\mathrm{I}}{u^a_{t_2},\Gamma'\vdash\Delta',u^a_{t_2}}\,\mathrm{W}\qquad \dfrac{\dfrac{\pi_2}{\psi^a_{t_2},\Gamma'\vdash\Delta'}}{\psi^a_{t_2},u^a_{t_2},\Gamma'\vdash\Delta'}\,\mathrm{W}}{u^a_{t_2},\delta(\psi^a_{t_2},u^a_{t_2}),\Gamma'\vdash\Delta'}\,\mathrm{L\delta\downarrow}}{u^a_{t_2},\delta(t_1,t_2),\delta(\psi^a_{t_1},u^a_{t_1}),\Gamma'\vdash\Delta'}\,\mathrm{S_2}\\ = \\ u^a_{t_2},\delta(t_1,t_2),\delta(\varphi^a_{t_1},t^a_{t_1}),\Gamma'\vdash\Delta'\end{array}}{u^a_{t_2},\delta(t_1,t_2),\delta(\varphi^a_{t_2},t^a_{t_2}),\Gamma'\vdash\Delta'}\,\mathrm{S_1}
$$

(We assume that $\varphi^a_{t_1} = \psi^a_{t_1}$ and $t^a_{t_1} = u^a_{t_1}$, as required for the S-axiom, and that π_2 does not contain the parameter a; if it does, we may first replace all occurrences of a by a fresh parameter.)

These transformations give us a proof in LK^*_δ which is strictly less complex than π, and by Claim 1 we may convert it into an S-reduced proof of no greater complexity.

2. π_1 ends in a structural rule: I, W, or one of the Cut-rules. In each case, it is routine to check that this rule commutes with the final cut of π, resulting in a decrease in the height of the proof above the cut of maximal complexity, and so a decrease in the complexity of the proof.

3. π_1 ends in an application of the P-rule. We assume that the parameter a does not occur in π_2; if it does, we must first replace every occurrence of a in π_1

with a fresh parameter.

$$\dfrac{\dfrac{\overset{\pi_1'}{a,\Gamma\vdash\Delta,\varphi_1}}{\Gamma\vdash\Delta,\varphi_1}\text{P}\qquad \overset{\pi_2}{\varphi_2,\Gamma'\vdash\Delta'}}{\varphi_3,\Gamma,\Gamma'\vdash\Delta,\Delta'}\quad\rightsquigarrow\quad\dfrac{\dfrac{\overset{\pi_1'}{a,\Gamma\vdash\Delta,\varphi_1}\qquad\overset{\pi_2}{\varphi_2,\Gamma'\vdash\Delta'}}{\varphi_3,a,\Gamma,\Gamma'\vdash\Delta,\Delta'}}{\Gamma,\Gamma'\vdash\Delta,\Delta'}\text{P}$$

4. π_1 ends in an application of one of the logical rules (L$\wedge$, R$\wedge$, L$\vee$, R$\vee$, L$\neg$, R$\neg$, L$\forall$, R$\forall$, L$\exists$, R$\exists$, L$\delta\downarrow$, R$\delta\downarrow$, L$\delta\uparrow$ or R$\delta\uparrow$) but φ_1 is *not* the principal formula of that rule. In each case, it is routine to check that each of these rules commute with the final cut of π. As before, the resulting proof is of lower complexity because of the decrease in height of the proofs of the left premises of maximal cuts. We shall illustrate the point with the $L\delta$-rule:

$$\dfrac{\dfrac{\overset{\pi_1'}{\Gamma\vdash\Delta,\varphi_1,t}\qquad\overset{\pi_1''}{\psi,\Gamma\vdash\Delta,\varphi_1}}{\delta(\psi,t),\Gamma\vdash\Delta,\varphi_1}\text{L}\delta\downarrow\qquad\overset{\pi_2}{\varphi_2,\Gamma'\vdash\Delta'}}{\varphi_3,\delta(\psi,t),\Gamma,\Gamma'\vdash\Delta,\Delta'}\text{cut}\quad\rightsquigarrow$$

$$\dfrac{\dfrac{\overset{\pi_1'}{\Gamma\vdash\Delta,\varphi_1,t}\qquad\overset{\pi_2}{\varphi_2,\Gamma'\vdash\Delta'}}{\varphi_3,\Gamma,\Gamma'\vdash\Delta,\Delta',t}\text{cut}\qquad\dfrac{\overset{\pi_1''}{\psi,\Gamma\vdash\Delta,\varphi_1}\qquad\overset{\pi_2}{\varphi_2,\Gamma'\vdash\Delta'}}{\varphi_3,\psi,\Gamma,\Gamma'\vdash\Delta,\Delta'}\text{cut}}{\varphi_3,\delta(\psi,t),\Gamma,\Gamma'\vdash\Delta,\Delta'}\text{L}\delta\downarrow$$

Note that π_1' and π_1'' are both shorter than π_1, and so the two cuts in the resulting proof are both less complex than the cut in π.

5. π_1 ends in an application of the T-rule *and the formulae in* φ_2,Γ,Δ *are all* δ*-formulae*. The cut must be an application of the C-rule ($\varphi_1=\varphi_2=\varphi$, $\varphi_3=\emptyset$) because the other Cut-rules require either φ_1 or φ_2 to be a non-δ-formula. In this case, the T-rule commutes with the cut:

$$\dfrac{\dfrac{\overset{\pi_1'}{t,\Gamma\vdash\Delta,\varphi}}{\Gamma\vdash\Delta,\varphi}\text{T}\qquad\overset{\pi_2}{\varphi,\Gamma'\vdash\Delta'}}{\Gamma,\Gamma'\vdash\Delta,\Delta'}\text{C}\quad\rightsquigarrow\quad\dfrac{\dfrac{\overset{\pi_1'}{t,\Gamma\vdash\Delta,\varphi}\qquad\overset{\pi_2}{\varphi,\Gamma'\vdash\Delta'}}{t,\Gamma,\Gamma'\vdash\Delta,\Delta'}}{\Gamma,\Gamma'\vdash\Delta,\Delta'}\text{T}$$

The condition that φ,Γ',Δ' are all δ-formulae is important because if one of these formulae was non-δ then the side-condition on the application of the T-rule in the resulting proof would be invalid.

6. – 10. The analogous cases for π_2 instead of π_1.

The above includes all cases except those in which either (i) φ_1 and φ_2 are principal formulae of the final steps of π_1 and π_2 respectively, or (ii) one of the proofs ends in an application of the T-rule, but the condition in 5. (or 10.) does not hold. We now turn to the second, somewhat tricky, case.

11. π_1 ends in an application of the T-rule but some of the formulae in $\varphi_2, \Gamma', \Delta'$ are non-δ-formulae. We consider the Cut-rules separately.

    (a) If the cut of π is an application of the CδL-rule then there is a formula ψ and a term u such that $\varphi_1 = \delta(\psi, u)$, $\varphi_2 = \psi$ and $\varphi_3 = u$, and so we are faced with a situation like this:

$$\dfrac{\dfrac{\begin{matrix}\pi_1'\\ t, \Gamma \vdash \Delta, \delta(\psi, u)\end{matrix}}{\Gamma \vdash \Delta, \delta(\psi, u)}\,\mathrm{T} \qquad \begin{matrix}\pi_2\\ \psi, \Gamma' \vdash \Delta'\end{matrix}}{u, \Gamma, \Gamma' \vdash \Delta, \Delta'}\,\mathrm{C}\delta\mathrm{L}$$

By repeated application of the δ-rules we may "protect" the non-δ-formulae of ψ, Γ', Δ' by embedding them inside δ-formulae. Now, let $\delta(\Gamma', u)$ be the set of formulae $\delta(\gamma, u)$ for each non-δ-formula γ of Γ', together with the δ-formulae of Γ'; and similarly for $\delta(\Delta', u)$. Now replace π by the following proof:

$$\begin{matrix}\dfrac{\dfrac{\begin{matrix}\pi_1'\\ t, \Gamma \vdash \Delta, \delta(\psi, u)\end{matrix}}{\Gamma \vdash \Delta, \delta(\psi, u)}\,\mathrm{T} \qquad \dfrac{\begin{matrix}\dfrac{\begin{matrix}\pi_2\\ \psi, \Gamma' \vdash \Delta'\end{matrix}}{u, \psi, \Gamma' \vdash \Delta'}\,\mathrm{W}\\ \vdots\ \ \mathrm{L}\delta{\downarrow} \text{ and } \mathrm{R}\delta{\downarrow}\\ u, \delta(\psi, u), \delta(\Gamma', u) \vdash \delta(\Delta', u)\end{matrix}}{\delta(\psi, u), \delta(\Gamma', u) \vdash \delta(\Delta', u)}\,\mathrm{T}}{\Gamma, \delta(\Gamma', u) \vdash \Delta, \delta(\Delta', u)}\,\mathrm{C}\\ \vdots\ \ \mathrm{L}\delta{\uparrow} \text{ and } \mathrm{R}\delta{\uparrow}\\ u, \Gamma, \Gamma' \vdash \Delta, \Delta'\end{matrix}$$

Call this proof π'. The maximal cut in π' is actually *more* complex than the maximal cuts in π because its cut-formula is $\delta(\psi, u)$ instead of ψ. However, the right premise of the cut is composed only of δ-formulae. Repeating the argument given cases 1.–5., we see that π' may be simplified until the cut-formula $\delta(\psi, u)$ is the principal formula of the last step in the proof of the left premise. We then have the following situation (possibly repeated many times):

$$
\dfrac{\dfrac{\begin{matrix}\pi_0' \\ \Gamma_0 \vdash \Delta_0, u\end{matrix} \qquad \begin{matrix}\pi_0'' \\ \Gamma_0 \vdash \Delta_0, \psi\end{matrix}}{\Gamma_0 \vdash \Delta_0, \delta(\psi, u)}\text{R}\delta{\downarrow} \qquad \dfrac{\begin{matrix}\pi_2 \\ \dfrac{\psi, \Gamma' \vdash \Delta'}{u, \psi, \Gamma' \vdash \Delta'}\text{W} \\ \vdots\ \text{L}\delta{\downarrow}\text{ and R}\delta{\downarrow} \\ u, \delta(\psi, u), \delta(\Gamma', u) \vdash \delta(\Delta', u)\end{matrix}}{\delta(\psi, u), \delta(\Gamma', u) \vdash \delta(\Delta', u)}\text{T}}{\Gamma_0, \delta(\Gamma', u) \vdash \Delta_0, \delta(\Delta', u)}\text{C}
$$

Each maximal cut in π' will be of this form, and may be replaced by the following:

$$
\begin{matrix}
\dfrac{\begin{matrix}\pi_0' \\ \Gamma_0 \vdash \Delta_0, u\end{matrix} \qquad \dfrac{\begin{matrix}\pi_0'' \\ \Gamma_0 \vdash \Delta_0, \psi\end{matrix} \qquad \begin{matrix}\pi_2 \\ \psi, \Gamma' \vdash \Delta'\end{matrix}}{\Gamma_0, \Gamma' \vdash \Delta_0, \Delta'}\text{C}}{\vdots}\text{L}\delta{\downarrow}\text{ and R}\delta{\downarrow} \\
\Gamma_0, \delta(\Gamma', u) \vdash \Delta_0, \delta(\Delta', u)
\end{matrix}
$$

Each of these cuts is less complex than the cut in π that we started with, because π_0'' is necessarily shorter than π_1, and so we have reduced the complexity of π, as required.

(b) If the cut of π is an application of the C-rule then $\varphi_1 = \varphi_2 = \varphi$ and $\varphi_3 = \emptyset$, and so we are faced with a situation like this:

$$
\dfrac{\dfrac{\begin{matrix}\pi_1' \\ t, \Gamma \vdash \Delta, \varphi\end{matrix}}{\Gamma \vdash \Delta, \varphi}\text{T} \qquad \begin{matrix}\pi_2 \\ \varphi, \Gamma' \vdash \Delta'\end{matrix}}{\Gamma, \Gamma' \vdash \Delta, \Delta'}\text{C}
$$

The side-condition on the T-rule requires φ to be a δ-formula, so $\varphi = \delta(\psi, u)$ for some formula ψ and term u. Moreover, we may assume that φ is the principal formula of the final step of π_2; otherwise this case would be subsumed by one of the cases 6.–10. (and if it is case 10. then we may be assured that the extra condition is satisfied, because each of the formulae Γ, Δ, φ is a δ-formula, by the side-condition on the T-rule). Thus we have the following:

$$
\dfrac{\dfrac{\begin{matrix}\pi_1' \\ t, \Gamma \vdash \Delta, \delta(\psi, u)\end{matrix}}{\Gamma \vdash \Delta, \delta(\psi, u)}\text{T} \qquad \dfrac{\begin{matrix}\pi_2' \\ \Gamma' \vdash \Delta', u\end{matrix} \qquad \begin{matrix}\pi_2'' \\ \psi, \Gamma' \vdash \Delta'\end{matrix}}{\delta(\psi, u), \Gamma' \vdash \Delta'}\text{L}\delta{\downarrow}}{\Gamma, \Gamma' \vdash \Delta, \Delta'}\text{C}
$$

which may be modified in the following way. Let a be a fresh parameter.

$$
\dfrac{
\dfrac{\begin{array}{c}\pi_1'\\ t, \Gamma \vdash \Delta, \delta(\psi, u)\end{array}}{\Gamma \vdash \Delta, \delta(\psi, u)}\,\text{T}
\qquad
\dfrac{\begin{array}{c}
\dfrac{\dfrac{\begin{array}{c}\pi_2' \\ \Gamma' \vdash \Delta', u\end{array} \qquad \begin{array}{c}\pi_2'' \\ \psi, \Gamma' \vdash \Delta'\end{array}}{\delta(\psi, u), \Gamma' \vdash \Delta'}\,\text{L}\delta{\downarrow}}{a, \delta(\psi, u), \Gamma' \vdash \Delta'}\,\text{W} \\
\vdots \quad \text{L}\delta{\downarrow} \text{ and } \text{R}\delta{\downarrow} \\
a, \delta(\psi, u), \delta(\Gamma', a) \vdash \delta(\Delta', a)
\end{array}}{\delta(\psi, u), \delta(\Gamma', a) \vdash \delta(\Delta', a)}\,\text{T}
}{
\begin{array}{c}
\Gamma, \delta(\Gamma', a) \vdash \Delta, \delta(\Delta', a) \\
\vdots \quad \text{L}\delta{\uparrow} \text{ and } \text{R}\delta{\uparrow} \\
\dfrac{a, \Gamma, \Gamma' \vdash \Delta, \Delta'}{\Gamma, \Gamma' \vdash \Delta, \Delta'}\,\text{P}
\end{array}
}\,\text{C}
$$

Again the result is a proof π' which is more complex than π, but which may be converted into a less complex proof by manoeuvres exactly analogous to those made in the previous case.[7]

(c) If the cut of π is an application of the CδR-rule then there is a non-δ-formula ψ and a term u such that $\varphi_1 = \psi$, $\varphi_2 = \delta(\psi, u)$ and $\varphi_3 = u$. This conflicts with the side-condition on the T-rule, which requires φ_1 to be a δ-formula, so this case is void.

12. π_2 ends in an application of the T-rule but some of the formulae in $\varphi_1, \Gamma, \Delta$ are non-δ-formulae. This case is handled in a way directly analogous to case 11.

13. Finally, the only possibility we are left with is that both π_1 and π_2 end in applications of rules whose principal formulae are φ_1 and φ_2 respectively. If the final cut of π is an application of the CδL-rule or CδR-rule then we use the following transformations.

$$
\dfrac{\dfrac{\begin{array}{c}\pi_1' \\ \Gamma \vdash \Delta, t\end{array} \quad \begin{array}{c}\pi_1'' \\ \Gamma \vdash \Delta, \psi\end{array}}{\Gamma \vdash \Delta, \delta(\psi, t)}\,\text{R}\delta{\downarrow} \qquad \begin{array}{c}\pi_2 \\ \psi, \Gamma' \vdash \Delta'\end{array}}{t, \Gamma, \Gamma' \vdash \Delta, \Delta'}\,\text{C}\delta\text{L}
\quad \rightsquigarrow \quad
\dfrac{\dfrac{\begin{array}{c}\pi_1'' \\ \Gamma \vdash \Delta, \psi\end{array} \quad \begin{array}{c}\pi_2 \\ \psi, \Gamma' \vdash \Delta'\end{array}}{\Gamma, \Gamma' \vdash \Delta, \Delta'}\,\text{C}}{t, \Gamma, \Gamma' \vdash \Delta, \Delta'}\,\text{W}
$$

$$
\dfrac{\begin{array}{c}\pi_1 \\ \Gamma \vdash \Delta, \psi\end{array} \qquad \dfrac{\begin{array}{c}\pi_2' \\ \Gamma' \vdash \Delta', t\end{array} \quad \begin{array}{c}\pi_2'' \\ \psi, \Gamma' \vdash \Delta'\end{array}}{\delta(\psi, t), \Gamma' \vdash \Delta'}\,\text{L}\delta{\downarrow}}{t, \Gamma, \Gamma' \vdash \Delta, \Delta'}\,\text{C}\delta\text{R}
\quad \rightsquigarrow \quad
\dfrac{\dfrac{\begin{array}{c}\pi_1 \\ \Gamma \vdash \Delta, \psi\end{array} \quad \begin{array}{c}\pi_2'' \\ \psi, \Gamma' \vdash \Delta'\end{array}}{\Gamma, \Gamma' \vdash \Delta, \Delta'}\,\text{C}}{t, \Gamma, \Gamma' \vdash \Delta, \Delta'}\,\text{W}
$$

If the cut of π is an application of the C-rule, we need to examine the various cases.

(a) π_1 ends in an application of the R$\delta{\uparrow}$-rule:

[7] When ψ is not a term, this whole case may be reduced to the previous case.

$$\dfrac{\dfrac{\begin{matrix}\pi_1' \\ \Gamma \vdash \Delta, \delta(\varphi, t)\end{matrix}}{t, \Gamma \vdash \Delta, \varphi}\text{R}\delta\uparrow \qquad \begin{matrix}\pi_2 \\ \varphi, \Gamma' \vdash \Delta'\end{matrix}}{t, \Gamma, \Gamma' \vdash \Delta, \Delta'}\text{C} \quad \leadsto \quad \dfrac{\begin{matrix}\pi_1' \\ \Gamma \vdash \Delta, \delta(\varphi, t)\end{matrix} \qquad \begin{matrix}\pi_2 \\ \varphi, \Gamma' \vdash \Delta'\end{matrix}}{t, \Gamma, \Gamma' \vdash \Delta, \Delta'}\text{C}\delta\text{L}$$

(b) π_2 ends in an application of the L$\delta\uparrow$-rule:

$$\dfrac{\begin{matrix}\pi_1 \\ \Gamma \vdash \Delta, \varphi\end{matrix} \qquad \dfrac{\begin{matrix}\pi_2' \\ \delta(\varphi, t), \Gamma' \vdash \Delta'\end{matrix}}{\varphi, t, \Gamma' \vdash \Delta'}\text{L}\delta\uparrow}{t, \Gamma, \Gamma' \vdash \Delta, \Delta'}\text{C} \quad \leadsto \quad \dfrac{\begin{matrix}\pi_1 \\ \Gamma \vdash \Delta, \varphi\end{matrix} \qquad \begin{matrix}\pi_2' \\ \delta(\varphi, t), \Gamma' \vdash \Delta'\end{matrix}}{t, \Gamma, \Gamma' \vdash \Delta, \Delta'}\text{C}\delta\text{R}$$

(c) π_1 and π_2 end in applications of the right and left rules, respectively, for $\wedge$, $\vee$, $\neg$, $\exists$ or $\forall$: use the classical transformations.

(d) π_1 ends in an application of R$\delta\downarrow$ and π_2 ends in an application of L$\delta\downarrow$:

$$\dfrac{\dfrac{\begin{matrix}\pi_1' \\ \Gamma \vdash \Delta, t\end{matrix} \quad \begin{matrix}\pi_1'' \\ \Gamma \vdash \Delta, \psi\end{matrix}}{\Gamma \vdash \Delta, \delta(\psi, t)}\text{R}\delta\downarrow \qquad \dfrac{\begin{matrix}\pi_2' \\ \Gamma' \vdash \Delta', t\end{matrix} \quad \begin{matrix}\pi_2'' \\ \psi, \Gamma' \vdash \Delta'\end{matrix}}{\delta(\psi, t), \Gamma' \vdash \Delta'}\text{L}\delta\downarrow}{\Gamma, \Gamma' \vdash \Delta, \Delta'}\text{C} \quad \leadsto \quad \dfrac{\begin{matrix}\pi_1'' \\ \Gamma \vdash \Delta, \psi\end{matrix} \quad \begin{matrix}\pi_2'' \\ \psi, \Gamma' \vdash \Delta'\end{matrix}}{\Gamma, \Gamma' \vdash \Delta, \Delta'}\text{C}$$

This completes the proof of Claim 2. ⊣

The theorem may now be proved. Each proof in LK_δ^* may be converted into an S-reduced proof of no greater complexity, by Claim 1. This proof has a finite number of cuts of maximal complexity. Claim 2 shows how a sub-proof ending in each one these maximal cuts may be replaced by a less complex proof. Apply this operation to each of the maximal cuts, and the resulting proof will be strictly less complex than the original. Thus, by induction on the complexity of proofs, we see that every LK_δ^* may be converted into a proof of minimal complexity. The only LK_δ^* proofs of minimal complexity are those of LK_δ augmented with S-axioms. Finally, the S-axioms are obviously derivable in LK_δ, using the S-rules, and so the theorem is proved.

Institute of Philosophy
National Chung Cheng University

REFERENCES

[Austin, 1950] J.L. Austin. *Truth.* Proceedings of the Aristotelian Society, 1950.

[Barwise, 1989] J. Barwise. *The Situation in Logic*. CSLI Publications, Stanford, CA, 1989.

[Barwise and Etchemendy, 1987] J. Barwise and J. Etchemendy. *The Liar: An Essay in Truth and Circularity*. Basil Blackwell, Oxford, 1987.

[Barwise and Perry, 1983] J. Barwise and J. Perry. *Situations and Attitudes*. The MIT Press, Cambridge, MA, 1983.

[Blackburn, 1993] P. Blackburn. Nominal tense logic. *Notre Dame Journal of Formal Logic*, 34:56–83, 1993.

[Cooper, 1991] R. Cooper. Persistence and structural determination. In J. Barwise, M.J. Gawron, G. Plotkin, and S. Tutiya, editors, *Situation Theory and its Applications, Vol. 2*, CSLI Publications, Stanford, CA, 1991.

[Cooper and Kamp, 1991] R. Cooper and H. Kamp. Negation in situation semantics and discourse representation theory. In J. Barwise, M.J. Gawron, G. Plotkin, and S. Tutiya, editors, *Situation Theory and its Applications, Vol. 2*, CSLI Publications, Stanford, CA, 1991.

[Fernando, 1990] T. Fernando. On the logic of situation theory. In R. Cooper, K. Mukai, and J. Perry, editors, *Situation Theory and its Applications, Vol. 1*, CSLI Publications, Stanford, CA, 1990.

[Gargov and Goranko, 1993] G. Gargov and V. Goranko. Modal logic with names. *Journal of Philosophical Logic*, 22:607–636, 1993.

[Kleene, 1952] S.C. Kleene. *Introduction to Metamathematics*. North-Holland, Amsterdam, 1952.

[Langholm, 1988] T. Langholm. *Partiality, Truth and Persistence*. CSLI Publications, Stanford, CA, 1988.

[Langholm, 1989] T. Langholm. Algorithms for partial logic. COSMOS Report 8, Dept. of Mathematics, Oslo, 1989.

[Prawitz, 1965] D. Prawitz. *Natural Deduction. A Proof-Theoretical Study*, volume 3 of *Acta Universitatis Stockholmiensis*. Almqvist & Wiksell, Stockholm, 1965.

[Prior, 1967] A. Prior. *Past, Present and Future*. Oxford University Press, 1967.

[Seligman, 1991] J. Seligman. A cut-free sequent calculus for elementary situated reasoning. Technical Report HCRC-RP 22, HCRC, Edinburgh, 1991.

[Seligman, 1992] J. Seligman. Situated consequence for elementary situation theory. Logic group preprint IULG-92-16, Indiana University, 1992.

[Seligman and ter Meulen, 1994] J. Seligman and A. ter Meulen. Dynamic aspect trees. In M. Masuch and L. Polos, editors, *Applied Logic: How, What and Why*, pages 287–320. Kluwer Academic Publishers, Dordrecht, 1994.

[Westerståhl, 1990] D. Westerståhl. Parametric types and propositions in first-order situation theory. In R. Cooper, K. Mukai, and J. Perry, editors, *Situation Theory and its Applications, Vol. 1*, CSLI Publications, Stanford, CA, 1990.

DIMITER VAKARELOV

MODAL LOGICS OF ARROWS

INTRODUCTION

There exist many formal schemes and tools for representing knowledge about different types of data. Sometimes we can gain a better understanding if our information has some graphical representation. In many cases arrows are very suitable visual objects for representing various data structures: different kinds of graphs, binary relations, mappings, categories and so on. An abstract form of this representation scheme is the notion of arrow structure, which, in this paper, is an algebraic version of the notion of directed multi-graph.

Simply speaking, an arrow structure is a two sorted algebraic system, consisting of a set of arrows Ar, a set of points Po and two functions $\mathbf{1}$ and $\mathbf{2}$ from arrows to points, assigning to each arrow x the point $\mathbf{1}(x)$ — the beginning of x, and the point $\mathbf{2}(x)$ — the end of x. Using the functions $\mathbf{1}$ and $\mathbf{2}$ we define four relations R_{ij}, $i, j = \mathbf{1}, \mathbf{2}$, such that $xR_{ij}y$ iff $i(x) = j(y)$. These relations define the four possibilities for two arrows to have a common point. So each arrow structure S determines a relational system $W(S) = (Ar, \{R_{ij} \mid i, j = 1, 2\})$ called *arrow frame*.

It is shown below that the information captured by an arrow structure S is contained in the arrow frame $W(S)$. Arrow frames as relational systems with binary relations are suitable for interpreting polymodal logics, having modal operations corresponding to each binary relation in the frame. So we introduce a modal language $\mathcal{L}$ with four boxes $[ij]$ with standard Kripke semantics in arrow frames. We show how different properties of arrow frames are modally definable by means of modal formulas of $\mathcal{L}$. The logic of all arrow frames is axiomatized and called BAL — the Basic Arrow Logic. This paper is mainly devoted to a study of BAL and some of its extensions.

The paper is organized as follows. Section 1 is devoted to arrow structures and arrow frames. In Section 2 we introduce the notion of arrow logic in a semantic way as the collection of all formulas true in a given class of arrow frames. Some definability and undefinability results are proved there; for instance, applying some special techniques, called 'copying,' we show that the logic of all arrow frames coincides with the logic of all normal arrow frames, which correspond to directed graphs, admitting no more than one arrow between an ordered pair of points.

In Section 3 we present an axiomatization of the logic of all arrow frames, and we prove several completeness theorems. Applying the filtration technique from

M. de Rijke (ed.), Advances in Intensional Logic, 137–171.

ordinary modal logic we prove in Section 4 that BAL and some other arrow logics possess the finite model property and are decidable.

In Section 5 we study an extension of BAL with a new connective interpreted by an equivalence relation between arrows, stating that two arrows are equivalent if they have common begins and common ends. In Section 6 we study another extension with a modal constant **Loop**, which is true in an arrow if it has common begin and end, i.e. if it forms a loop.

Section 7 is devoted to a short survey of possible directions for further development, including extensions with different polyadic modalities, corresponding to some typical relations between arrows such as $\mathbf{Path}_n$, $\mathbf{Path}_\infty$, $\mathbf{Loop}_n$, $\mathbf{Trapezium}_n$, **Triangle** and others. There are some natural generalizations of modal logics of binary relations and the Lambek Calculus. Finally, we exhibit a many-dimensional generalization of arrow structures; this enables one to consider n-ary relations on a set as an n-dimensional arrow structure. Among the logics based on n-dimensional arrow frames are some natural generalizations of the so-called Cylindric Modal Logics.

The idea to consider logics based on two sorted structures having points and arrows, was suggested to me by Johan van Benthem (personal communication). The first results were included in the manuscript [Vakarelov, 1990], and the many-dimensional generalization were announced in the abstracts [Vakarelov, 1991a, 1992a]. The present paper is an extended and revised version of [Vakarelov, 1992b]. The terms *arrow frame* and *arrow logic* were introduced by [van Benthem, 1989] in connection with some generalizations of the modal logic of algebras of relations. Van Benthem's arrow frames consist of a set of objects called arrows, with composition as a ternary relation, converse as a binary relation and a set of identity arrows. Although these relational structures are different from ours, we adopt van Benthem's terminology, because it fits very well with the subject of this paper.

1 ARROW STRUCTURES AND ARROW FRAMES

By an *arrow structure* we shall mean any system $S = (Ar, Po, \mathbf{1}, \mathbf{2})$, where

- Ar is a non-empty set, whose elements are called arrows,
- Po is a non-empty set, whose elements are called points. We assume also that $Ar \cap Po = \emptyset$.
- $\mathbf{1}$ and $\mathbf{2}$ are total functions from Ar to Po associating to each arrow x the following two points:
  - $\mathbf{1}(x)$ — the first point of x (beginning, source, domain), and

- $\mathbf{2}(x)$ — the last point of x (end, target, codomain).

Graphically:

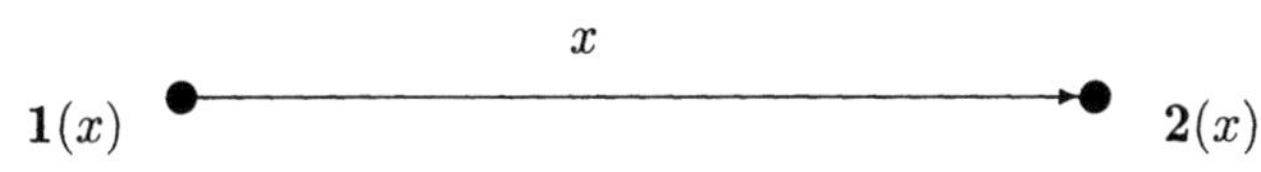

If $A = \mathbf{1}(x)$ and $B = \mathbf{2}(x)$, we say that x *connects* A with B, or, that (A, B) is a *connected* pair of points. It is possible for a pair of points (A, B) to be connected by different arrows.

- For technical reasons we assume the following axiom for arrow structures:

  (Ax) For each point A there exists an arrow x such that $A = \mathbf{1}(x)$ or $A = \mathbf{2}(x)$. In other words, each point is either the first or the last point of some arrow.

- An arrow structure S is called *normal* if it satisfies the following condition of normality

  (Nor) If $\mathbf{1}(x) = \mathbf{1}(y)$ and $\mathbf{2}(x) = \mathbf{2}(y)$, then $x = y$.

- To denote that Ar, Po, $\mathbf{1}$, $\mathbf{2}$ belong to a given arrow structure S, we will sometimes write Ar_S, Po_S, $\mathbf{1}_S$ and $\mathbf{2}_S$.

The main examples of arrow structures are directed multi-graphs, and for normal arrow structures — directed graphs without isolated points. These are notions studied in Graph Theory where graphs are visualized, or sometimes defined, by geometrical notions of a point and arrow. According to graph-theoretic intuitions, an arrow is a part of a line with some direction, connecting two points. Formally, the notion of an arrow structure coincides with the notion of directed multi-graph without isolated points. We will prefer, however, the term ‘arrow structure’, as it is more neutral, since it has models that are not only connected with graph-theoretic intuitions such as categories and binary relations.

The example of arrow structures constructed from a binary relation can be defined as follows. Let R be a nonempty binary relation over a non-empty set W. Define $Ar = R$, $Po = \{x \in W \mid (\exists y \in W)(xRy \text{ or } yRx)\}$, and for $(x, y) \in Ar$ define $\mathbf{1}((x, y)) = x$ and $\mathbf{2}((x, y)) = y$. Then, obviously $(Ar, Po, \mathbf{1}, \mathbf{2})$ is a normal arrow structure. In some sense this example is typical, because each normal arrow structure can be represented as an arrow structure determined by a non-empty binary relation.

Let S be an arrow structure. The following binary relation $\rho = \rho_S$ can be defined on the set Po_S. For each $A, B \in Po_S$:

$$A\rho B \text{ iff } (\exists x \in Ar_S)(\mathbf{1}(x) = A \text{ and } \mathbf{2}(x) = B).$$

We will consider different kinds of arrow structures, depending on the properties of the relation ρ_S:

- S is a *serial* arrow structure if ρ_S is serial (i.e. if $\forall A \exists B\ A\rho B$),
- S is *reflexive* if ρ_S is a reflexive relation,
- S is *symmetric* if ρ_S is a symmetric relation,
- S is *transitive* if ρ_S is a transitive relation,
- S is *total* if ρ_S is a total relation, i.e. if $\rho_S = Po_S \times Po_S$.

Let S be a given arrow structure. The following four relations $R_{ij} = R_{ij}^S$, (i, $j \in \{1,2\}$) in Ar_S (called *incidence relations* in S), will play a fundamental role in this paper:

$$xR_{ij}y \text{ iff } i(x) = j(y).$$

The following pictures illustrate the above relations:

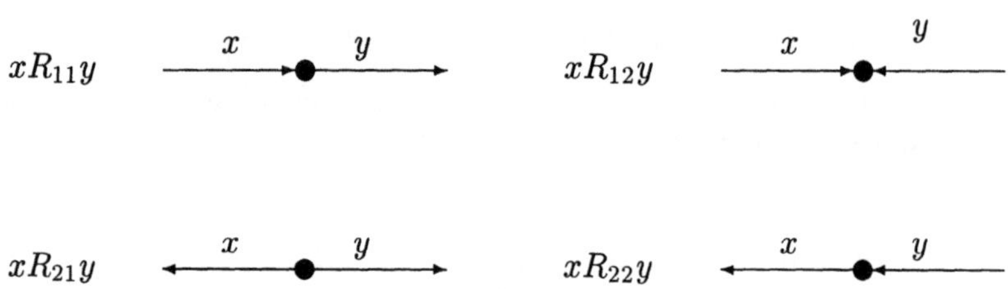

LEMMA 1.1 *The relations R_{ij} satisfy the following conditions for any x, y, $z \in Ar_S$ and i, j, $k \in \{1,2\}$:*

(ρii) $\quad xR_{ii}x$,

(σij) $\quad$ *If* $xR_{ij}y$ *then* $yR_{ji}x$,

(τijk) $\quad$ *If* $xR_{ij}y$ *and* $yR_{jk}z$, *then* $xR_{ik}z$.

Proof. By an easy verification. $\dashv$

Let $\underline{W} = (W, R_{11}, R_{22}, R_{12}, R_{21})$, $W \neq \emptyset$, be a relational system. $\underline{W}$ will be called an *arrow frame* if it satisfies the axioms (ρii), (σij) and (τijk) for any i, j, $k \in \{1,2\}$ and $x, y, z \in W$. The class of all arrow frames will be denoted by ARROW. If S is an arrow structure, then the arrow frame

$$\text{SAF}(S) = (Ar_S, R_{11}^S, R_{22}^S, R_{12}^S, R_{21}^S)$$

will be called a *standard arrow frame* over S. The class of all standard arrow frames will be denoted by (standard)ARROW. One of the main results of this section will be the proof that each arrow frame is a standard arrow frame over some arrow structure, i.e. (standard)ARROW = ARROW.

LEMMA 1.2 *Let S be an arrow structure. Then the following equivalences are true, where x, y, z range over Ar_S:*

(i) *S is a normal arrow structure iff* $\forall xy(xR_{11}y \wedge xR_{22}y \rightarrow x = y)$,

(ii) *S is a serial arrow structure iff* $\forall x \exists y xR_{21}y$,

(iii) *S is a reflexive arrow structure iff* $\forall x \exists y(xR_{11}y \wedge yR_{21}x)$ *and* $\forall x \exists y(xR_{21}y \wedge yR_{22}x)$,

(iv) *S is a symmetric arrow structure iff* $\forall x \exists y(xR_{12}y \wedge yR_{12}x)$,

(v) *S is a transitive arrow structure iff* $\forall xy \exists z(xR_{21}y \rightarrow xR_{11}z \wedge yR_{22}z)$.

(vi) *S is a total arrow structure iff* $\forall ij \in \{1, 2\} \forall xy \exists z(xR_{i1}z \wedge zR_{2j}y)$.

Proof. As an example we shall prove (ii).

($\Rightarrow$) Suppose that S is serial, and let $x \in Ar_S$ and $\mathbf{2}(x) = A$. By seriality there exists $B \in Pos_S$ such that $A\rho B$. Then for some $y \in Ar_S$ we have $\mathbf{1}(y) = A$ and $\mathbf{2}(y) = B$, so $\mathbf{2}(x) = \mathbf{1}(y)$, which yields $xR_{21}y$. Thus $\forall x \exists y xR_{21}y$.

($\Leftarrow$) Suppose that $\forall x \exists y xR_{21}y$ and let $A \in Pos_S$. Then by (Ax) there exists $x \in Ar_S$ such that $A = \mathbf{1}(x)$ or $A = \mathbf{2}(x)$. In the first case, let $B = \mathbf{2}(x)$, then $A\rho B$. In the second case, take y such that $xR_{21}y$. From this we get $\mathbf{2}(x) = \mathbf{1}(y)$ and $A = \mathbf{1}(y)$. Take $B = \mathbf{2}(y)$, then we get $A\rho B$. So in both cases $\forall A \exists B\ A\rho B$.

The remaining conditions can be proved in a similar way. ⊣

The above lemma suggests the following definitions for arrow frames. Let

$$\underline{W} = (W, R_{11}, R_{22}, R_{12}, R_{21})$$

be an arrow frame.

- $\underline{W}$ is a *normal* arrow frame iff $\forall xy(xR_{11}y \wedge xR_{22}y \rightarrow x = y)$,
- $\underline{W}$ is a *serial* arrow frame iff $\forall x \exists y\, xR_{21}y$,
- $\underline{W}$ is a *reflexive* arrow frame iff

$$\forall x \exists y(xR_{21}y \wedge xR_{22}y) \text{ and } \forall x \exists y(xR_{12}y \wedge xR_{11}y),$$

- $\underline{W}$ is a *symmetric* arrow frame iff $\forall x \exists y(xR_{12}y \wedge xR_{21}y)$,
- $\underline{W}$ is *transitive* arrow frame iff $\forall xy \exists z(xR_{21}y \rightarrow xR_{11}z \wedge yR_{22}z)$,

- $\underline{W}$ is *total* iff $\forall ij \in \{1, 2\}\forall xy\exists z(xR_{i1}z \wedge zR_{2j}y)$, where the variables x, y, z range over the set W.

The class of all normal arrow frames will be denoted by (nor)ARROW. Analogously, we introduce the notations (ser)ARROW, (ref)ARROW, (sym)ARROW, (tr)ARROW and (total)ARROW for the classes of all serial, reflexive, symmetric, transitive, and total arrow frames respectively. We will use also notations like (ref)(sym)ARROW to denote the class of all reflexive and symmetric arrow frames.

Obviously, if $\underline{W}$ is total arrow frame, then $\underline{W}$ is a reflexive, symmetric and transitive arrow frame. An arrow frame is called *pretotal* if it is reflexive, symmetric and transitive. The class of all pretotal arrow frames is denoted by (pretotal)ARROW. Using our combined notations we have that (pretotal)ARROW = $(ref)(sym)(tr)ARROW$.

Let $\underline{W} = (W, R_{11}, R_{22}, R_{12}, R_{21})$ be an arrow frame, $W' \subseteq W$, and $W' \neq \emptyset$. We write R'_{ij} for the relations R_{ij} restricted to W'. Then obviously the system $\underline{W'} = (W', R'_{11}, R'_{22}, R'_{12}, R'_{21})$ is an arrow frame; it is called a *subframe* of $\underline{W}$. The frame $\underline{W'}$ is called *generated subframe* of $\underline{W}$ if $\forall ij \in \{1, 2\}\forall x \in W'\forall y \in W(xR_{ij}y \rightarrow y \in W')$.

If $a \in W$, then by $\underline{W}_a$ we denote the smallest generated subframe of $\underline{W}$ that contains a. $\underline{W}_a$ is called an *arrow subframe* of $\underline{W}$ generated by a. If $\underline{W}$ is an arrow frame and there exists an $a \in W$ such that $\underline{W} = \underline{W}_a$, then $\underline{W}$ is called a *generated arrow frame*. If Σ is a class of arrow frames, then by Σ_{gen} we denote the class of all generated frames of Σ.

LEMMA 1.3 $((\text{pretotal})\text{ARROW})_{\text{gen}} = (\text{total})\text{ARROW}$.

Proof. By an easy verification. ⊣

Let S be an arrow structure. For $i \in \{1, 2\}$ and $A \in Pos$ define

$$i(A) = \{x \in Ar_S \mid i(x) = A\}, \quad g(A) = (\mathbf{1}(A), \mathbf{2}(A)).$$

LEMMA 1.4 *The following is true for each x, $y \in Ar_S$ and i, $j \in \{1,2\}$:*

(1) If $x \in i(A)$ and $y \in j(A)$, then $xR^S_{ij}y$,

(2) If $xR^S_{ij}y$ then $x \subset i(A)$ iff $y \in j(A)$,

(3) $\mathbf{1}(A) \cup \mathbf{2}(A) \neq \emptyset$.

Proof. By an easy verification. ⊣

Lemma 1.4 suggests the following definition. Let $\underline{W} = (W, R_{11}, R_{22}, R_{12}, R_{21})$ be an arrow frame, and let α_1 and α_2 be subsets of W. The pair (α_1, α_2) will be called a *generalized point* in $\underline{W}$ if it satisfies the following conditions for each $x, y \in W$ and $i, j \in \{1, 2\}$:

(1) If $x \in \alpha_i$ and $y \in \alpha_j$ then $xR_{ij}y$,

(2) If $xR_{ij}y$ then $x \in \alpha_i$ iff $y \in \alpha_j$,

(3) $\alpha_1 \cup \alpha_2 \neq \emptyset$.

The set of generalized points of an arrow frame $\underline{W}$ will be denoted by $Po(\underline{W})$. Lemma 1.4 now says that $g(A) = (\mathbf{1}(A), \mathbf{2}(A))$ is a generalized point in the standard arrow frame SAF(S) over S.

For a binary relation R in W and $x \in W$, we define $R(x) = \{y \in W \mid xRy\}$.

LEMMA 1.5 *Let* $\underline{U} = (U, R_{11}, R_{22}, R_{12}, R_{21})$ *be an arrow frame. Then for any* x, $y \in U$ *and* i, $j \in \{1,2\}$ *we have* $xR_{ij}y$ *iff* $R_{i1}(x) = R_{j1}(y)$ *and* $R_{i2}(x) = R_{j2}(y)$.

Proof. By an easy calculation, using the axioms of arrow frames. ⊣

LEMMA 1.6 *Let* $\underline{W}$ *be an arrow frame. Then for any* x, y, $z \in W$ *and* i, j, $k \in \{1,2\}$:

(i) The pair $k(z) = (R_{k1}(z), R_{k2}(z))$ *is a generalized point in* $\underline{W}$.

(ii) For each generalized point (α_1, α_2) *there exists* $z \in W$ *and* $k \in \{1,2\}$ *such that* $k(z) = (\alpha_1, \alpha_2)$.

(iii) $xR_{ij}y$ *iff* $i(x) = j(y)$.

Proof. (i) Let $i, j \in \{1,2\}$ and $x \in R_{ki}(z)$ and $y \in R_{kj}(z)$. Then we have $zR_{ki}x$ and $zR_{kj}y$. Then by (σki) we obtain $xR_{ik}z$ and by (τikj) we get $xR_{ij}y$. This proves condition (1) from the definition of generalized point. In a similar way one can verify condition (2). By (ρkk) we have $xR_{kk}x$, so $R_{kk}(x) \neq \emptyset$. This shows that $R_{k1}(x) \cup R_{k2}(x) \neq \emptyset$, which proves condition (3).

(ii) Let (α_1, α_2) be a generalized point in $\underline{W}$. Then there exists $z \in W$ such that $z \in \alpha_1 \cup \alpha_2$.

Case 1. $z \in \alpha_1$. In this case we will show that $k = 1$, i.e. that $(\alpha_1, \alpha_2) = \mathbf{1}(z) = (R_{11}(z), R_{12}(z))$ i.e. that $\alpha_1 = R_{11}(z)$ and that $\alpha_2 = R_{12}(z)$. Let $x \in \alpha_1$. Since $z \in \alpha_1$, then by item (1) of the definition of a generalized point we get $xR_{11}z$. So by $(\sigma 11)$ we obtain $zR_{11}x$, which shows that $x \in R_{11}(z)$. Now let $x \in R_{11}(z)$. Then $zR_{11}x$, and since $z \in \alpha_1$, by item (2) of the definition of a generalized point we get $x \in \alpha_1$. This proves the equality $\alpha_1 = R_{11}(z)$. In a similar way one can prove that $\alpha_2 = R_{12}(z)$.

Case 2: $z \in \alpha_2$. In this case $k = 2$ and we can proceed as in Case 1.

(iii) By Lemma 1.5 we have that $xR_{ij}y$ iff $R_{i1}(x) = R_{j1}(y)$, and $R_{i2}(x) = R_{j2}(y)$ iff $(R_{i1}(x), R_{i2}(x)) = (R_{j1}(y),_{j2}(y))$ iff $i(x) = j(y)$. ⊣

We will now give a construction of arrow structures from arrow frames. Let

$$\underline{W} = (W, R_{11}, R_{22}, R_{12}, R_{21}).$$

Define a system $S = S(\underline{W})$ as follows: $Ar_S = W$, $Po_S = Po(\underline{W})$ — the set of generalized points of $\underline{W}$ —; for $k = 1, 2$ and $z \in W$, let $k_S(z) = k(z) = (R_{k1}(z), R_{k2}(z))$ be as in Lemma 1.6. In the next theorem we shall show that $S(\underline{W})$ is an arrow structure called the *arrow structure over* $\underline{W}$.

THEOREM 1.7

(i) The system $S(\underline{W})$ defined above is an arrow structure. Moreover:

(ii) The standard arrow frame $\mathrm{SAF}(S(\underline{W}))$ *over $S(\underline{W})$ coincides with $\underline{W}$.*

(iii) $S(\underline{W})$ is a normal (serial, reflexive, symmetric, transitive, total) arrow structure iff $\underline{W}$ is a normal (serial, reflexive, and so on) arrow frame.

Proof. (i) By Lemma 1.6.(i) and (ii) we obtain that the system $S(\underline{W})$ is an arrow structure.

(ii) By Lemma 1.1 and lemma 1.6.(iii) $\mathrm{SAF}(S(\underline{W}))$ is a standard arrow frame such that for any $x, y \in W$ and $i, j \in \{1, 2\}$: $xR_{ij}y$ iff $i(x) = j(y)$ iff $xR^S_{ij}y$, which shows that $\mathrm{SAF}(S(\underline{W})) = \underline{W}$.

(iii) By Lemma 1.2 $S(\underline{W})$ is a normal (serial, ...) arrow structure iff the corresponding standard arrow frame $\mathrm{SAF}(S(\underline{W}))$ over $S(\underline{W})$ is normal (serial, ...). By (ii) $\mathrm{SAF}(S(\underline{W})) = \underline{W}$, which proves the assertion. ⊣

COROLLARY 1.8 (standard)ARROW = ARROW.

Proof. Immediate from Theorem 1.7. ⊣

Let S and S' be two arrow structures. A pair (f, g) of one-one functions $f : Ar_S \to Ar_{S'}$ and $g : Po_S \to Po_{S'}$ is called an *isomorphism from S onto S'* if for any $x \in Ar_S$ and $i = \mathbf{1}, \mathbf{2}$ we have $g(i_S(x)) = i_{S'}(f(x))$.

LEMMA 1.9 *Let S be an arrow structure, $\underline{W} = \mathrm{SAF}(S)$ the standard arrow frame over S, $Po(\underline{W})$ the set of generalized points of $\underline{W}$, and $S' = S(\underline{W})$ the arrow structure over $\underline{W}$. Assume furthermore that for $A \in Po_S$ the function $g(A) = (\mathbf{1}(A), \mathbf{2}(A))$ is defined as before Lemma 1.4, and that for $x \in Ar_S$ and $i = \mathbf{1}, \mathbf{2}$ the function $i_{S'}(x) = (R^{S'}_{i1}(x), R^{S'}_{i2}(x))$ is defined as in Lemma 1.6.(i). Then:*

(i) g is a one-one function from Po_S onto $Po(\underline{W})$.

(ii) For any $x \in Ar_S$ and $i = \mathbf{1}, \mathbf{2}$ we have $g(i(x)) = i_{S'}(x)$.

Proof. Obviously $g(A)$ is a generalized point in $\underline{W}$. Let $g(A) = g(B)$. Then $\mathbf{1}(A) = \mathbf{1}(B)$ and $\mathbf{2}(A) = \mathbf{2}(B)$. For A we can find $x \in Ar_S$ such that $\mathbf{1}(x) = A$ or $\mathbf{2}(x) = A$. Then $x \in \mathbf{1}(A)$ or $x \in \mathbf{2}(A)$. Suppose $x \in \mathbf{1}(A)$. Then $x \in \mathbf{1}(B)$, so $\mathbf{1}(x) = B$. From $\mathbf{1}(x) = A$ and $\mathbf{1}(x) = B$ we get $A = B$. In the case where $x \in \mathbf{2}(A)$ we proceed in the same way and get $A = B$. This shows that the mapping is injective.

To show that it is onto suppose that (α_1, α_2) is a generalized point in $\underline{W}$. We will show that for some $A \in Pos$ it holds that $g(A) = (\mathbf{1}(A), \mathbf{2}(A)) = (\alpha_1, \alpha_2)$. Since $\alpha_1 \cup \alpha_2 \neq \emptyset$ there exists $z \in \alpha_1$ or $z \in \alpha_2$.
Case 1. $z \in \alpha_1$. Let $\mathbf{1}(z) = A$, so $z \in \mathbf{1}(A)$. We will show that $\mathbf{1}(A) = \alpha_1$, and that $\mathbf{2}(A) = \alpha_2$. Suppose $x \in \mathbf{1}(A)$. Then $\mathbf{1}(x) = A$, so $\mathbf{1}(x) = \mathbf{1}(z)$, which yields $xR^S_{11}z$. Since $z \in \alpha_1$, then, by the properties of generalized points, we get $x \in \alpha_1$, so $\mathbf{1}(A) \subseteq \alpha_1$. Suppose now that $x \in \alpha_1$. Then, since $z \in \alpha_1$, we get $xR_{11}z$, so $\mathbf{1}(x) = \mathbf{1}(z) = A$. Then $x \in \mathbf{1}(A)$, so $\alpha_1 \subseteq \mathbf{1}(A)$. Consequently, $\mathbf{1}(A) = \alpha_1$. In a similar way one can show that $\mathbf{2}(A) = \alpha_2$. Hence, in this case $g(A) = (\alpha_1, \alpha_2)$.
Case 2. $z \in \alpha_2$. The proof is similar to that of Case 1.

(ii) Let $x \in Ar_S$ and $i = 1, 2$. Since $g(i(x)) = (\mathbf{1}(\mathbf{i}(x)), \mathbf{2}(\mathbf{i}(x))$ and $i_{S'}(x) = (R^S_{i1}(x), R^S_{i2}(x))$, to show that $g(\mathbf{i}(x)) = \mathbf{i}_{S'}(x)$, we have to prove that $\mathbf{1}(i(x)) = R^S_{i1}(x)$ and that $\mathbf{2}(i(x)) = R^S_{i1}(x)$. Suppose to this end that $y \in \mathbf{1}((i(x))$, so $\mathbf{1}(y) = i(x)$. Thus $xR^S_{i1}y$, which yields $y \in R^S_{i1}(x)$. Consequently, $\mathbf{1}(i(x)) \subseteq R^S_{i1}(x)$. The converse inclusion and the second equality can be proved in a similar way. ⊣

THEOREM 1.10 *Let S be an arrow structure, $\underline{W} = \mathrm{SAF}(S)$ the standard arrow frame over S, and $S(\underline{W})$ the arrow structure over $\underline{W}$. Then S is isomorphic to $S(\underline{W})$.*

Proof. For $x \in Ar_S$ let $f(x) = x$, and for $A \in Pos$ let $g(A) = (\mathbf{1}(A), \mathbf{2}(A))$. Lemma 1.9 shows that the pair (f, g) is the required isomorphism. ⊣

Theorems 1.10 and 1.7 show that all the information of an arrow structure S is contained in the standard arrow frame $\mathrm{SAF}(S)$ over S, and it can be expressed in terms of arrows and the relations R^S_{ij}.

An example of such a correspondence is provided by Lemma 1.2. As far as first-order conditions on the relation ρ are concerned, this correspondence can be defined in an effective way by means of a translation. The intuitive idea of this translation is the following. By the axiom (Ax), for each point A there exists $i \in \{\mathbf{1}, \mathbf{2}\}$ such that $A = i(x)$. So each variable A for a point is translated into a pair (i, x), where x denotes an arrow and i denotes one of the numbers 1 and 2. Suppose now that we have $A\rho B$, $A = i(x)$ and $B = j(y)$. Then by the definition of ρ we have: $(\exists u)(\mathbf{1}(u) = i(x) \wedge \mathbf{2}(u) = j(y))$ which is equivalent to $(\exists u)(xR_{i1}u \wedge uR_{2j}y)$. So if A is translated by (i, x) and B by

(j, y), then the corresponding translation of $A\rho B$ will be the formula $\varphi = xS_{ij}y = (\exists u)(xR_{i1}u \wedge uR_{2j}y)$. Here obviously $S_{ij} = R_{i1} \circ R_{2j}$. The parameters i and j in φ can be eliminated depending on the quantifiers in whose scope A and B occur. If for example A is under the scope of $(\forall A)$, we replace this quantifier by $(\forall i)(\forall x)$, and similarly for $(\exists A)$. Then quantifiers of the type $(\forall i)$ and $(\exists i)$ can be eliminated in a standard way by conjunctions and disjunctions of formulas, substituting 1 and 2 for i.

As an example, let us take the formula $(\forall A)(A\rho A)$. First, this formula is translated into $(\forall i)(\forall x)xS_{ii}x$. Eliminating the first quantifier, we obtain $(\forall x)(xS_{11}x) \wedge (\forall x)(xS_{22}x)$, which is equivalent to

$$(\forall x)(\exists y)(xR_{11}y \wedge yR_{21}x) \wedge (\forall x)(\exists y)(xR_{21}y \wedge yR_{22}x),$$

which is exactly the condition of reflexivity of ρ from Lemma 1.2. The translation of the formula $(\forall A)(\exists B)(A\rho B)$ is the following:

$$(\forall i)(\forall x)(\exists j)(\exists y)(xS_{ij}y).$$

Eliminating $(\forall i)$, we obtain the conjunction of the following two formulas:

$$\varphi_{1j} = (\forall x)(\exists j)(\exists y)(xS_{1j}y), \quad \varphi_{2j} = (\forall x)(\exists j)(\exists y)xS_{2j}y).$$

Eliminating $(\exists j)$ from φ_{1j} and φ_{2j}, we obtain the following formulas φ_1 and φ_2:

$$\begin{aligned} \varphi_1 &= (\forall x)((\exists y)(xS_{11}y) \vee (\exists y)(xS_{12}y)), \\ \varphi_2 &= (\forall x)((\exists y)(xS_{21}y) \vee (\exists y)(xS_{22}y)). \end{aligned}$$

Substituting S_{ij} here we obtain

$$\begin{aligned} \varphi_1 &= (\forall x)((\exists y)(\exists z)(xR_{11}z \wedge zR_{21}y) \vee (\exists y)(\exists z)(xR_{11}z \& zR_{22}y)), \\ \varphi_2 &= (\forall x)((\exists y)(\exists z)(xR_{21}z \wedge zR_{21}y) \vee (\exists y)(\exists z)(xR_{21}z \& zR_{22}y). \end{aligned}$$

The formula φ_1 is always true in arrow structures because in the second disjunct we can take $y = z = x$. Furthermore, the formula $\varphi = (\forall x)(\exists z)(xR_{21}z)$ follows from φ_2, and φ is exactly the condition of seriality from Lemma 1.2. It is easy to see that, in turn, φ implies the formula φ_2.

The intuitive idea of translating first-order sentences about points in terms of ρ and $=$ in arrow structures into equivalent sentences about arrows in terms of the relations R_{ij} can be stated in precise terms, but we will refrain from doing so in this paper.

2 ARROW LOGICS — SEMANTIC DEFINITIONS AND SOME DEFINABILITY AND NONDEFINABILITY RESULTS

In this section we will give a semantic definition of a class of modal logics called arrow logics. To this end we introduce the following modal language $\mathcal{L}$. It contains the following symbols:

- VAR, a denumerable set of proposition variables,
- $\neg$, $\wedge$, $\vee$, classical propositional connectives,
- $[ij]$, $i, j = 1, 2$, four modal operations,
- (,), parentheses.

The definition of the set of all formulas FOR for $\mathcal{L}$ is defined in the usual way. We use the following abbreviations: $A \Rightarrow B = \neg A \vee B$, $A \Leftrightarrow B = (A \Rightarrow B) \wedge (B \Rightarrow A)$, $1 = A \vee \neg A$, $0 = \neg 1$, $\langle ij \rangle A = \neg [ij] \neg A$.

The general semantics of $\mathcal{L}$ is a Kripke semantics over relational structures of the type $\underline{W} = (W, R_{11}, R_{22}, R_{12}, R_{21})$ with $W \neq \emptyset$, called *frames*. The standard semantics of $\mathcal{L}$ is over the class ARROW of all arrow frames. We will now introduce the basic semantic definitions and notations that we will use (for more details about Kripke semantics and related notions we refer the reader to [Segerberg, 1971; Hughes and Cresswell, 1984; van Benthem, 1985]).

Let $\underline{W} = (W, R_{11}, R_{22}, R_{12}, R_{21})$ be a frame. A function $v : VAR \rightarrow 2^W$ assigning to each variable $p \in VAR$ a subset $v(p)$ of W is called a *valuation*, and the pair $M = (\underline{W}, v)$ is called a *model* over $\underline{W}$. For $x \in W$ and $A \in$ FOR we define a satisfiability relation $x \Vdash_v A$ in M (to be read 'A is true in x under the valuation v') by induction on the complexity of the formula A as in the usual Kripke definition:

$$
\begin{array}{rll}
x \Vdash_v A & \text{iff} & x \in v(A) \text{ for } A \in VAR \\
x \Vdash_v \neg A & \text{iff} & x \nVdash_v A \\
x \Vdash_v A \wedge B & \text{iff} & x \Vdash_v A \text{ and } x \Vdash_v B \\
x \Vdash_v A \vee B & \text{iff} & x \Vdash_v A \text{ or } x \Vdash_v B \\
x \Vdash_v [ij]A & \text{iff} & (\forall y \in W)(x R_{ij} y \rightarrow y \Vdash_v A).
\end{array}
$$

We say that A is *true* in the model $M = (\underline{W}, v)$, or that M is a *model* for A, if for any $x \in W$ we have $x \Vdash_v A$. A is *true* in the frame $\underline{W}$, or $\underline{W}$ is a *frame* for A, if A is true in any model on $\underline{W}$. A is *true* in a class Σ of frames if A is true in any member of Σ. A class of formulas L is *true* in a model M, or M is a *model* for L, if any member of L is true in M. L is *true* in a class of frames Σ if any formula from L is true in Σ. L is called the *logic* of Σ, denoted by $L(\Sigma)$, if it contains all formulas true in Σ. Obviously, this operation of assigning sets of formulas to classes of frames is anti-monotonic in the following sense: if $\Sigma \subseteq \Sigma'$, then $L(\Sigma') \subseteq L(\Sigma)$.

Below we will study the logics

- L((standard)ARROW),
- L(ARROW),

- $L(\text{(nor)ARROW})$,
- $L(\text{(ser)ARROW})$,
- $L(\text{(ref)ARROW})$,
- $L(\text{(sym)ARROW})$,
- $L(\text{(tr)ARROW})$,
- $L(\text{(pretotal)ARROW})$, and
- $L(\text{(total)ARROW})$.

The most important logic on this list is $L(\text{(standard)ARROW})$. The first result that can be stated for $L(\text{(standard)ARROW})$, and that follows immediately from Corollary 1.8, is that $L(\text{(standard)ARROW}) = L(\text{ARROW})$.

We say that a condition φ for R_{ij} is *modally definable* in a class Σ of frames if there exists a formula A such that for any frame $\underline{W} \in \Sigma$: A is true in $\underline{W}$ iff φ holds in $\underline{W}$. If a class of frames is characterized by a condition φ which is modally definable in the class of all frames, then we say that Σ is a *modally definable* class of frames. The following lemma is a standard result in modal definability theory.

LEMMA 2.1 (Definability of arrow frames) *Let Σ be the class of all frames, and $A \in VAR$. The conditions on the left-hand side in the table below are modally definable in Σ by the formulas on the right-hand side* $(i, j, k = 1, 2)$*:*

(ρii)	$(\forall x) x R_{ii} x$	$(\mathrm{R}ii)$	$[ii]A \Rightarrow A$
(σij)	$(\forall xy)(x R_{ij} y \rightarrow y R_{ji} x)$	(Σij)	$A \vee [ij]\neg[ji]A$
(τijk)	$(\forall xyz)(x R_{ij} y \text{ and } y R_{jk} z \rightarrow x R_{ik} z)$	$(\mathrm{T}ijk)$	$[ik]A \Rightarrow [ij][jk]A$.

COROLLARY 2.2 *The class* ARROW *is modally definable.*

LEMMA 2.3 *Let* $\Sigma =$ ARROW *and* $A \in VAR$. *The conditions on arrow frames listed on the left-hand side in the table below are modally definable in Σ by the corresponding formulas on the right-hand side:*

seriality	*(ser)*	$\langle 21 \rangle 1$,
reflexivity	*(ref)*	$([11][21]A \Rightarrow A) \wedge ([21][22]A \Rightarrow A)$,
symmetry	*(sym)*	$[12][12]A \Rightarrow A$,
transitivity	*(tr)*	$[11][22]A \Rightarrow [21]A$.

Proof. As an example we will show the validity of (tr) in an arrow frame $\underline{W}$ implies that $\underline{W}$ is a transitive. For the sake of contradiction, suppose that (tr) is true in $\underline{W}$, but that $\underline{W}$ is not transitive. Then, for some x, $y \in W$ we have $x R_{21} y$ and not $(\exists z \in W)(x R_{11} z \wedge z R_{22} y)$. Define $v(A) = W \setminus \{y\}$. Then $y \not\Vdash_v A$, and since $x R_{21} y$ we get $x \not\Vdash_v [21]A$. We will show that $x \Vdash_v [11][22]A$. Suppose that this is not true. Then for some $z, t \in W$ we have $x R_{11} z$, $z R_{22} t$ and $t \not\Vdash_v A$, hence $t = y$. So $(\exists z)(x R_{11} z \& z R_{22} y)$, which is a contradiction. $\dashv$

COROLLARY 2.4 *The following classes are modally definable:*

- (ser)ARROW,
- (ref)ARROW,
- (sym)ARROW,
- (tr)ARROW, *and*
- (pretotal)ARROW.

We will show that the condition of normality of an arrow frame is not modally definable, and consequently that the class (nor)ARROW is not modally definable. We will show first that the logic L((nor)ARROW) coincides with the logic L(ARROW). To this end we will use a special construction called *copying*, which is adapted here for relational structures in the type of arrow frames.

Let $\underline{W} = (W, R_{11}, R_{22}, R_{12}, R_{21})$ and $\underline{W'} = (W', R'_{11}, R'_{22}, R'_{12}, R'_{21})$ be two frames, and $M = (\underline{W}, v)$, $M' = (\underline{W'}, v')$ models over $\underline{W}$ and $\underline{W'}$, respectively. Let I be a non-empty set of mappings from W into W'. We say that I is a *copying from* $\underline{W}$ *to* $\underline{W'}$ if the following conditions are satisfied for any i, $j \in \{1, 2\}$, x, $y \in W$ and $f, g \in I$:

$(I1)$ $(\forall y' \in W')(\exists y \in W)(\exists g \in I)g(y) = y'$

$(I2)$ If $f(x) = g(y)$, then $x = y$,

$(R_{ij}1)$ If $xR_{ij}y$ then $(\forall f \in I)(\exists g \in I)f(x)R'_{ij}g(y)$,

$(R_{ij}2)$ If $f(x)R'_{ij}g(y)$ then $xR_{ij}y$.

We say that I is a *copying from* M *to* M' if in addition the following condition is satisfied for any $p \in VAR$, $x \in W$ and $f \in I$:

(V) $x \in v(p)$ iff $f(x) \in v'(p)$.

For $x \in W$ and $f \in I$ we call $f(x)$ the *f-th copy of* x, and $f(W) = \{f(x) \mid x \in W\}$ is called the *f-th copy of* W. By $(I1)$ we obtain that $W' = \bigcup\{f(W) \mid f \in I\}$, so W' is a sum of its copies. If I is one element set $\{f\}$ then f is an isomorphism from $\underline{W}$ onto $\underline{W'}$.

The importance of the copying construction is in the following

LEMMA 2.5 (Copying Lemma)

(i) Let I be a copying from the model M to the model M'. Then for any formula $A \in \mathcal{L}$, $x \in W$ and $f \in I$ the following holds: $x \Vdash_v A$ iff $f(x) \Vdash_{v'} A$.

(ii) *If I is a copying from the frame $\underline{W}$ to the frame $\underline{W'}$, and v is a valuation, then there exists a valuation v' such that I is a copying from the model $M = (\underline{W}, v)$ to the model $M' = (\underline{W'}, v')$.*

Proof. (i) The proof is by induction on the complexity of the formula A. For $A \in VAR$ the statement holds by condition (V) of copying. If A is a Boolean combination of formulas the proof is straightforward. Let $A = [ij]B$, and assume as induction hypothesis that the statement holds for B.

($\Rightarrow$) Suppose $x \Vdash_v [ij]B$ and $f \in I$. To show that $f(x) \Vdash_{v'} [ij]B$, we suppose that $f(x)R'_{ij}y'$ and proceed to show that $y' \Vdash_v B$. By $(I1)$ $(\exists y \in W)(\exists g \in I)g(y) = y'$, so $f(x)R'_{ij}g(y)$, and by $(R_{ij}2)$ we get $xR_{ij}y$. From $xR_{ij}y$ and $x \Vdash_v [ij]B$ we get $y \Vdash_v B$. Then, by the induction hypothesis we get $g(y) \Vdash_{v'} B$, so $y' \Vdash_{v'} B$.

($\Leftarrow$) Suppose $f(x) \Vdash_{v'} [ij]B$. To show that $x \Vdash_v [ij]B$, we suppose that $xR_{ij}y$ and proceed to show that $y \Vdash_v B$. From $xR_{ij}y$ we obtain by $(R_{ij}1)$ that there exists $g \in W$ such that $f(x)R'_{ij}g(y)$. Then, since $f(x) \Vdash_{v'} [ij]B$, we get $g(y) \Vdash_{v'} B$ and, by the induction hypothesis, $y \Vdash_v B$.

(ii) For $p \in VAR$ define:

$$v'(p) = \{x' \in W' \mid (\exists x \in W)(\exists f \in I)f(x) = x' \text{ and } x \in v(p)\}.$$

We will show that the copying condition (V) is fulfilled. Let $x \in W$ and $f \in I$, and suppose $x \in v(p)$. Then, by the definition of v' we have $f(x) \in v'(p)$. For the converse implication suppose $f(x) \in v'(p)$. Then there exists $y \in W$ and $g \in I$ such that $f(x) = g(y)$ and $y \in v(p)$. By $(I2)$ we get $x = y$, so $x \in v(p)$. ⊣

LEMMA 2.6 *Let $\underline{W} = (W, R_{11}, R_{22}, R_{12}, R_{21})$ be an arrow frame. Then there exists a normal arrow frame $\underline{W'} = (W', R'_{11}, R'_{22}, R'_{12}, R'_{21})$ and a copying I from $\underline{W}$ to $\underline{W'}$. Moreover, if $\underline{W}$ is finite, we can take $\underline{W'}$ to be finite as well.*

Proof. Let $\underline{B}(W) = (B(W), 0, 1, +, .)$ be the Boolean ring over the set W, namely $B(W)$ is the set of all subsets of W, $0 = \emptyset$, $1 = W$, $A + B = (A \setminus B) \cup (B \setminus A)$ and $A \cdot B = AB = A \cap B$. Note that in Boolean rings $a - b = a + b$.

We put $W' = W \times B(W)$, and $I = B(W)$. The elements of $I = B(W)$ are viewed as functions from W in W' in the following manner: for $f \in I$ and $x \in W$ we put $f(x) = (x, f)$. Obviously, the conditions $(I1)$ and $(I2)$ from the definition of copying are fulfilled, and each element of W' has the form $f(x)$ for some $f \in I$ and $x \in$W. For the relations R'_{ij} we have the following definition:

$$f(x)R'_{ij}g(y) \text{ iff } xR_{ij}y \wedge (f + i \cdot \{x\} = g + j \cdot \{y\}).$$

Here the indices $i, j \in \{1, 2\}$ are considered as elements of $B(W)$: 1 is the unit of $B(W)$ and $2 = 1 + 1 = 1 - 1 = 0$.

To verify the condition $(R_{ij}1)$, suppose $xR_{ij}y$ and $f \in$I. Put $g = f + i \cdot \{x\} - j \cdot \{y\}$. Then $f + i \cdot \{x\} = g + j \cdot \{y\}$, which implies $f(x)R'_{ij}g(y)$. Condition $(R_{ij}2)$ follows immediately from the definition of R'_{ij}. So I is a copying.

The proof that W' together with the relations R'_{ij} forms an arrow frame is straightforward. As to the condition of normality, suppose that both $f(x)R'_{11}g(y)$ and $f(x)R'_{22}g(y)$. Then we obtain $xR_{11}y \wedge (f + 1 \cdot \{x\} = g + 1 \cdot \{y\})$ and $xR_{22}y \wedge (f + 2 \cdot \{x\} = g + 2 \cdot \{y\})$. From this and the fact that $2 = 0$, we get $f = g$ and $f + \{x\} = g + \{y\}$, which implies $\{x\} = \{y\}$, hence $x = y$ and $f(x) = g(x)$. Thus W' is a normal arrow frame.

Suppose now that $\underline{W}$ is a finite arrow frame. Then the Boolean ring over W is finite too and hence $\underline{W'}$ is a finite arrow frame. $\dashv$

If Σ is a class of arrow frames, then the class of all finite arrow frames from Σ is denoted by Σ_{fin}.

THEOREM 2.7

(i) $L(\text{(nor)ARROW}) = L(\text{ARROW})$.

(ii) $L(\text{((nor)ARROW)}_{\text{fin}}) = L(\text{ARROW}_{\text{fin}})$.

Proof. (i) Since $\text{(nor)ARROW} \subseteq \text{ARROW}$ we get

$$L(\text{ARROW}) \subseteq L(\text{(nor)ARROW}).$$

To prove that $L(\text{(nor)ARROW}) \subseteq L(\text{ARROW})$, suppose $A \notin L(\text{ARROW})$. Then there exists an arrow frame $\underline{W}$, $x \in W$ and a valuation v such that $x \not\Vdash_v A$. By Lemma 2.6 there exists a normal arrow frame $\underline{W'}$ and a copying I from $\underline{W}$ to $\underline{W'}$. By Lemma 2.5.(ii) there exists a valuation v' in W' such that I is a copying from the model $(\underline{W}, v)$ to the model $(\underline{W'}, v')$. By the copying lemma we get for any $f \in I$ that $f(x) \not\Vdash_{v'} A$. So A is not true in $\underline{W'}$, and hence $A \notin L(\text{(nor)ARROW})$. This implies $L(\text{(nor)ARROW}) \subseteq L(\text{ARROW})$.

(ii) The proof is the same as the proof of (i), using the fact that Lemma 2.6 guarantees that $\underline{W'}$ is a finite arrow frame. $\dashv$

COROLLARY 2.8 *Normality of an arrow frame is not modally definable.*

Proof. Suppose that there exists a formula φ such that for any arrow frame $\underline{W}$: φ is true in $\underline{W}$ iff $\underline{W}$ is normal. So $\varphi \in L(\text{(nor)ARROW})$. Let $\underline{W}_0$ be a non-normal arrow frame. Then φ is not true in $\underline{W}_0$, so $\varphi \notin L(\text{ARROW})$, hence by Theorem 2.7 $\varphi \notin L(\text{(nor)ARROW})$ — a contradiction. $\dashv$

Another example of a modally undefinable condition is totality. First we need the following standard result from modal logic.

LEMMA 2.9 *Let Σ be a nonempty class of arrow frames closed under subframes, and let Σ_{gen} be the class of generated frames of Σ. Then $L(\Sigma) = L(\Sigma_{\text{gen}})$.*

COROLLARY 2.10 *The following logics coincide:*

(i) – $L(\text{(pretotal)ARROW})$,
– $L((\text{(pretotal)ARROW})_{\text{gen}})$, *and*
– $L(\text{(total)ARROW})$.

(ii) – $L((\text{(pretotal)ARROW})_{\text{fin}})$,
– $L(((\text{(pretotal)ARROW})_{\text{gen}})_{\text{fin}})$, *and*
– $L((\text{(total)-ARROW})_{\text{fin}})$.

Proof. (i) Use Lemma 2.9 and Lemma 1.3. (ii) Use the fact that a generated frame of a finite frame is a finite frame too. ⊣

COROLLARY 2.11 *Totality of an arrow frame is not modally definable.*

Proof. Suppose that there exists a formula φ such that for any arrow frame $\underline{W}$: φ is true in $\underline{W}$ iff $\underline{W}$ is total arrow frame. Then $\varphi \in L(\text{(total)ARROW})$ and by Corollary 2.10 $\varphi \in L(\text{(pretotal)ARROW})$. Let $\underline{W_0}$ be a pretotal arrow frame which is not total (such frames obviously exist). Then φ is not true in $\underline{W_0}$, so $\varphi \notin L(\text{(pretotal)ARROW})$ — a contradiction. ⊣

3 AXIOMATIZATION OF SOME ARROW LOGICS

In this section we introduce a syntactical definition of arrow logic as sets of formulas containing some formulas as axioms and closed under some rules. The minimal set of axioms which we shall use contains axioms from the minimal modal logic for each modality $[ij]$; in addition it has formulas which modally define arrow frames. The formal system obtained in this way is denoted by BAL and called *Basic Arrow Logic.*

Axioms and Rules for BAL.

$(Bool)$ All or enough Boolean tautologies,

$(K[ij])$ $[ij](A \Rightarrow B) \Rightarrow ([ij]A \Rightarrow [ij]B)$,

(Rii) $[ii]A \Rightarrow A$,

(Σij) $A \vee [ij]\neg[ji]A$,

$(Tijk)$ $[ik]A \Rightarrow [ij][jk]A$,

(MP) $\dfrac{A,\ A \Rightarrow B}{B}$

$(N[ij])$ $\dfrac{A}{[ij]A}$, where $i, j \in \{1, 2\}$ and A and B are arbitrary formulas.

We identify BAL with the set of its theorems.

By an *arrow logic* we mean any set L of formulas containing BAL and closed under the rules (MP), $(N[ij])$ and the rule of substitution of propositional variables. So BAL is the smallest arrow logic. We adopt the following notation. If X is a finite sequence of formulas (viewed as new axioms), then by BAL $+ X$ we denote the smallest arrow logic containing all formulas from X. We will us the following formulas as additional axioms:

(ser) $\langle 21\rangle 1$,

(ref) $([11][21]A \Rightarrow A) \wedge ([21][22]A \Rightarrow A)$,

(sym) $[12][12]A \Rightarrow A$,

(tr) $[11][22]A \Rightarrow [21]A$.

Let $X \subseteq \{\text{ser, ref, sym, tr}\}$, and let for instance $X = \{\text{ser, tr}\}$. Then BAL $+ X =$ BAL $+$ ser $+$ tr. We will use also the notation (X)ARROW.

Let L be an arrow logic, and Σ a class of arrow frames. We say that L is *sound* for Σ if $L \subseteq L(\Sigma)$; L is *complete* for Σ if $L(\Sigma) \subseteq L$; and that L is *characterized* by Σ, or that $L(\Sigma)$ is *axiomatized* by L, if L is sound and complete for Σ, i.e. if $L = L(\Sigma)$.

In the completeness proofs we shall use the standard method of canonical models. We will give a brief description of the method. For more details and some definitions we refer the reader to [Segerberg, 1971] or [Hughes and Cresswell, 1984].

Let L be an arrow logic. The frame $\underline{W}_L = (W_L, R^L_{11}, R^L_{22}, R^L_{12}, R^L_{21})$ will be called a *canonical frame* for the logic L if W_L is the set of all maximal consistent sets in L, and the relations R^L_{ij} are defined on W_L as follows:

$$xR^L_{ij}y \text{ iff } \{A \in \text{FOR} \mid [ij]A \in x\} \subseteq y.$$

For $p \in VAR$ the function $v_L(p) = \{x \in W_L \mid p \in x\}$ is called the *canonical valuation* and the pair $M_L = (\underline{W}_L, v_L)$ is called the *canonical model* for L. The following is a standard result from modal logic.

LEMMA 3.1

(i) (Truth lemma for the canonical model for L.) The following is true for any formula A and $x \in W_L$: $x \Vdash_{v_L} A$ iff $A \in x$.

(ii) If $A \notin L$ then there exists $x \in W_L$ such that $A \notin x$.

LEMMA 3.2 *Let L be an arrow logic. Then the canonical frame $\underline{W}_L$ of L is an arrow frame.*

Proof. It is a well-known fact from standard modal logic that the axiom (Rii) yields the condition (ρii) for the canonical frame. Likewise, the axioms (Σij) and (Tijk) yield the conditions (σij) and (τijk) for $\underline{W}_L$. Thus, $\underline{W}_L$ is an arrow frame. ⊣

THEOREM 3.3 BAL *is sound and complete for the class of all arrow frames.*

Proof. Soundness follows by Lemma 2.1, and completeness can be proved by the method of canonical models. Let L = BAL. By Lemma 3.2 the canonical frame for L is an arrow frame. To show that $L(\mathsf{ARROW}) \subseteq L$ suppose that $A \notin L$. Then by Lemma 3.1.(ii) there exists $x \in W_L$ such that $A \notin x$. Then by the truth lemma we have $x \not\Vdash_{v_L} A$, so A is not true in the arrow frame $\underline{W}_L$. Then $A \notin L(\mathsf{ARROW})$, which proves the theorem. ⊣

COROLLARY 3.4 BAL $= L(\mathsf{ARROW}) = L((\mathsf{nor})\mathsf{ARROW})$.

Proof. Immediate from Theorem 3.3 and Theorem 2.7. ⊣

LEMMA 3.5 *Let L be an arrow logic. Then the following conditions are true:*

(i) (ser) $\in L$ *iff* $\underline{W}_L$ *is a serial arrow frame,*

(ii) (ref) $\in L$ *iff* $\underline{W}_L$ *is a reflexive arrow frame,*

(iii) (sym) $\in L$ *iff* $\underline{W}_L$ *is a symmetric arrow frame,*

(iv) (tr) $\in L$ *iff* $\underline{W}_L$ *is a transitive arrow frame.*

Proof. As an example we will show (iv) ($\Rightarrow$). Suppose (tr) $= [11][22]A \Rightarrow [21]A \in L$. We need to establish the transitivity of $\underline{W}_L$: $(\forall xy \in W_L)(\exists z \in W_L)(xR^L_{21}y \rightarrow xR^L_{11}z \wedge zR^L_{22}y)$. Let $M_1 = \{A \mid [11]A \in x\}$, $M_2 = \{A \mid (\exists B \in y)(\neg A \Rightarrow [22]\neg B \in L)\}$ and $M = M_1 \cup M_2$. Then the following claim is true:

CLAIM

(i) *If* $A_1, \ldots, A_n \vdash M_i$, *then* $A_1 \wedge \ldots \wedge A_n \in M_i$, $L \subseteq M_i$, $i = 1, 2$,

(ii) *If* $A \in M_i$ *and* $A \Rightarrow B \in L$, *then* $B \in M_i$, $i = 1, 2$,

(iii) $M_1 \cup M_2$ *is L-inconsistent set iff* $\exists A \in$ FOR*:* $A \in M_1$ *and* $\neg A \in M_2$,

(iv) *If* $xR^L_{21}y$, *then* M *is L-consistent set of formulas.*

(v) *Let z be a maximal consistent set. Then* $M_2 \subseteq z$ *implies* $zR^L_{22}y$, *and* $M_1 \subseteq z$ *implies* $xR^L_{11}z$.

(vi) If $xR_{21}^L y$ *then* $(\exists z \in W_L)(xR_{11}^L z \wedge zR_{22}^L y)$.

Proof of Claim. The proofs of (i) and (ii) are straightforward. Let us prove (iii). The right-to-left implication is obvious. For the left-to-right implication suppose that M is L-inconsistent. Then there exist $A_1, \ldots, A_n \in M$ such that $\neg(A_1 \wedge \ldots \wedge A_n) \in L$. Let $\{A_1, \ldots, A_i\} \in M_1$ and $\{A_{i+1}, \ldots, A_n\} \in M_2$. Put $A = A_1 \wedge \ldots \wedge A_i$, and $B = A_{i+1} \wedge \ldots \wedge A_n$. If $\{A_1, \ldots, A_i\} = \emptyset$ we put $A = T$, and similarly for B, where T is a fixed element of L. Then by (i) $A \in M_1$, $B \in M_2$ and $\neg(A \wedge B) \in L$, hence $B \Rightarrow \neg A \in L$. By (ii) we obtain that $\neg A \in M_2$. So there exists $A \in$ FOR such that $A \in M_1$ and $\neg A \in M_2$.

(iv) Suppose that $xR_{21}^L y$, and that M is not L-consistent. Then by (iii) there exists a formula A such that $A \in M_1$ and $\neg A \in M_2$. Then $[11]A \in x$ and there exists $B \in y$ such that $\neg\neg A \Rightarrow [22]\neg B \in L$, hence $A \Rightarrow [22]\neg B \in L$. Then by the rule $(N[11])$, we get $[11](A \Rightarrow [22]\neg B) \in L$, and by axiom $(K[11])$ and (MP) we obtain that $[11]A \Rightarrow [11][22]\neg B \in L$. But $[11]A \in x$, so $[11][22]\neg B \in x$. Now, by the axiom (tr): $[11][22]\neg B \Rightarrow [21]\neg B$ and (MP) we get $[21]\neg B \in x$ and since $xR_{21}^L y$ we get $\neg B \in y$. Since $B \in y$ we obtain a contradiction.

(v) Suppose $M_2 \subseteq z \in W_L$. Suppose that $zR_{22}^L y$ does not hold. Then for some formula A we have: $[22]B \in z$ and $B \notin y$, so $\neg B \in y$. Since $\neg\neg[22]B \Rightarrow [22]\neg\neg B \in L$, then by the definition of M_2 we get that $\neg[22]B \in M_2$, hence $\neg[22]B \in z$ — a contradiction. The second part of (v) follows by the definition of R_{11}^L.

(vi) Suppose that $xR_{21}^L y$. Then by (iv) M is an L-consistent set. Then there exists a maximal consistent set z such that $M \subseteq z$, and by (v) we have $xR_{11}^L z$ and $zR_{22}^L y$. ⊣

Now the proof of (iv) $(\Rightarrow)$ follows directly from claim (vi). ⊣

THEOREM 3.6 *If* $X \subseteq \{\text{ser}, \text{ref}, \text{sym}, \text{tr}\}$, *then* $\text{BAL} + X = L((X)\text{ARROW})$.

Proof. The soundness part of the theorem follows from Lemma 2.3, and the completeness part can be obtained from Lemma 3.5 as in the proof of Theorem 3.3. ⊣

COROLLARY 3.7

(i) $\text{BAL} + \text{ref} + \text{sym} + \text{tr} = L((\text{pretotal})\text{ARROW})$,

(ii) $\text{BAL} + \text{ref} + \text{sym} + \text{tr} = L((\text{total})\text{ARROW})$.

Proof. Part (i) is a direct consequence of Theorem 3.6, and part (ii) follows from Corollary 2.10. ⊣

4 FILTRATION AND FINITE MODEL PROPERTY

In this section we will show that BAL and some of its extensions enjoy the finite model property and are decidable, by applying the filtration techniques from classical modal logic. We adopt the Segerberg's definition of filtration, adapted for the language $\mathcal{L}$ of arrow logics (see [Segerberg, 1971]).

Let $\underline{W} = (W, R_{11}, R_{22}, R_{12}, R_{21})$ be an arrow frame, and $M = (\underline{W}, v)$ a model over $\underline{W}$. Let Ψ be a finite set of formulas, closed under subformulas. For $x, y \in W$ define:

$$\begin{array}{rcl} x \sim y & \text{iff} & (\forall A \in \Psi)(x \Vdash_v A \text{ iff } y \Vdash_v A) \\ |x| & = & \{y \in W \mid x \sim y\} \\ W' & = & \{|x| \mid x \in W\} \\ v'(p) & = & \{|x| \mid x \in v(p)\} \text{ for } p \in VAR. \end{array}$$

Let R'_{ij} $(i, j = 1, 2)$ be binary relations in W' such that $\underline{W'} = (W', R'_{11}, R'_{22}, R'_{12}, R'_{21})$ be an arrow frame. We say that the model $M' = (W', v')$ is a *filtration* of the model M through Ψ if the following conditions are satisfied for any $i, j = 1, 2$ and $x, y \in W$:

$(FR_{ij}1)$ If $xR_{ij}y$, then $|x|R'_{ij}|y|$

$(FR_{ij}2)$ If $|x|R_{ij}|y|$, then $(\forall[ij]A \in \Psi)\ (x \Vdash_v [ij]A \rightarrow y \Vdash_v A)$.

The following lemma is a standard result in filtration theory.

LEMMA 4.1 (Filtration Lemma, [Segerberg, 1971])

(i) *For any formula $A \in \Psi$ and $x \in W$ the following is true: $x \Vdash_v A$ iff $|x| \Vdash_{v'} A$.*

(ii) *$Card\, W' \leq 2^n$, where $n = Card\, \Psi$.*

Let L be an arrow logic. We say that L *admits a filtration* if for any frame $\underline{W}$ for L and a model $M = (\underline{W}, v)$ over $\underline{W}$ and for any formula A, there exist a finite set of formulas Ψ containing A and closed under subformulas and a filtration $M' = (\underline{W'}, v')$ of M through Ψ such that $\underline{W'}$ is a frame for L.

COROLLARY 4.2

(i) *Let Σ be a class of arrow frames, let Σ_{fin} be the class of all finite arrow frames from Σ and let $L(\Sigma)$ admit a filtration and the filtrations be in Σ. Then $L(\Sigma) = L(\Sigma_{\text{fin}})$.*

(ii) *Let $L(\Sigma) = L(\Sigma_{\text{fin}})$. If $L(\Sigma)$ is finitely axiomatizable then it is decidable.*

LEMMA 4.3 *Let $\underline{W}$ be an arrow frame, $M = (\underline{W}, v)$ a model over $\underline{W}$, and $M' = (\underline{W'}, v,)$ a filtration of M through Ψ. Then:*

(i) If $\underline{W}$ is a serial arrow frame, then $\underline{W'}$ is a serial arrow frame,

(ii) If $\underline{W}$ is a reflexive arrow frame, then $\underline{W'}$ is a reflexive arrow frame,

(iii) If $\underline{W}$ is a symmetric arrow frame, the $\underline{W'}$ is a symmetric arrow frame,

(iv) If $\underline{W}$ is a total arrow frame, then $\underline{W'}$ is a total arrow frame.

Proof. As an example we shall prove (iii). We have to show that

$$(\forall|x| \in W')(\exists|y| \in W')(|x|R'_{12}|y| \wedge |y|R'_{12}|x|).$$

Suppose $|x| \in W'$. Then there exists $y \in W$ such that $xR_{12}y \wedge yR_{12}x$. Then by the condition $(FR_{12}1)$ of filtrations we obtain $|x|R'_{12}|y| \wedge |y|R'_{12}|x|$. ⊣

THEOREM 4.4 *The logic $L(\mathsf{ARROW})$ admits a filtration.*

Proof. Let A be a formula and let Ψ be the smallest set of formulas containing A, closed under subformulas and satisfying the following condition

(∗) If for some $i, j = 1, 2$ $[ij]A \in \Psi$, then for any $i, j = 1, 2$ we have $[ij]A \in \Psi$.

It is easy to see that Ψ is finite and if n is the number of subformulas of A then $Card\,\Psi \leq 4n$. Then define W' and v' as in the definition of filtration. We define the relations R'_{ij} in W' as follows:

$$|x|R'_{ij}|y| \text{ iff } (\forall[ij]A \in \Psi)(\forall k \in \{1,2\})(x \Vdash_v [ik]A \leftrightarrow y \Vdash_v [jk]A).$$

First we will show that the frame $\underline{W'}$ is an arrow frame. The conditions (ρii) and (σij) follow directly from the definition of R'_{ij}. For the condition (τijk) suppose $|x|R'_{ij}|y|$ and $|y|R'_{jk}|z|$. To prove $|x|R'_{ik}|z|$ suppose $[ik]A \in \Psi$ $(l \in \{1,2\})$, and for the direction $(\rightarrow)$ suppose $x \Vdash_v [il]A$. We will show that $z \Vdash_v [kl]A$. From $[ik]A \in \Psi$ we get $[ij]A, [jl]A \in \Psi$. Now $|x|R'_{ij}|y|$, $[ij]A \in \Psi$ and $x \Vdash_v [il]A$ imply $y \Vdash_v [jl]A$. This together with $[jl]A \in \Psi$ and $|y|R'_{jk}|z|$ imply $z \Vdash_v [kl]A$.

The converse direction $(\leftarrow)$ can be proved in a similar way.

It remains to show that the conditions $(FR_{ij}1)$ and $(FR_{ij}2)$ of the definition of a filtration are satisfied. For the condition $(FR_{ij}1)$ suppose $xR_{ij}y$, $[ij]A \in \Psi$, $k \in \{1,2\}$ and for the direction $(\rightarrow)$ suppose $x \Vdash_v [ik]A$, $yR_{jk}z$; we will show that $z \Vdash_v A$. From $xR_{ij}y$ and $yR_{jk}z$ we get $xR_{ik}z$, and since $x \Vdash_v [ik]A$, we get $z \Vdash_v A$. For the direction $(\leftarrow)$ suppose $y \Vdash_v [jk]A$, and $xR_{ik}z$; we will show that $z \Vdash_v A$. From $xR_{ij}y$ we get $yR_{ji}x$, and by $xR_{ik}z$ we get $yR_{jk}z$. From this and $y \Vdash_v [jk]A$ we obtain $z \Vdash_v A$. This ends the proof of $(FR_{ij}1)$.

For the condition $(FR_{ij}2)$ suppose $|x|R'_{ij}|y|$, $[ij]A \in \Psi$ and $x \Vdash_v [ij]A$. From this we obtain $y \Vdash_v [jj]A$, and since $yR_{jj}y$ we get $y \Vdash_v A$. This completes the proof of the theorem. ⊣

COROLLARY 4.5

(i) $\mathrm{BAL} = L(\mathsf{ARROW}) = L(\mathsf{ARROW}_{\mathsf{fin}}) = L(((\mathsf{nor})\mathsf{ARROW})_{\mathsf{fin}})$.

(ii) BAL *is a decidable logic.*

Proof. (i) The first two equalities follow from Corollary 3.4 and Theorem 4.4. The last equality follows from Theorem 2.7.

(ii) This is a consequence of Corollaries 4.2, 3.4, and Theorem 4.4. ⊣

THEOREM 4.6 *If* $X \subseteq \{\mathrm{ser}, \mathrm{ref}, \mathrm{sym}\}$, *then the logic* $L((X)\mathsf{ARROW})$ *admits a filtration.*

Proof. Use the same filtration as in Theorem 4.4 and apply Lemma 4.3. ⊣

COROLLARY 4.7 *Let* $X \subseteq \{\mathrm{ser}, \mathrm{ref}, \mathrm{sym}\}$. *Then*

(i) $\mathsf{BAL} + X = L((X)\mathsf{ARROW}) = L(((X)\mathsf{ARROW})_{\mathsf{fin}})$

(ii) $\mathrm{BAL} + X$ *is a decidable logic.*

THEOREM 4.8 *The logic* $L((\mathsf{total})\mathsf{ARROW})$ *admits a filtration.*

Proof. Use the same filtration as in Theorem 4.4 and apply Lemma 4.3. ⊣

COROLLARY 4.9

(i) The following logics coincide:

- $\mathrm{BAL} + \mathrm{ref} + \mathrm{sym} + \mathrm{tr}$,
- $L((\mathsf{pretotal})\mathsf{ARROW})$,
- $L((\mathsf{total})\mathsf{ARROW})$,
- $L(((\mathsf{total})\mathsf{ARROW})_{\mathsf{fin}})$, *and*
- $L(((\mathsf{pretotal})\mathsf{ARROW})_{\mathsf{fin}})$

(ii) $\mathrm{BAL} + \mathrm{ref} + \mathrm{sym} + \mathrm{tr}$ *is a decidable logic.*

5 AN EXTENSION OF BAL WITH A MODALITY FOR EQUIVALENT ARROWS

We have seen that the condition of normality is not modally definable. This means that the language $\mathcal{L}$ is not strong enough to distinguish between normal and non normal arrow frames. In this section we will show that there exists a natural extension of the language $\mathcal{L}$ in which normality becomes modally definable.

Let $\underline{W}$ be an arrow frame; for $x, y \in W$ define

$$(\equiv) \qquad x \equiv y \text{ iff } xR_{11}y \wedge xR_{22}y.$$

Graphically:

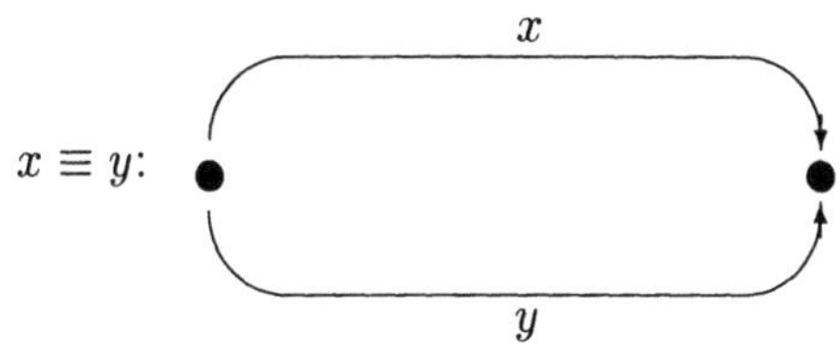

The relation $\equiv$ is called an equivalence of two arrows. In terms of the relation $\equiv$ the normality condition is equivalent to the following one:

$$(\mathrm{Nor}') \qquad (\forall xy \in W)\,(x \equiv y \to x = y).$$

If we extend our language $\mathcal{L}$ with a new modality $[\equiv]$, which is interpreted on arrow frames using the relation $\equiv$, then (Nor') is definable by means of the formula $p \Rightarrow [\equiv]p$.

Let $\mathcal{L}([\equiv])$ denote the extension of $\mathcal{L}$ with the operator $[\equiv]$. The *general semantics* of $\mathcal{L}([\equiv])$ is given by the class of all relational structures (called also *frames*) of the form

$$\underline{W} = (W, R_{11}, R_{22}, R_{12}, R_{21}, \equiv).$$

The *standard semantics* of $\mathcal{L}([\equiv])$ is given by the class of arrow frames in which the relation $\equiv$ is defined by $(\equiv)$. We shall show that the condition $(\equiv)$ is not modally definable. To this end we introduce the following nonstandard semantics for $\mathcal{L}([\equiv])$.

By a nonstandard $\equiv$-arrow frame we mean any system $\underline{W} = (W, R_{11}, R_{22}, R_{12}, R_{21}, \equiv)$ satisfying the following conditions for any $x, y, z \in W$ and $i, j, k \in \{1,2\}$:

(ρii), (σij) and (τijk),

$(\equiv \rho) \qquad x \equiv x,$

$(\equiv \sigma) \qquad x \equiv y \to y \equiv x,$

$(\equiv \tau) \qquad x \equiv y \wedge y \equiv z \to x \equiv z,$

$(\equiv \subseteq R_{ii}) \quad x \equiv y \to xR_{ii}y.$

If a nonstandard $\equiv$-arrow frame satisfies the condition

$(R_{11} \cap R_{22} \subseteq \equiv) \quad xR_{11}y \wedge xR_{22}y \to x \equiv y,$

then it is called a *standard* $\equiv$-arrow frame. It is easily seen that in any standard $\equiv$-arrow frame we have

$$x \equiv y \leftrightarrow xR_{11}y \wedge xR_{22}y.$$

We use Nonstandard-$\equiv$-ARROW and Standard-$\equiv$-ARROW to denote the classes of nonstandard and standard $\equiv$-arrow frames, respectively.

The conditions from the definition of nonstandard $\equiv$-arrow frame are modally definable by the following formulas respectively:

$(\equiv R)$ $[\equiv]A \Rightarrow A$,

$(\equiv \Sigma)$ $A \vee [\equiv]\neg[\equiv]A$,

$(\equiv T)$ $[\equiv]A \Rightarrow [\equiv][\equiv]A$,

$(\subseteq_{ii})$ $[ii]A \Rightarrow [\equiv]A, i = 1, 2$.

We will show that the condition $(R_{11} \cap R_{22} \subseteq \equiv)$ is not modally definable. To this end we will first prove that

$$L(\text{Standard-}\equiv\text{-ARROW}) = L(\text{Nonstandard-}\equiv\text{-ARROW}),$$

using the copying construction. Let us note that the definition of copying should be extended by the relevant copying conditions $(\equiv 1)$ and $(\equiv 2)$, which together guarantee the copying lemma.

LEMMA 5.1 *Let* $\underline{W} = (W, R_{11}, R_{22}, R_{12}, R_{21}, \equiv)$ *be a nonstandard* $\equiv$*-arrow frame. Then there exists a standard* $\equiv$*-arrow frame* $\underline{W'} = (W', R'_{11}, R'_{22}, R'_{12}, R'_{21}, \equiv')$, *and a copying from* $\underline{W}$ *to* $\underline{W'}$. *Moreover, if* $\underline{W}$ *is a finite arrow frame then* $\underline{W'}$ *is a finite arrow frame too.*

Proof. Use the same construction as in the Lemma 2.6 with the following modification. Let $\equiv(x) = \{y \in W \mid x \equiv y\}$. Since $\equiv$ is an equivalence relation, $\equiv(x) = \equiv(y)$ implies $x \equiv y$. The definitions of R'_{ij} and $\equiv'$ are the following:

$$\begin{array}{lll} f(x)R'_{ij}g(y) & \text{iff} & xR_{ij}y \wedge (f + i \cdot \equiv(x) = g + j \cdot \equiv(y)) \\ f(x) \equiv' g(y) & \text{iff} & x \equiv y \wedge f = g. \end{array}$$

The details that this works are left to the reader. ⊣

COROLLARY 5.2 $L(\text{Standard-}\equiv\text{-ARROW}) = L(\text{Nonstandard-}\equiv\text{-ARROW})$.

Now the axiomatization of L(Nonstandard-$\equiv$-ARROW) is easy.

Axioms and Rules for $[\equiv]$**BAL.** Let $[\equiv]$BAL denote the following axiomatic system.

(I) All axioms and rules of BAL,

(II) The following new axioms:

$(\equiv \text{R})$ $[\equiv]A \Rightarrow A$,
$(\equiv \Sigma)$ $A \vee [\equiv]\neg[\equiv]A$,
$(\equiv \text{T})$ $[\equiv]A \Rightarrow [\equiv][\equiv]A$,
$(\subseteq ii)$ $[ii]A \Rightarrow [\equiv]A, i = 1, 2$.

THEOREM 5.3 $[\equiv]$BAL *is sound and complete for the class* Nonstandard-$\equiv$-ARROW.

Proof. By the canonical model construction. ⊣

COROLLARY 5.4 *The following logics coincide:*

- $[\equiv]$BAL,
- L(Nonstandard-$\equiv$-ARROW), *and*
- L(Standard-$\equiv$-ARROW).

THEOREM 5.5

(i) The following logics coincide:

- $[\equiv]$BAL,
- L(Nonstandard-$\equiv$-ARROW), *and*
- L((Nonstandard-$\equiv$-ARROW)$_{\mathsf{fin}}$).

(ii) $[\equiv]$BAL *is a decidable logic.*

Proof. Apply the filtration technique with the following modification: the definition of the relations R'_{ij} is the same as in Theorem 4.4, the definition of $\equiv'$ is the following

$$|x| \equiv' |y| \text{ iff}$$
$$(\forall [\equiv]A \in \Psi)(x \Vdash_v [\equiv]A \leftrightarrow y \Vdash_v [\equiv]A) \wedge |x|R'_{11}|y| \wedge |x|R'_{22}|y|.$$

We leave it to the reader to verify that this definition will do. ⊣

6 EXTENSIONS OF BAL WITH THE PROPOSITIONAL CONSTANT LOOP

We say that an arrow x *forms a loop* if $xR_{12}x$. Graphically

x

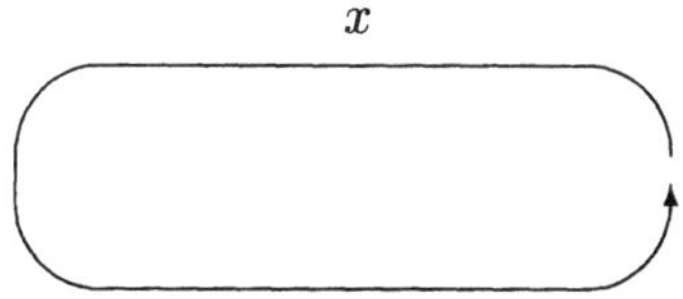

Let $\underline{W}$ be an arrow frame. We let $\text{Loop}_{\underline{W}} = \{x \in W \mid xR_{12}x\}$.

LEMMA 6.1 *In the language* $\mathcal{L}$ Loop *is not expressible in the sense that there is no formula* A *in* $\mathcal{L}$ *such that for any arrow frame* $\underline{W}$*, valuation* v *and* $x \in W$*:* $x \Vdash_v A$ *iff* $x \in \text{Loop}_{\underline{W}}$.

Proof. Let $W = \{a, b, c\}$, $R_{11} = R_{22} = \{(a,a), (b,b), (c,c)\}$, $R_{12} = R_{21} = \{(a,a), (b,c), (c,b)\}$. It is easy to see that W with the relations R_{ij} is an arrow frame. Let v be a valuation in W such that for any $p \in VAR$, $v(p) = \emptyset$. Then by induction on the complexity of formulas it follows that for any formula A the set $v(A) = \{x \in W \mid x \Vdash_v A\}$ is either W or $\emptyset$.

Suppose now that there exists a formula A such that $x \Vdash_v A$ iff $x \in$Loop. Then $v(A) = \{a\}$ which contradicts the previous remark. ⊣

Let $\mathcal{L}$(Loop) extend the language $\mathcal{L}$ with a new propositional constant Loop with the following standard semantics: for any arrow frame $\underline{W}$, valuation v and $x \in W$: $x \Vdash_v$ Loop iff $x \in \text{Loop}_{\underline{W}}$. Loop also has a nonstandard semantics which can be defined as follows. By a *nonstandard* Loop *arrow frame* we mean any system $\underline{W} = (W, R_{11}, R_{22}, R_{12}, R_{21}, \delta)$ such that $(W, R_{11}, R_{22}, R_{12}, R_{21})$ is an arrow frame, and δ (sometimes denoted by $\delta_{\underline{W}}$) is a subset of W. Then the interpretation of Loop in a nonstandard Loop arrow frame $\underline{W}$ is: $x \Vdash_v$ Loop iff $x \in \delta_{\underline{W}}$. A nonstandard Loop arrow frame $\underline{W}$ is called a *standard* one if the following two conditions are satisfied; together they state that $\text{Loop}_{\underline{W}} = \delta_{\underline{W}}$:

(Loop 1) $(\forall x \in W)(x \in \delta_{\underline{W}} \rightarrow x \in \text{Loop}_{\underline{W}})$,

(Loop 2) $(\forall x \in W)(x \in \text{Loop}_{\underline{W}} \rightarrow x \in \delta_{\underline{W}})$.

We write NonstandardLoopARROW to denote the class of all nonstandard Loop arrow frames. Likewise, the class of all standard Loop arrow frames is denoted by StandardLoopARROW. It can easily be shown that the condition (Loop 1) is modally definable in NonstandardLoopARROW by the following formula

(Loop) Loop$\Rightarrow$ ($[12]A \Rightarrow A$).

If a nonstandard Loop arrow frame satisfies (Loop 1) we call it a *general Loop arrow frame*. We use GeneralLoopARROW to denote the class of general Loop arrow frames. We will show that condition (Loop 2) is not modally definable in GeneralLoopARROW. To this end we will again use the copying construction, which contains an additional condition to deal with frames equipped with δ:

(δ) For any $x \in W$ and $f \in I$: $x \in \delta$ iff $f(x) \in \delta'$.

The copying lemma for this version of copying is also true.

LEMMA 6.2 *Let* $\underline{W} = (W, R_{11}, R_{22}, R_{12}, R_{21}, \delta)$ *be a general Loop arrow frame. Then there exists a standard Loop arrow frame* $\underline{W'} = (W', R'_{11}, R'_{22}, R'_{12}, R'_{21}, \delta')$*, and a copying* I *from* $\underline{W}$ *to* $\underline{W'}$*. Moreover, if* $\underline{W}$ *is finite then* $\underline{W'}$ *is a finite frame too.*

Proof. Define I and W' as in the proof of Lemma 2.6. To define R'_{ij} we first define the function

$$\delta(x) = \begin{cases} 0 & \text{if } x \in \delta \\ 1 & \text{if } x \notin \delta, \end{cases}$$

where 0 and 1 are considered as zero and unit of the Boolean ring. Then:

$$\begin{array}{rcl} f(x)R'_{ij}g(y) & \text{iff} & xR_{ij}y \wedge (f + i \cdot \delta(x) = g + j \cdot \delta(y)) \\ \delta' & = & \{x' \mid \exists x \in \delta \exists f \in I\, f(x) = x'\}. \end{array}$$

The proof that this is a copying and that $\underline{W}$ is an arrow frame is the same as in Lemma 2.6. Let us show that $\underline{W'}$ is a standard Loop arrow frame.

For the condition (Loop 1) suppose $x' \in \delta'$. Then $x' = f(x)$ for some $x \in \delta$ and $f \in I$. So we have $xR_{12}x$, $\delta(x) = 0$, and hence $f + 1 \cdot \delta(x) = f + 2 \cdot \delta(x)$. This shows that $f(x)R'_{12}f(x)$, hence $x'R'_{12}x'$.

For the condition (Loop 2) suppose $x'R'_{12}x'$. Then for some $x \in W$ and $f \in I$ we have $f(x) = x'$ and $f(x)R'_{12}f(x)$. So $xR_{12}x$ and $f + \delta(x) = f$, and therefore $\delta(x) = 0$, which yields $x \in \delta$. Thus $x' \in \delta'$. ⊣

Lemma 6.2 implies the following

THEOREM 6.3 $L(\mathsf{LoopARROW}) = L(\mathsf{GeneralLoopARROW})$.

COROLLARY 6.4 *Condition* (Loop 2) *is not modally definable.*

Using the above results we axiomatizatize the logic $L(\mathsf{LoopARROW})$ as follows: we simply axiomatize $L(\mathsf{GeneralLoopARROW})$ by adding to the axioms of BAL the following axiom

(Loop) Loop $\Rightarrow$ ($[12]A \Rightarrow A$).

We use LoopBAL to denote the system thus obtained. Using the canonical construction one can prove the following

THEOREM 6.5 LoopBAL *is sound and complete for* GeneralLoopARROW.

COROLLARY 6.6 *The following logics coincide:*

- LoopBAL,
- $L(\mathsf{GeneralLoopARROW})$, *and*
- $L(\mathsf{LoopARROW})$.

The constant Loop enables us to distinguish the logics $L(\mathsf{LoopARROW})$ and $L((\mathsf{nor})\mathsf{LoopARROW})$. Namely, we have

LEMMA 6.7 *Let* $\varphi = A \wedge \text{Loop} \Rightarrow [12](\text{Loop} \Rightarrow A)$. *Then:*

(i) $\varphi \notin L(\mathsf{LoopARROW})$,

(ii) $\varphi \in L((\mathsf{nor})\mathsf{LoopARROW})$,

(iii) $L(\mathsf{LoopARROW}) \neq L((\mathsf{nor})\mathsf{LoopARROW})$.

Proof. This is straightforward by the completeness theorem. ⊣

The formula φ from Lemma 6.7 defines the following condition in General-LoopARROW:

$$(\text{Nor}_0) \qquad (\forall x\, y)(xR_{11}y \wedge x \in \delta \wedge y \in \delta \rightarrow x = y)$$

Let $\underline{W}$ be a general Loop arrow frame. We call $\underline{W}$ *quasi-normal* if it satisfies the condition (Nor_0).

LEMMA 6.8 *Let* $\underline{W} = (W, R_{11}, R_{22}, R_{12}, R_{21}, \delta)$ *be a quasi-normal general Loop arrow frame. Then there exists a normal Loop arrow frame* $\underline{W'} = (W', R'_{11}, R'_{22}, R'_{12}, R'_{21}, \delta')$, *and a copying* I *from* $\underline{W}$ *to* $\underline{W'}$. *Moreover, if* $\underline{W}$ *is finite, then so is* $\underline{W'}$.

Proof. The construction of I, W' and R'_{ij} is the same as in Lemma 6.2 with the following modification of the function $\delta(x)$:

$$\delta(x) = \begin{cases} 0 & \text{if } x \in \delta \\ \{x\} & \text{if } x \notin \delta. \end{cases}$$

The proof that $\underline{W'}$ is a standard Loop arrow frame is the same as in Lemma 6.2. Let us show the condition of normality. Suppose $f(x)R'_{11}g(y)$ and $f(x)R'_{22}g(y)$. Then we have $xR_{11}y \wedge (f + \delta(x) = g + \delta(y))$, and $xR_{22}y \wedge (f = g)$. From this we get $\delta(x) = \delta(y)$.
Case 1. $\delta(x) = \emptyset$. Then $\delta(y) = \emptyset$ and hence $x, y \in \delta_{\underline{W}}$. By $xR_{11}y$ and $x, y \in \delta_{\underline{W}}$ we get by (Nor_0) $x = y$, and by $f = g$ we obtain $f(x) = f(y)$.
Case 2. $\delta(x) \neq \emptyset$. Then $\delta(y) \neq \emptyset$ and hence $\{x\} = \{y\}$, so $x = y$ and consequently $f(x) = g(y)$. This proves the condition of normality. ⊣

From Lemma 6.8 we obtain the following

THEOREM 6.9 $L((\mathsf{nor}_0)\mathsf{GeneralLoopARROW}) = L((\mathsf{nor})\mathsf{LoopARROW})$.

Define NorLoopBAL = LoopBAL + $A \wedge \text{Loop} \Rightarrow [12](\text{Loop} \Rightarrow A)$. Using the canonical method we can easily prove the following

THEOREM 6.10 NorLoopBAL *is sound and complete for the class of frames* $(\mathsf{nor}_0)\mathsf{GeneralLoopARROW}$.

COROLLARY 6.11 NorLoopBAL = $L(\text{(nor)LoopARROW})$.

LEMMA 6.12 *The following logics admit a filtration and are decidable:*

- $L(\text{GeneralLoopARROW})$ *and*
- $L((\text{nor}_0)\text{GeneralLoopARROW})$.

Proof. For the logic $L(\text{generalLoopARROW})$ use the same filtration as for the logic $L(\text{ARROW})$ with the following definition for δ: $\delta' = \{|x| \mid x \in \delta\}$. We have to show that the filtered frame satisfies the condition (Loop 1). Suppose $|x| \in \delta'$. Then $x \in \delta$, and by (Loop 1) we get $xR_{12}x$. By the properties of filtration it follows that $|x|R'_{12}|y|$.

For the logic $L((\text{nor}_0)\text{GeneralLoopARROW})$ we modify the definition of R'_{ij} as follows:

$$|x|R'_{ij}|y| \text{ iff}$$
$$(\forall [ij]A \in \Psi)(\forall k \in \{1,2\})(x \Vdash_v [ik]A \leftrightarrow x \Vdash_v [jk]A) \wedge$$
$$(x,\, y \in \delta' \rightarrow |x| = |y|).$$

The proof that this definition works is left to the reader. ⊣

COROLLARY 6.13 LoopBAL *and* NorLoopBAL *enjoy the finite model property and are decidable.*

The language $\mathcal{L}([\equiv], \text{Loop})$ is an extension of the language $\mathcal{L}([\equiv])$ with the constant Loop. The standard semantics of this language is a combination of the standard semantics of $\mathcal{L}([\equiv])$ and $\mathcal{L}(\text{Loop})$. This semantics is also modally undefinable. To axiomatize it we introduce a general semantics for $\mathcal{L}([\equiv], \text{Loop})$ as follows.

A frame $\underline{W} = (W, R_{11}, R_{22}, R_{12}, R_{21}, \equiv, \delta)$ is called a *general* Loop-$\equiv$ *arrow frame* if $(W, R_{11}, R_{22}, R_{12}, R_{21})$ is an arrow frame, and $\equiv$ and δ satisfy the conditions in table below:

$(\rho \equiv)$	$x \equiv x,$
$(\sigma \equiv)$	$x \equiv y \rightarrow y \equiv x$
$(\tau \equiv)$	$x \equiv y \wedge y \equiv z \rightarrow x \equiv z,$
$(\equiv \subseteq R_{ii})$	$x \equiv y \rightarrow xR_{ii}y,\ i = 1, 2$
(Loop 1)	$x \in \delta \rightarrow xR_{12}x,$
$(\equiv \delta)$	$x \equiv y \& x \in \delta \rightarrow y \in \delta,$
$(\equiv R_{11}\delta)$	$xR_{11}y \& x \in \delta \& y \in \delta \rightarrow x \equiv y.$

If, in addition, $\underline{W}$ satisfies the condition $(R_{11} \cap R_{22} \subseteq \equiv)$ and (Loop 2), then it is called *standard* Loop-$\equiv$ *arrow frame*. We use GeneralLoop-$\equiv$-ARROW and StandardLoop-$\equiv$-ARROW to denote the classes of general Loop-$\equiv$ arrow frames and standard Loop-$\equiv$-arrow frames, respectively.

The conditions in the above list are modally definable by the formulas in the list below.

($\equiv$ R) $[\equiv]A \Rightarrow A$,

($\equiv \Sigma$) $A \vee [\equiv]\neg[\equiv]A$,

($\equiv$ T) $[\equiv]A \Rightarrow [\equiv][\equiv]A$,

($\subseteq ii$) $[ii]A \Rightarrow [\equiv]A,\ i = 1, 2$,

(Loop) Loop $\Rightarrow ([\equiv]A \Rightarrow A)$,

($\equiv$ Loop) Loop $\Rightarrow [\equiv]$Loop,

($\equiv$ 11 Loop) Loop $\wedge [\equiv]A \Rightarrow [11]$(Loop $\Rightarrow A$).

We axiomatize the logic $\mathcal{L}([\equiv], \text{Loop})$ by adding the above formulas as axioms to the logic BAL. The resulting system is called $[\equiv]$LoopBAL.

THEOREM 6.14 *The logic* $[\equiv]$LoopBAL *is sound and complete for the class of arrow frames* GeneralLoop-$\equiv$-ARROW.

Proof. Use the canonical model construction. ⊣

LEMMA 6.15 *Let* $\underline{W} = (W, R_{11}, R_{22}, R_{12}, R_{21}, \equiv, \delta)$ *be a general* Loop-$\equiv$ *arrow frame. Then there exist a standard* Loop-$\equiv$ *arrow frame* $\underline{W'} = (W', R'_{11}, R'_{22}, R'_{12}, R'_{21}, \equiv', \delta')$, *and a copying I from* $\underline{W}$ *to* $\underline{W'}$.

Proof. The sets W', I, δ' and R'_{ij} are defined as in Lemma 6.8 with the following modification of the function $\delta(x)$:

$$\delta(x) = \begin{cases} 0 & \text{if } x \in \delta \\ \equiv(x) & \text{if } x \notin \delta \end{cases}$$

The relation $\equiv'$ is defined as in Lemma 5.1. The proof that this construction works is left to the reader. ⊣

COROLLARY 6.16 *The following logics coincide:*

- $[\equiv]$LoopBAL,
- L(GeneralLoop-$\equiv$-ARROW), *and*
- L(StandardLoop-$\equiv$-ARROW).

7 FURTHER PERSPECTIVES

A. Extensions of BAL with Additional Connectives

In Sections 5 and 6 we considered examples of possible extensions of the language $\mathcal{L}$ with operators that have their standard semantics in terms of arrow frames. There are many possibilities for such extensions, depending on what kind of relations between arrows we want to describe in a modal setting. The main scheme is the following: to each $n+1$-ary relation $R(x_0, x_1, \ldots, x_n)$ we associate an n-place modal box operation $[R](A_1, \ldots, A_n)$ with the following semantics, coming from the representation theory of Boolean algebras with operators [Jónsson and Tarski, 1951]; see also [Vakarelov, 1991b]:

$$x_0 \Vdash_v [R](A_1, \ldots, A_n) \text{ iff}$$
$$(\forall x_1, \ldots, x_n \in W)(R(x_0, x_1, \ldots, x_n) \rightarrow$$
$$x_1 \Vdash_v A_1 \text{ or } \ldots \text{ or } x \Vdash_v A_n).$$

The dual operator $\langle R\rangle(A_1, \ldots, A_n)$ is defined as $\neg[R](\neg A_1, \ldots, \neg A_n)$.

Below we list some natural relations between arrows which are candidates for a modal study:

- $\text{Path}_n(x_1, \ldots, x_n)$ iff $x_1 R_{21} x_2 \wedge x_2 R_{21} x_3 \wedge \ldots \wedge x_{n-1} R_{21} x_n$, $n \geq 2$

- $\text{Path}_\infty(x_1, x_2, x_3, \ldots)$ iff $(\forall n)\text{Path}_n(x_1, \ldots, x_n)$
- $\text{Loop}_n(x_1, x_2, \ldots, x_n)$ iff $\text{Path}_{n+1}(x_1, \ldots, x_n, x_1)$

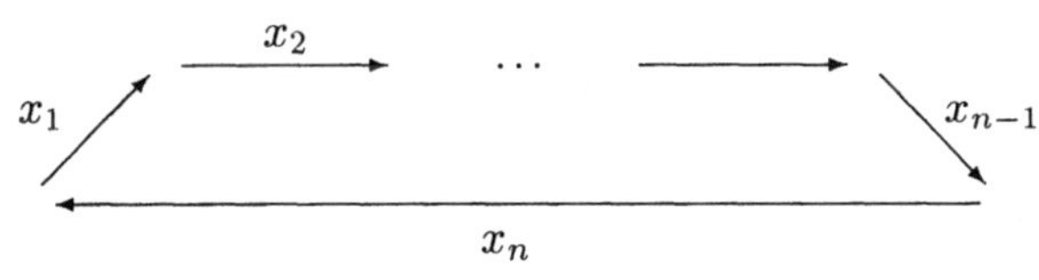

- Converse: xSy iff $\text{Loop}_2(x, y)$

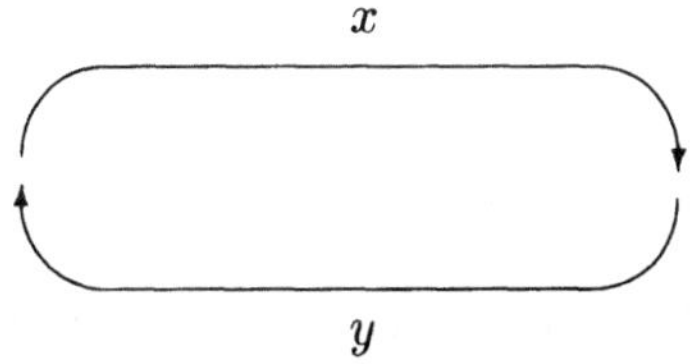

- $\text{Trapezium}_n(x_1, \ldots, x_n, y)$ iff $\text{Path}_n(x_1, \ldots, x_n) \wedge x_1 R_{11} y \wedge x_n R_{22} y$

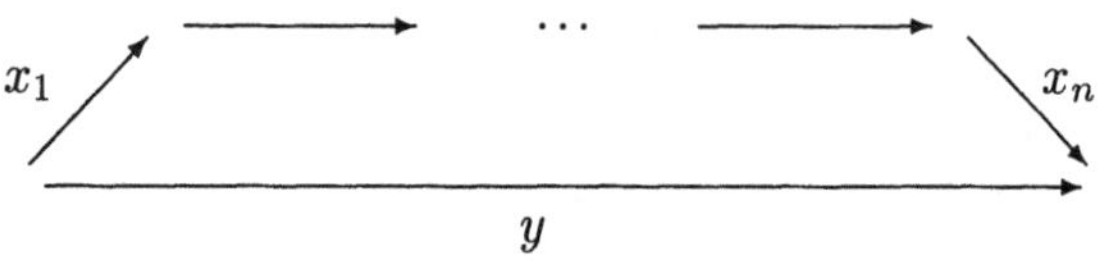

- $\text{Triangle}(x, y, z)$ iff $\text{Trapezium}_2(x, y, z)$

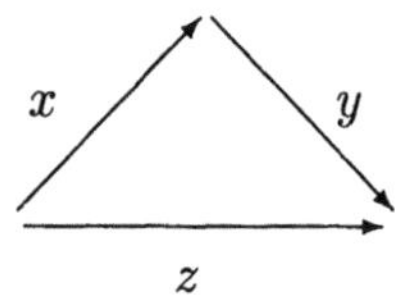

- Connectedness: $\text{Con}(x, y)$ iff $\exists n \geq 2 \exists x_1 \ldots x_n$: $x = x_1 \wedge x_n = y \wedge \text{Path}_n(x_1, \ldots, x_n)$

- Double sided connectedness: $\text{Dcon}(x, y)$ iff $\text{Con}(x, y)$ and $\text{Con}(y, x)$.

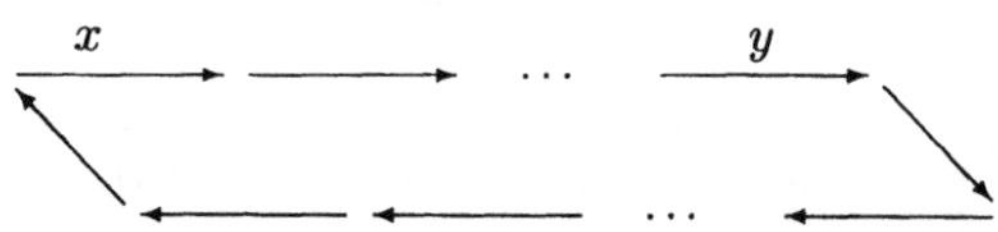

The relations Path_n, Path_∞ , Loop_n can be also used to define semantics for suitable propositional constants such as $\mathbf{Path}_n$, $\mathbf{Path}_\infty$, $\mathbf{Loop}_n$, **Loop** in the following manner:

$$\begin{array}{rcl} x_1 \Vdash_v \mathbf{Path}_n & \text{iff} & (\exists x_2, \ldots x_n \in W)\, \text{Path}_n(x_1, x_2, \ldots, x_n), \\ x_1 \Vdash_v \mathbf{Path}_\infty & \text{iff} & (\exists x_2, x_3, \ldots)\, \text{Path}_\infty(x_1, x_2, x_3, \ldots), \\ x_1 \Vdash_v \mathbf{Loop}_n & \text{iff} & (\exists x_2, \ldots, x_n)\, \text{Loop}_n(x_1, x_2, \ldots, x_n), \\ x \Vdash_v \mathbf{Loop} & \text{iff} & \exists n\, x \Vdash_v \mathbf{Loop}_n. \end{array}$$

These considerations motivate the following general problem: develop a modal theory (axiomatization, definability, (un-) decidability) of some extensions of BAL with modal operations corresponding to the relations on arrow structures defined above.

For example, the extension of BAL with the modal operations

- $A \bullet B = \langle \text{Triangle} \rangle (A, B)$,
- $A^{-1} = [\text{Converse}]A$, and
- the propositional constant Id = **Loop**

is a natural generalization of the modal logic of binary relations (van Benthem 1989; Venema 1989, 1991). This logic is closely related to various versions of representable relativized relational algebras [Kramer, 1989; Maddux, 1982; Németi, 1991]. An axiomatization is given by [Arsov, 1994].

B. Arrow Semantics of Lambek Calculus and its Generalizations

Let $A \,/\, B$ and $A \setminus B$ be the 'duals' of $A \bullet B$ with the following semantics:

$$x \Vdash_v B \,/\, A \quad \text{iff} \quad (\forall y\, z \in W)(\text{Triangle}(x, y, z) \wedge y \Vdash_v A \rightarrow z \Vdash_v B)$$
$$y \Vdash_v A \setminus B \quad \text{iff} \quad (\forall x\, z \in W)(\text{Triangle}(x, y, z) \wedge x \Vdash_v A \rightarrow z \Vdash_v B).$$

The modal connectives $A \bullet B$, $A \setminus B$, and $A \,/\, B$ can be considered as the operations of the Lambek Calculus. [Andréka and Mikulás, 1994] prove a completeness theorem for the Lambek Calculus with respect to a relational semantics of the above type over transitive normal arrow frames (this is an equivalent reformulation of the result by Andréka and Mikulás in our 'arrow' terminology). [Roorda, 1991, 1993)] studies a modal version of Lambek Calculus extended with classical Boolean operations. So it is natural to study an extension of BAL with the above dyadic modal operations, which will give another intuition for the operations in the Lambek Calculus.

C. Arrow Logics and Point Logics over Arrow Systems

To each arrow structure $S = (Ar, Po, \mathbf{1}, \mathbf{2})$ we can associate the following two relational systems: the arrow frame $(Ar, R_{11}, R_{22}, R_{12}, R_{21})$, and the point frame (Po, ρ). The first system is used as a semantic base of the logic BAL and the later can be used as a semantic base of an ordinary modal language with a modal operator $\blacksquare$. So each class Σ of arrow systems determines a class of arrow frames $\text{Ar}(\Sigma)$ and a class $\text{Po}(\Sigma)$ of point frames. A general question which arises is the problem of comparative study of the corresponding logics $L(\text{Ar}(\Sigma))$ and $L(\text{Po}(\Sigma))$. A kind of a correspondence between first-order properties of $\text{Po}(\Sigma)$ and $\text{Ar}(\Sigma)$ was shown in Section 1.

D. Many-dimensional Generalizations of Arrow Systems

The arrow structures introduced in this paper can be generalized to so-called n-*dimensional arrow structures* [Vakarelov, 1991a] in the following way. Let $S =$

$(Ar, Po, 1, \ldots, n)$ be a two sorted algebraic system. S is called an *n-dimensional arrow structure* if for any $i = 1, \ldots, n$, i is a function from Ar to Po satisfying the axiom

$$(\forall A \in Po)(\exists i 1 \leq i \leq n)(\exists x \in Ar)(i(x) = A).$$

The arrows in an n-dimensional arrow structure look as follows:

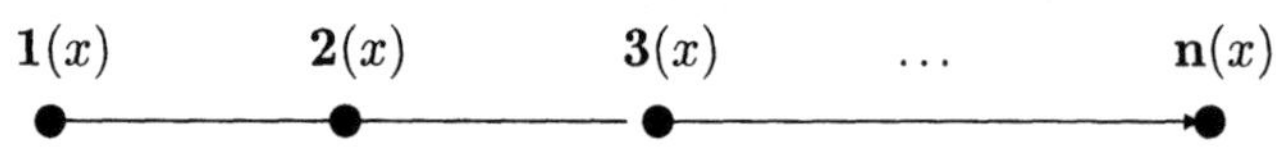

A natural example of an n-dimensional arrow structure is the set of all n-tuples of a given n-ary relation. Among the logics based on n-dimensional arrow frames are some generalizations of the so-called Cylindric Modal Logics [Venema, 1989; Venema, 1991]. These logics also have a very close connection with some versions of representable relativized cylindric algebras [Németi, 1991].

Department of Mathematical Logic with Laboratory for Applied Logic
Sofia University

Acknowledgments. Thanks are due to Johan van Benthem for drawing my attention to arrows and arrow logics, and to Hajnal Andréka and István Németi for many stimulating discussions and pointing out the connection of the 'arrow approach' to some problems in algebraic logic. I am much obliged to Tsutomu Hosoi and Catsumi Sasaki, who very carefully read the manuscript and helped me to correct the typos. I also want to thank the anonymous referee for valuable remarks that helped improve the quality of the paper.

REFERENCES

[Andréka and Mikulás, 1994] H. Andréka and Sz. Mikulás. Lambek Calculus and its relational semantics: completeness and incompleteness. *Journal of Logic, Language and Information*, 3:1–37, 1994.

[Arsov, 1994] A. Arsov. Basic arrow logic with relation algebraic operators. ILLC Research Report LP-94-02, University of Amsterdam, 1994.

[van Benthem, 1989] J. van Benthem. Modal logic and relational algebra. Manuscript, ILLC, University of Amsterdam, 1989.

[van Benthem, 1985] J. van Benthem. *Modal Logic and Classical Logic*. Bibliopolis, 1985.

[Hughes and Cresswell, 1984] G. Hughes and M. Cresswell. *A Companion to Modal Logic*. Methuen, London, 1984.

[Jónsson and Tarski, 1951] B. Jónsson and A. Tarski. Boolean algebras with operators, Part I. *American Journal of Mathematics*, 73:891–939, 1951.

[Kramer, 1989] R.L. Kramer. Relativized relational algebras. In H. Andréka, D. Monk, and I. Németi, editors, *Algebraic Logic*. North-Holland, Amsterdam, 1989.

[Maddux, 1982] R. Maddux. Some varieties containing relation algebras. *Transactions of the American Mathematical Society*, 272:501–526, 1982.

[Németi, 1991] I. Németi. Algebraizations of quantifier logics. *Studia Logica*, 51:485–570, 1991.

[Roorda, 1991] D. Roorda. *Resource Logics*. PhD thesis, Department of Mathematics and Computer Science, University of Amsterdam, 1991.

[Roorda, 1993] D. Roorda. Dyadic modalities and lambek calculus. In M. de Rijke, editor, *Diamonds and Defaults*, pages 215–253. Kluwer, Dordrecht, 1993.

[Segerberg, 1971] K. Segerberg. *An Essay in Classical Modal Logic*. Filosofiska Studier 13. University of Uppsala, 1971.

[Vakarelov, 1990] D. Vakarelov. Arrow logics. Manuscript, University of Sofia, 1990.

[Vakarelov, 1991a] D. Vakarelov. Modal logics for reasoning about arrows: arrow logics. In *Conference Proc. 9th International Congress of Logic, Methodology, and Philosophy of Sciences*, Uppsala, 1991.

[Vakarelov, 1991b] D. Vakarelov. Rough polyadic modal logics. *Journal of Applied Non-Classical Logics*, 1:9–35, 1991.

[Vakarelov, 1992a] D. Vakarelov. Arrow logics with cylindric operators. Abstract of a paper presented at the 1992 European Summer Meeting of the ASL, 1992.

[Vakarelov, 1992b] D. Vakarelov. A modal theory of arrows: arrow logics I. In D. Pearce and G. Wagner, editors, *Logics in AI, European Workshop '92*, pages 1–24. LNAI 633, Springer-Verlag, 1992.

[Venema, 1989] Y. Venema. Two-dimensional modal logics for relation algebras and temporal logic of intervals. Technical Report LP-89-03, Institute for Language, Logic and Information, 1989.

[Venema, 1991] Y. Venema. *Many-Dimensional Modal Logic*. PhD thesis, Department of Mathematics and Computer Science, University of Amsterdam, 1991.

HEINRICH WANSING

A FULL-CIRCLE THEOREM FOR SIMPLE TENSE LOGIC

A *full-circle theorem*[1] for a given logical system $\mathcal{L}$ says that certain proof systems $S_1, \ldots, S_4$ for $\mathcal{L}$ of the four most important types of inference systems (Hilbert-style, natural deduction, tableaux, sequent calculi) are all equivalent in the following sense (cf. Figure 1):

- Every proof of a wff A from wffs $A_1, \ldots, A_k$ in S_1 can be transformed into a proof of A from $A_1, \ldots, A_k$ in S_2;

$$\vdots$$

- every proof of A from $A_1, \ldots, A_n$ in S_4 can be transformed into a proof of A from $A_1, \ldots, A_k$ in S_1.

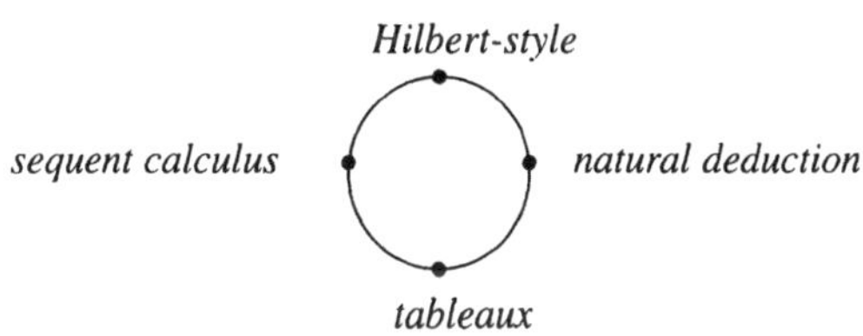

Figure 1. A full circle

In order to establish such a full circle, one has to make sure that in each case the data $A_1, \ldots, A_k$ are structured in the same way. If one, for instance, considers proofs from sets of assumptions, then these proofs have to be transformed into proofs from (representations of) the same data structures. Moreover, in the case of simple tense logic[2] there is a whole bunch of logics rather than one designated logical system. The standard syntactic presentation of these systems is in Hilbert-style and is thus *modular*. If one wants to establish a full circle for the most important systems of simple tense logic, the problem is that up to now modular natural deduction, tableaux, or sequent calculus presentations of these systems,

[1] This term was pointed out to me by Tijn Borghuis, who found it in [Barth and Krabbe, 1982, Chapter XI].

[2] That is, (propositional) tense logic without 'since' and 'until'.

M. de Rijke (ed.), Advances in Intensional Logic, 173–193.

which are not obtained by simply adding *axiom schemes* to presentations of the minimal normal tense logical system K_t, seem to be unknown. We shall therefore first of all extend Tijn Borghuis' generalization of *Fitch-style natural deduction* [Borghuis, 1993] and Nuel Belnap's *display logic* [Belnap, 1982, 1990] (as presented by [Wansing, 1994]) to modular proof systems for simple tense logic. It will turn out that the display calculi are not only modular, but in addition satisfy *Došen's principle*: the axiomatic extensions of K_t under consideration are captured by *structural rules*, so that the proof-theoretic meaning of the connectives, as codified in the operational rules, remains unaltered.

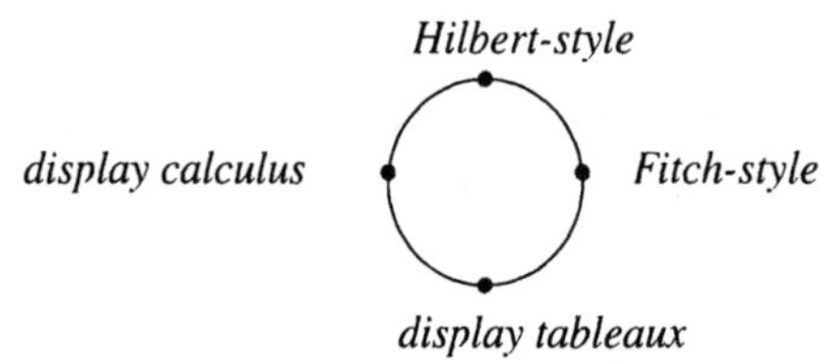

Figure 2. Another full circle.

Although the proof of the full-circle theorem to be given is neither complicated nor particularly deep, it may be of some interest as a contribution to the emergence of a general proof theory of intensional logic.

1 HILBERT-STYLE

Consider the language of simple tense logic in the vocabulary of classical propositional logic CPL (including **1** 'truth' and **0** 'falsity') together with the unary connectives **G** 'always in the future', **F** 'sometimes in the future', **H** 'always in the past' and **P** 'sometimes in the past'. The minimal normal propositional tense logic K_t can be axiomatized as follows:

*A*1 a set of convenient axiom schemes for CPL, including **1** and $\neg\mathbf{0}$

*A*2 $\mathbf{F}A \supset \neg\mathbf{G}\neg A, \quad \neg\mathbf{G}\neg A \supset \mathbf{F}A$
$\mathbf{P}A \supset \neg\mathbf{H}\neg A, \quad \neg\mathbf{H}\neg A \supset \mathbf{P}A$

*A*3 $\mathbf{G}(A \supset B) \supset (\mathbf{G}A \supset \mathbf{G}B), \quad \mathbf{H}(A \supset B) \supset (\mathbf{H}A \supset \mathbf{H}B)$

*A*4 $A \supset \mathbf{GP}A, \quad A \supset \mathbf{HF}A$

(MP) $A, A \supset B \;/\; B$

(MN) $\vdash A \;/\; \vdash \mathbf{G}A, \quad \vdash A \;/\; \vdash \mathbf{H}A$

Table 1. Additional axiom schemes

$\langle A1 \rangle$	$\mathbf{G}A \supset \mathbf{GG}A$	$\forall x \forall y \forall z((x < y \wedge y < z) \supset x < z)$
$\langle A2 \rangle$	$\mathbf{GG}A \supset \mathbf{G}A$	$\forall x \forall y(x < y \supset \exists z(z < y \wedge x < z))$
$\langle A3 \rangle$	$A \supset \mathbf{FP}A$	$\forall x \exists y(x < y)$
$\langle A4 \rangle$	$A \supset \mathbf{PF}A$	$\forall x \exists y(y < x)$
$\langle A5 \rangle$	$\mathbf{F}A \supset \mathbf{G}(\mathbf{F}A \vee A \vee \mathbf{P}A)$	$\forall x \forall y \forall z((x < y \wedge x < z) \supset (z < y \vee y < z \vee z = y))$
$\langle A6 \rangle$	$\mathbf{P}A \supset \mathbf{G}(\mathbf{P}A \vee A \vee \mathbf{F}A)$	$\forall x \forall y \forall z((y < x \wedge z < x) \supset (z < y \vee y < z \vee z = y))$

We consider a number of further axiom schemes, which are presented in Table 1 along with their defining first-order properties on Kripke frames.

A (Hilbert-style) *axiomatic proof* of A from a finite set $\Delta = \{A_1, \ldots, A_k\}$ is a finite sequence of wffs $B_1, \ldots, B_n$, where $B_n = A$ and for each B_j $(1 \leq j \leq n)$, either

1. $B_j \in \Delta$, or
2. B_j is an instantiation of an axiom scheme, or
3. B_j is derivable from earlier items in the sequence by means of (MP) or (MN).

We write $\mathcal{P}_H(\Pi, A, \Delta)$, if Π is a (Hilbert-style) axiomatic proof of A from Δ.

2 FITCH-STYLE

Tijn Borghuis [1993] obtains a modular natural deduction proof-theoretic framework for the most important normal modal propositional logics by adding both modal *import* and *export* rules to the Fitch-style natural deduction system for *CPL*. This method can be extended to the above systems of simple tense logic.[3] A characteristic feature of Fitch-style natural deduction is that the extent of proofs and parts of proofs which are called 'subordinate proofs' is indicated by vertical lines. If $\Delta = \{A_1, \ldots, A_n\}$ and $n > 0$, then $\mathbf{G}\Delta = \{\mathbf{G}A_1, \ldots, \mathbf{G}A_n\}$; if $n = 0$, then $\mathbf{G}\Delta = \emptyset$. We adopt analogous conventions for $\mathbf{H}$, $\mathbf{F}$, and $\mathbf{P}$. For K_t the notion $\mathcal{P}_N(\Pi, A, \Delta)$ ('Π is a proof in natural deduction of A from the finite set of assumptions Δ') is inductively defined as follows:

$F1$ $\mathcal{P}_N(| A, A, \{A\})$

[3]Note that the rule $\langle F1 \rangle$ below does not exactly reproduce Borghuis' 4-import rule.

$$F2 \quad \mathcal{P}_N(\mid \mathbf{1}, \mathbf{1}, \emptyset)$$

$$F3 \quad \mathcal{P}_N\left(\left|\begin{array}{l}\mathbf{0}\\ A\end{array}\right. , A, \{\mathbf{0}\}\right)$$

$$F4 \quad \mathcal{P}_N(\mid \Pi_1, A_1, \Delta_1), \mathcal{P}_N(\mid \Pi_2, A_2, \Delta_2) \;\Rightarrow\; \mathcal{P}_N\left(\left|\begin{array}{l}\Pi_1\\ \Pi_2\\ A_1 \wedge A_2\end{array}\right. , A_1 \wedge A_2, \Delta_1 \cup \Delta_2\right)$$

$$F5 \quad \mathcal{P}_N(\mid \Pi, A \wedge B, \Delta) \;\Rightarrow\; \mathcal{P}_N\left(\left|\begin{array}{l}\Pi\\ A\end{array}\right. , A, \Delta\right)$$

$$\mathcal{P}_N(\mid \Pi, A \wedge B, \Delta) \;\Rightarrow\; \mathcal{P}_N\left(\left|\begin{array}{l}\Pi\\ B\end{array}\right. , B, \Delta\right)$$

$$F6 \quad \mathcal{P}_N(\mid \Pi, A, \Delta) \;\Rightarrow\; \mathcal{P}_N\left(\left|\begin{array}{l}\Pi\\ A \vee B\end{array}\right. , A \vee B, \Delta\right)$$

$$\mathcal{P}_N(\mid \Pi, B, \Delta) \;\Rightarrow\; \mathcal{P}_N\left(\left|\begin{array}{l}\Pi\\ A \vee B\end{array}\right. , A \vee B, \Delta\right)$$

$$F7 \quad \mathcal{P}_N(\mid \Pi_1, C, \Delta_1),\; \mathcal{P}_N(\mid \Pi_2, C, \Delta_2),\; \mathcal{P}_N(\mid \Pi_3, A \vee B, \Delta_3) \;\Rightarrow\; \mathcal{P}_N\left(\left|\begin{array}{l}\Pi_3\\ \left|\begin{array}{l}\Pi_1\\ \Pi_2\end{array}\right.\\ C\end{array}\right. , C, \Delta_3 \cup (\Delta_1 \setminus \{A\}) \cup (\Delta_2 \setminus \{B\})\right)$$

$$F8 \quad \mathcal{P}_N(\mid \Pi_1, A, \Delta_1),\; \mathcal{P}_N(\mid \Pi_2, A \supset B, \Delta_2) \;\Rightarrow\; \mathcal{P}_N\left(\left|\begin{array}{l}\Pi_1\\ \Pi_2\\ B\end{array}\right. , B, \Delta_1 \cup \Delta_2\right)$$

$$F9 \quad \mathcal{P}_N(\mid \Pi, B, \Delta) \;\Rightarrow\; \mathcal{P}_N\left(\left|\begin{array}{l}\mid \Pi\\ A \supset B\end{array}\right. , A \supset B, \Delta \setminus \{A\}\right)$$

$$F10 \quad \mathcal{P}_N(\mid \Pi_1, A, \Delta_1),\; \mathcal{P}_N(\mid \Pi_2, \neg A, \Delta_2) \;\Rightarrow\; \mathcal{P}_N\left(\left|\begin{array}{l}\Pi_1\\ \Pi_2\\ B\end{array}\right. , B, \Delta_1 \cup \Delta_2\right)$$

$$F11 \quad \mathcal{P}_N(\mid \Pi_1, A, \Delta_1),\; \mathcal{P}_N(\mid \Pi_2, \neg A, \Delta_2) \;\Rightarrow\; \mathcal{P}_N\left(\left|\begin{array}{l}\Pi_1\\ \Pi_2\\ \neg B\end{array}\right. , \neg B, (\Delta_1 \cup \Delta_2) \setminus \{B\}\right)$$

$$F12 \quad \mathcal{P}_N(\mid \Pi_1, A, \Delta_1),\; \mathcal{P}_N(\mid \Pi_2, \neg A, \Delta_2) \;\Rightarrow\; \mathcal{P}_N\left(\left|\begin{array}{l}\Pi_1\\ \Pi_2\\ B\end{array}\right. , B, (\Delta_1 \cup \Delta_2) \setminus \{\neg B\}\right)$$

$$F13\ \mathcal{P}_N(|\,\Pi, A, \Delta) \Rightarrow \mathcal{P}_N\left(\left|\begin{array}{l} A_1 \\ \vdots \\ A_n \\ \Pi \end{array}\right., A, \Delta \cup \{A_1, \ldots, A_n\}\right)$$

$$F14\ \mathcal{P}_N(|\,\Pi, A, \Delta) \Rightarrow \mathcal{P}_N\left(\left|\begin{array}{l} \quad\mathbf{G} \mid \Pi \\ \mathbf{G}A \end{array}\right., \mathbf{G}A, \mathbf{G}\Delta\right)$$

$$F15\ \mathcal{P}_N(|\,\Pi, A, \Delta) \Rightarrow \mathcal{P}_N\left(\left|\begin{array}{l} \quad\mathbf{H} \mid \Pi \\ \mathbf{H}A \end{array}\right., \mathbf{H}A, \mathbf{H}\Delta\right)$$

$$F16\ \mathcal{P}_N(|\,\Pi_1, B, \Delta_1),\ \mathcal{P}_N(|\,\Pi_2, \neg B, \Delta_2) \Rightarrow \mathcal{P}_N\left(\left|\begin{array}{l} \Pi_1 \\ \Pi_2 \\ \mathbf{F}A \end{array}\right., \mathbf{F}A, (\Delta_1 \cup \Delta_2) \setminus \{\mathbf{G}\neg A\}\right)$$

$$F17\ \mathcal{P}_N(|\,\Pi_1, B, \Delta_1),\ \mathcal{P}_N(|\,\Pi_2, \neg B, \Delta_2) \Rightarrow \mathcal{P}_N\left(\left|\begin{array}{l} \Pi_1 \\ \Pi_2 \\ \mathbf{P}A \end{array}\right., \mathbf{P}A, (\Delta_1 \cup \Delta_2) \setminus \{\mathbf{H}\neg A\}\right)$$

$$F18\ \mathcal{P}_N(|\,\Pi_1, B, \Delta_1),\ \mathcal{P}_N(|\,\Pi_2, \neg B, \Delta_2) \Rightarrow \mathcal{P}_N\left(\left|\begin{array}{l} \Pi_1 \\ \Pi_2 \\ \neg\mathbf{G}\neg A \end{array}\right., \neg\mathbf{G}\neg A, (\Delta_1 \cup \Delta_2) \setminus \{\neg\mathbf{F}A\}\right)$$

$$F19\ \mathcal{P}_N(|\,\Pi_1, B, \Delta_1),\ \mathcal{P}_N(|\,\Pi_2, \neg B, \Delta_2) \Rightarrow \mathcal{P}_N\left(\left|\begin{array}{l} \Pi_1 \\ \Pi_2 \\ \neg\mathbf{H}\neg A \end{array}\right., \neg\mathbf{H}\neg A, (\Delta_1 \cup \Delta_2) \setminus \{\neg\mathbf{P}A\}\right)$$

$$F20\ \mathcal{P}_N(|\,\Pi, A, \Delta) \Rightarrow \mathcal{P}_N\left(\left|\begin{array}{l} \Pi \\ \quad\mathbf{H} \mid \mathbf{F}A \\ \mathbf{HF}A \end{array}\right., \mathbf{HF}A, \Delta\right)$$

$$F21\ \mathcal{P}_N(|\,\Pi, A, \Delta) \Rightarrow \mathcal{P}_N\left(\left|\begin{array}{l} \Pi \\ \quad\mathbf{G} \mid \mathbf{P}A \\ \mathbf{GP}A \end{array}\right., \mathbf{GP}A, \Delta\right)$$

The additional axiom schemes are taken into account by the following rules:

$$\langle F1\rangle\ \ \mathcal{P}_N(|\,\Pi, A, \Delta) \Rightarrow \mathcal{P}_N\left(\left|\begin{array}{l} \quad\mathbf{G} \mid \Pi \\ \mathbf{G}A \end{array}\right., \mathbf{G}A, \mathbf{GG}\Delta\right)$$

$$\langle F2\rangle\ \ \mathcal{P}_N(|\,\Pi, A, \Delta) \Rightarrow \mathcal{P}_N\left(\left|\begin{array}{l} \quad\mathbf{G} \mid \Pi \\ \mathbf{GG}A \end{array}\right., \mathbf{GG}A, \mathbf{G}\Delta\right)$$

$$\langle F3\rangle \quad \mathcal{P}_N(\mid \Pi, A, \Delta) \Rightarrow \mathcal{P}_N\left(\left|\begin{matrix}\Pi\\ \mathbf{FP}A\end{matrix}\right., \mathbf{FP}A, \Delta\right)$$

$$\langle F4\rangle \quad \mathcal{P}_N(\mid \Pi, A, \Delta) \Rightarrow \mathcal{P}_N\left(\left|\begin{matrix}\Pi\\ \mathbf{PF}A\end{matrix}\right., \mathbf{PF}A, \Delta\right)$$

$$\langle F5\rangle \quad \mathcal{P}_N(\mid \Pi, \mathbf{F}A, \Delta) \Rightarrow \mathcal{P}_N\left(\left|\begin{matrix}\Pi\\ \mathbf{G}(\mathbf{F}A \vee A \vee \mathbf{P}A)\end{matrix}\right., \mathbf{G}(\mathbf{F}A \vee A \vee \mathbf{P}A), \Delta\right)$$

$$\langle F6\rangle \quad \mathcal{P}_N(\mid \Pi, \mathbf{P}A, \Delta) \Rightarrow \mathcal{P}_N\left(\left|\begin{matrix}\Pi\\ \mathbf{G}(\mathbf{P}A \vee A \vee \mathbf{F}A)\end{matrix}\right., \mathbf{G}(\mathbf{P}A \vee A \vee \mathbf{F}A), \Delta\right)$$

3 DISPLAY TABLEAUX

Tableau calculi are refutation methods consisting of certain rules that manipulate *sequents*, i.e., derivability assertions. The intuition behind tableaux is semantic in nature: if the premise sequent of a tableau rule has a countermodel, then so has at least one of its conclusion sequents. The language of *display* tableau (and sequent) calculi extends the language of ordinary tableau (and sequent) systems (cf. [Belnap, 1982, 1990; Wansing, 1994]. In addition to a certain logical language there is also a *structural* language, which describes the structure of the data (assumptions) and the goals (conclusions). This structural language plays an important role in display logic, because properties of the structural connectives have an impact on the set of provable sequents. The vocabulary of the structural language to be considered consists of

- the wffs,
- the constant **I**,
- the unary operations * and $\bullet$,
- the binary operation $\circ$, and
- brackets (,).

The structural operations have a context sensitive reading, depending on whether they occur in the data (i.e., on the left side of the sequent arrow $\rightarrow$) or in the goals (i.e., on the right side of $\rightarrow$). In data position, **I** is to be interpreted as **1**; in goal position, **I** is to be interpreted as **0**. The meaning of the tense logical operations **G**, **F**, **H**, and **P** depends on the modal structuring by means of $\bullet$ (and *). In data position $\bullet$ is to be understood as **P**, whereas in goal position $\bullet$ is to be understood as **G**. On both sides of $\rightarrow$, * is to be read as negation $\neg$. In data position, however, a structure in the scope of an odd number of *s is a goal context, whereas in goal position it forms a data context. Eventually, in the presence of suitable structural assumptions, $\circ$ is to be interpreted as Boolean conjunction $\wedge$ on the left hand side of $\rightarrow$ and as Boolean disjunction $\vee$ on the right hand side of $\rightarrow$. A sequent then is an expression $X \rightarrow Y$, where X and Y are structures. If s_1 and s_2 are sequents, then

$$\frac{s_1}{\overline{s_2}}$$

abbreviates the rules

$$\frac{s_1}{s_2} \text{ and } \frac{s_2}{s_1}.$$

One can easily verify that, under the intended interpretation of the structural connectives, the following tableau rules are sound with respect to the Kripkean possible worlds semantics:

$$T1 \quad \frac{X \circ Y \rightarrow Z}{\overline{X \rightarrow Z \circ Y^*}} \qquad \frac{X \circ Y \rightarrow Z}{\overline{Y \rightarrow X^* \circ Z}}$$

$$T2 \quad \frac{X \rightarrow Y \circ Z}{\overline{X \circ Z^* \rightarrow Y}} \qquad \frac{X \rightarrow Y \circ Z}{\overline{Y^* \circ X \rightarrow Z}}$$

$$T3 \quad \frac{X \rightarrow Y}{\overline{Y^* \rightarrow X^*}} \qquad \frac{X \rightarrow Y}{\overline{X \rightarrow Y^{**}}}$$

$$T4 \quad \frac{X \rightarrow \bullet Y}{\overline{\bullet X \rightarrow Y}}$$

$$T5 \quad \frac{X \rightarrow Y}{X \rightarrow A \quad A \rightarrow Y}$$

$$T6 \quad \frac{\mathbf{1} \rightarrow X}{\mathbf{I} \rightarrow X} \qquad \frac{X \rightarrow \mathbf{0}}{X \rightarrow \mathbf{I}}$$

$$T7 \quad \frac{X \rightarrow \neg A}{X \rightarrow A^*} \qquad \frac{\neg A \rightarrow X}{A^* \rightarrow X}$$

$$T8 \quad \frac{X \circ Y \rightarrow A \wedge B}{X \rightarrow A \quad Y \rightarrow B} \qquad \frac{A \wedge B \rightarrow X}{A \circ B \rightarrow X}$$

$$T9 \quad \frac{X \rightarrow A \vee B}{X \rightarrow A \circ B} \qquad \frac{A \vee B \rightarrow X \circ Y}{A \rightarrow X \quad B \rightarrow Y}$$

$$T10 \quad \frac{X \rightarrow A \supset B}{X \circ A \rightarrow B} \qquad \frac{A \supset B \rightarrow X^* \circ Y}{X \rightarrow A \quad B \rightarrow Y}$$

$$T11 \quad \frac{X \to \mathbf{G}A}{\bullet X \to A} \qquad \frac{\mathbf{G}A \to \bullet X}{A \to X}$$

$$T12 \quad \frac{\bullet X \to \mathbf{P}A}{X \to A} \qquad \frac{\mathbf{P}A \to X}{A \to \bullet X}$$

$$T13 \quad \frac{(\bullet(X^*))^* \to \mathbf{F}A}{X \to A} \qquad \frac{\mathbf{F}A \to X}{X^* \to \bullet(A^*)}$$

$$T14 \quad \frac{X \to \mathbf{H}A}{\bullet(A^*) \to X^*} \qquad \frac{\mathbf{H}A \to (\bullet(X^*))^*}{A \to X}$$

$$T15 \quad \frac{\mathbf{I} \circ X \to Z}{X \to Z} \qquad \frac{X \circ \mathbf{I} \to Z}{X \to Z}$$

$$T16 \quad \frac{X \to Z}{X \to \mathbf{I}} \qquad \frac{Z \to X}{\mathbf{I} \to X}$$

$$T17 \quad \frac{(X_1 \circ X_2) \circ X_3 \to Z}{X_1 \circ (X_2 \circ X_3) \to Z} \qquad \frac{X_2 \circ X_1 \to Z}{X_1 \circ X_2 \to Z} \qquad \frac{X \to Z}{X \circ X \to Z}$$

$$T18 \quad \frac{X_1 \circ X_2 \to Z}{X_1 \to Z} \qquad \frac{X_2 \circ X_1 \to Z}{X_1 \to Z}$$

$$T19 \quad \frac{\bullet \mathbf{I} \to X}{\mathbf{I} \to X} \qquad \frac{\bullet(X^*) \to \mathbf{I}^*}{\mathbf{I} \to X}$$

$$\langle T1 \rangle \quad \frac{X \to \bullet \bullet Y}{X \to \bullet Y}$$

$$\langle T2 \rangle \quad \frac{\bullet X \to Y}{\bullet \bullet X \to Y}$$

$$\langle T3 \rangle \quad \frac{X \to Y}{(\bullet((\bullet X)^*))^* \to Y}$$

$$\langle T4 \rangle \quad \frac{X \to Y}{\bullet((\bullet(X^*))^*) \to Y}$$

$$\langle T5 \rangle \quad \frac{\bullet((\bullet(X^*))^*) \to Y}{(\bullet(X^*))^* \to Y \quad X \to Y \quad \bullet X \to Y}$$

$$\langle T6\rangle^4 \quad \frac{(\bullet((\bullet X)^*))^* \to Y}{(\bullet(X^*))^* \to Y \quad X \to Y \quad \bullet X \to Y}$$

A *tableau* for a sequent $X \to Y$ is a tree with root $X \to Y$. The *nodes* of the tree are sequents, and every branching is an instantiation of one of the reduction rules. A *closed tableau* is a finite tableau such that every leaf is of the form $A \to A$, $\mathbf{I} \to \mathbf{1}$, or $\mathbf{0} \to \mathbf{I}$. We write $\mathcal{CT}(\Pi, X \to Y)$, if Π is a closed tableau for $X \to Y$.

4 DISPLAY SEQUENT CALCULI

Display sequent calculi are the proof systems of their corresponding tableau systems. The display sequent calculi are obtained by reading the tableau rules 'bottom up' and adding axioms (i.e., axiomatic sequent rules), viz. $\vdash A \to A$, $\vdash \mathbf{I} \to \mathbf{1}$, and $\vdash \mathbf{0} \to \mathbf{I}$. Thus, every proof in a display sequent calculus amounts to a closed display tableau. For the sequent systems under consideration, it follows from Belnap's [1982] general cut-elimination theorem that applications of the cut rule, i.e., the converse of $T5$, can be eliminated.

5 THE TRANSFORMATIONS

In this section we define the transformations needed to form the circle.

5.1 From Axioms to Natural Deduction

We define the mapping $(\cdot)^H$ of axiomatic proofs into natural deduction proofs. Suppose $\mathcal{P}_H(\Pi, A, \Delta = \{A_1, \ldots, A_n\})$.

Case 1: $A = A_j \in \Delta$. Take any enumeration $B_1 \ldots B_k$ of $\Delta \setminus \{A_j\}$ and set

$$\Pi^N = \left| \begin{array}{l} B_1 \\ \vdots \\ B_k \\ A_j. \end{array} \right.$$

Case 2: $A \notin \Delta$, and A is an instantiation of one of the axiom schemes. Then

$$\Pi^N = \left| \begin{array}{l} A_1 \\ \vdots \\ A_n \\ \Pi', \end{array} \right.$$

[4] Note that $\langle T1\rangle$–$\langle T6\rangle$ are purely structural rules.

where Π' is a certain natural deduction proof of A from $\emptyset$. We shall present Π' here only for (a characteristic sample of) the tense logical axiom schemes:

$$\begin{array}{|l}\begin{array}{|l}\mathbf{F}A\\ \neg\mathbf{F}A\\ \neg\mathbf{G}\neg A\end{array}\\ \mathbf{F}A\supset\neg\mathbf{G}\neg A\end{array}\qquad\begin{array}{|l}\begin{array}{|l}\neg\mathbf{G}\neg A\\ \mathbf{G}\neg A\\ \mathbf{F}A\end{array}\\ \neg\mathbf{G}\neg A\supset\mathbf{F}A\end{array}$$

$$\begin{array}{|l}\begin{array}{|l}A\\ \quad\mathbf{G}\mid\mathbf{P}A\\ \mathbf{GP}A\end{array}\\ A\supset\mathbf{GP}A\end{array}\qquad\begin{array}{|l}\begin{array}{|l}\begin{array}{|l}\mathbf{G}\begin{array}{|l}A\supset B\\ A\\ B\end{array}\\ \mathbf{G}B\end{array}\\ \mathbf{G}A\supset\mathbf{G}B\end{array}\\ \mathbf{G}(A\supset B)\supset(\mathbf{G}A\supset\mathbf{G}B)\end{array}$$

$$\begin{array}{|l}\begin{array}{|l}\quad\mathbf{G}\mid A\\ \mathbf{GG}A\end{array}\\ \mathbf{G}A\supset\mathbf{GG}A\end{array}\qquad\begin{array}{|l}\begin{array}{|l}\quad\mathbf{G}\mid A\\ \mathbf{G}A\end{array}\\ \mathbf{GG}A\supset\mathbf{G}A\end{array}$$

$$\begin{array}{|l}\begin{array}{|l}A\\ \mathbf{FP}A\end{array}\\ A\supset\mathbf{FP}A\end{array}\qquad\begin{array}{|l}\begin{array}{|l}\mathbf{F}A\\ \mathbf{G}(\mathbf{F}A\vee A\vee\mathbf{P}A)\end{array}\\ \mathbf{F}A\supset\mathbf{G}(\mathbf{F}A\vee A\vee\mathbf{P}A)\end{array}$$

Case 3: $A\not\in\Delta$, A is not an instantiation of one of the axiom schemes, and A is obtained from earlier items in Π by means of (i) (MP) or (ii) (MN). (i): In this case there is a formula C and there are proofs Π_1, Π_2, such that

$$\mathcal{P}_N(\Pi_1, C, \Delta_1\subseteq\Delta)$$

and

$$\mathcal{P}_N(\Pi_2, C\supset A, \Delta_2\subseteq\Delta).$$

Take any enumeration $B_1\ldots B_k$ of $\Delta\setminus(\Delta_1\cup\Delta_2)$ and set

$$\Pi^N=\begin{array}{|l}B_1\\ \vdots\\ B_k\\ \Pi_1\\ \Pi_2\\ A.\end{array}$$

(ii): In this case $A = \mathbf{G}B$ or $A = \mathbf{H}B$, and there is a proof Π_1 such that $\mathcal{P}_N(\Pi_1, B, \emptyset)$. Then $\Pi^N =$

$$\left|\begin{array}{l} A_1 \\ \vdots \\ A_n \\ \quad \mathbf{G} \mid \Pi_1 \\ A \end{array}\right. \qquad \text{or} \qquad \left|\begin{array}{l} A_1 \\ \vdots \\ A_n \\ \quad \mathbf{H} \mid \Pi_1 \\ A \end{array}\right.$$

The effect of $(\cdot)^H$ can be summarized as

LEMMA 5.1 $\mathcal{P}_H(\Pi, A, \{A_1, \ldots, A_n\})$ *implies* $\mathcal{P}_N(\Pi^N, A, \{A_1, \ldots, A_n\})$.

5.2 *From Natural Dedution to Closed Tableaux*

We inductively define the mapping $(\cdot)^T$ of natural deduction proofs into closed tableaux. If $\Delta = \{A_1, \ldots, A_n\}$, then $\circ\Delta = A_1 \circ \ldots \circ A_n$; if $n = 1$, then $\circ\Delta = A_1$; and if $\Delta = \emptyset$, then $\circ\Delta = \mathbf{I}$. A proof of A from $\Delta = \{A_1, \ldots, A_n\}$ will be mapped to a closed tableau for $\circ\Delta \to A$.[5]

$F1$: $\Pi^T = A \to A$

$F2$: $\Pi^T = \mathbf{I} \to \mathbf{1}$

$F3$: $\Pi^T = \dfrac{\mathbf{0} \to A}{\mathbf{0} \to \mathbf{I}}$

$F4$: Suppose $\mathcal{CT}(\Pi_1, \circ\Delta_1 \to A_1)$ and $\mathcal{CT}(\Pi_2, \circ\Delta_2 \to A_2)$.

$$\Pi^T = \begin{array}{c} \circ\Delta_1 \cup \Delta_2 \to A_1 \wedge A_2 \\ \vdots \\ \circ\Delta_1 \circ \circ\Delta_2 \to A_1 \wedge A_2 \\ \hline \Pi_1 \qquad\qquad \Pi_2 \end{array}$$

$F5$: We treat only one rule. Suppose $\mathcal{CT}(\Pi_1, \circ\Delta \to A \wedge B)$.

$$\Pi^T = \dfrac{\Pi_1 \quad \dfrac{\dfrac{\overline{\circ\Delta \to A}}{A \wedge B \to A}}{A \circ B \to A}}{A \to A}$$

[5]This is justified, since, by virtue of the tableau (and sequent) rules, $A_1 \circ \ldots \circ A_n$ can be viewed as the set $\{A_1, \ldots, A_n\}$, and $\mathbf{I}$ can be viewed as $\emptyset$.

$F6$: We treat only one rule. Suppose $\mathcal{CT}(\Pi_1, \circ\Delta \to A)$.

$$\Pi^T = \begin{array}{l} \circ\Delta \to A \lor B \\ \hline \circ\Delta \to A \circ B \\ \hline \circ\Delta \circ B^* \to A \\ \hline \quad \Pi_1 \end{array}$$

$F7$: Suppose $\mathcal{CT}(\Pi_1, \circ\Delta_1 \to C)$, $\mathcal{CT}(\Pi_2, \circ\Delta_2 \to C)$, and $\mathcal{CT}(\Pi_3, \circ\Delta_3 \to A \lor B)$.

$$\Pi^T = \begin{array}{l} \circ\Delta_3 \cup (\Delta_1 \setminus \{A\}) \cup (\Delta_2 \setminus \{B\}) \to C \\ \hline \quad \vdots \\ \circ\Delta_3 \circ \circ(\Delta_1 \setminus \{A\}) \circ \circ(\Delta_2 \setminus \{B\}) \circ \mathbf{I} \to C \\ \hline \circ\Delta_3 \circ \circ(\Delta_1 \setminus \{A\}) \circ \circ(\Delta_2 \setminus \{B\}) \to C \circ \mathbf{I}^* \\ \hline C^* \circ \circ\Delta_3 \circ \circ(\Delta_1 \setminus \{A\}) \circ \circ(\Delta_2 \setminus \{B\}) \to \mathbf{I}^* \\ \hline \quad \vdots \\ \circ\Delta_3 \to (C \circ (\circ\Delta_1 \setminus \{A\})^*) \circ (C \circ (\circ\Delta_2 \setminus \{B\})^*) \\ \hline \Pi_3 \quad A \lor B \to (C \circ (\circ\Delta_1 \setminus \{A\})^*) \circ (C \circ (\circ\Delta_2 \setminus \{B\})^*) \\ \hline \qquad A \to C \circ (\circ\Delta_1 \setminus \{A\})^* \quad B \to C \circ (\circ\Delta_2 \setminus \{B\})^* \\ \hline \qquad\qquad \vdots \qquad\qquad\qquad\qquad \vdots \\ \qquad\qquad \Pi_1 \qquad\qquad\qquad\qquad \Pi_2 \end{array}$$

$F8$: Suppose $\mathcal{CT}(\Pi_1, \circ\Delta_1 \to A)$ and $\mathcal{CT}(\Pi_2, \circ\Delta_2 \to A \supset B)$.

$$\Pi^T = \begin{array}{r} \circ\Delta_1 \cup \Delta_2 \to B \\ \hline \vdots \\ \circ\Delta_1 \circ \circ\Delta_2 \to B \\ \hline \circ\Delta_1 \to B \circ (\circ\Delta_2)^* \\ \hline \Pi_1 \quad A \to B \circ (\circ\Delta_2)^* \\ \hline A \circ \circ\Delta_2 \to B \\ \hline \circ\Delta_2 \to A^* \circ B \\ \hline \Pi_2 \quad A \supset B \to A^* \circ B \\ \hline A \to A \quad B \to B \end{array}$$

$F9$: Suppose $\mathcal{CT}(\Pi_1, \circ\Delta \to B)$.

$$\Pi^T = \begin{array}{c} \circ\Delta \setminus \{A\} \to A \supset B \\ \hline \vdots \\ \Pi_1 \end{array}$$

$F10$: Suppose $\mathcal{CT}(\Pi_1, \circ\Delta_1 \to A)$ and $\mathcal{CT}(\Pi_2, \circ\Delta_2 \to \neg A)$.

$$
\Pi^T = \begin{array}{r}
\underline{\circ\Delta_1 \cup \Delta_2 \to B} \\
\vdots \\
\underline{\circ\Delta_1 \circ \circ\Delta_2 \to B} \\
\underline{\circ\Delta_1 \to B \circ (\circ\Delta_2)^*} \\
\Pi_1 \quad \underline{A \to B \circ (\circ\Delta_2)^*} \\
\underline{A \circ \Delta_2 \to B} \\
\underline{\circ\Delta_2 \to A^* \circ B} \\
\Pi_2 \quad \underline{\neg A \to A^* \circ B} \\
\underline{A^* \to A^* \circ B} \\
\underline{A \circ A^* \to B} \\
\underline{A^* \to A^* \circ B} \\
\underline{A^* \circ B^* \to A^*} \\
\underline{A^* \to A^*} \\
A \to A
\end{array}
$$

$F11$: Suppose $\mathcal{CT}(\Pi_1, \circ\Delta_1 \to A)$ and $\mathcal{CT}(\Pi_2, \circ\Delta_2 \to \neg A)$. Then, by the previous construction, there is a tableau Π_3 such that

$$\mathcal{CT}(\Pi_3, \circ\Delta_1 \cup \Delta_2 \to \mathbf{0}).$$

Then

$$
\Pi^T = \begin{array}{rr}
 & \underline{\circ(\Delta_1 \setminus \{B\}) \cup (\Delta_2 \setminus \{B\}) \to \neg B} \\
 & \vdots \\
 & \underline{\circ(\Delta_1 \setminus \{B\}) \circ \circ(\Delta_2 \setminus \{B\}) \to \neg B} \\
\circ(\Delta_1 \setminus \{B\}) \circ \circ(\Delta_2 \setminus \{B\}) \to (B \supset \mathbf{0}) & \underline{(B \supset \mathbf{0}) \to \neg B} \\
\underline{\circ(\Delta_1 \setminus \{B\}) \cup (\Delta_2 \setminus \{B\}) \circ B \to \mathbf{0}} & \underline{(B \supset \mathbf{0}) \to B^*} \\
\vdots & \underline{(B \supset \mathbf{0}) \circ \mathbf{I} \to B^*} \\
\Pi_3 & \underline{(B \supset \mathbf{0}) \to B^* \circ \mathbf{I}^*} \\
 & B \to B \quad \underline{\mathbf{0} \to \mathbf{I}^*} \\
 & \mathbf{0} \to \mathbf{I}
\end{array}
$$

$F12$: Suppose $\mathcal{CT}(\Pi_1, \circ\Delta_1 \to A)$ and $\mathcal{CT}(\Pi_2, \circ\Delta_2 \to \neg A)$. Then there is a tableau Π_3 such that $\mathcal{CT}(\Pi_3, \circ\Delta_1 \cup \Delta_2 \rightarrow \mathbf{0})$.

$$
\Pi^T = \begin{array}{rr}
 & \circ(\Delta_1 \setminus \{\neg B\}) \cup (\Delta_2 \setminus \{\neg B\}) \rightarrow B \\ \hline
 & \vdots \\
 & \circ(\Delta_1 \setminus \{\neg B\}) \circ \circ(\Delta_2 \setminus \{\neg B\}) \rightarrow B \\ \hline
\circ(\Delta_1 \setminus \{\neg B\}) \circ \circ(\Delta_2 \setminus \{\neg B\}) \rightarrow \neg B \supset \mathbf{0} & \neg B \supset \mathbf{0} \rightarrow B \\ \hline
\circ(\Delta_1 \setminus \{\neg B\}) \circ \circ(\Delta_2 \setminus \{\neg B\}) \circ \neg B \rightarrow \mathbf{0} & (\neg B \supset \mathbf{0}) \circ \mathbf{I} \rightarrow B \\ \hline
\vdots & (\neg B \supset \mathbf{0}) \rightarrow B \circ \mathbf{I}^* \\ \hline
\Pi_3 & B^* \circ (\neg B \supset \mathbf{0}) \rightarrow \mathbf{I}^* \\ \hline
 & B^* \circ (\neg B \supset \mathbf{0}) \rightarrow \mathbf{I} \\ \hline
 & (\neg B \supset \mathbf{0}) \rightarrow B^{**} \circ \mathbf{I} \\ \hline
 & B^* \rightarrow \neg B \quad \mathbf{0} \rightarrow \mathbf{I} \\ \hline
 & B^* \rightarrow B^* \\ \hline
 & B \rightarrow B
\end{array}
$$

$F13$: Suppose $\mathcal{CT}(\Pi_1, \circ\Delta \rightarrow A)$.

$$
\Pi^T = \begin{array}{c}
\circ\Delta \cup \{A_1, \ldots, A_n\} \rightarrow A \\ \hline
\vdots \\
\circ\Delta \circ A_1 \rightarrow A \\ \hline
\Pi_1
\end{array}
$$

$F14$: Suppose $\mathcal{CT}(\Pi_1, \circ\Delta \rightarrow A)$, $\Delta = \{A_1, \ldots, A_n\} \neq \emptyset$.

$$
\Pi^T = \begin{array}{ll}
\circ\mathbf{G}\Delta \rightarrow \mathbf{G}A & \\ \hline
\circ\mathbf{G}\Delta \rightarrow \mathbf{G} \circ \Delta & \mathbf{G} \circ \Delta \rightarrow \mathbf{G}A \\ \hline
\bullet \circ \mathbf{G}\Delta \rightarrow \circ\Delta & \bullet\mathbf{G} \circ \Delta \rightarrow A \\ \hline
\vdots & \mathbf{G} \circ \Delta \rightarrow \bullet A \\ \hline
 & \Pi_1 \\
\bullet \circ \mathbf{G}\Delta \circ (A_n^* \circ \ldots \circ A_2^*) \rightarrow A_1 & \\ \hline
\bullet \circ \mathbf{G}\Delta \rightarrow A_1 & \\ \hline
\circ\mathbf{G}\Delta \rightarrow \bullet A_1 & \\ \hline
\vdots & \\
\mathbf{G}A_1 \rightarrow \bullet A_1 & \\ \hline
A_1 \rightarrow A_1 &
\end{array}
$$

Suppose $\mathcal{CT}(\Pi_1, \mathbf{I} \rightarrow A)$. Then

$$
\Pi^T = \begin{array}{c}
\mathbf{I} \rightarrow \mathbf{G}A \\ \hline
\bullet\mathbf{I} \rightarrow A \\ \hline
\Pi_1.
\end{array}
$$

$F15$: Suppose $\mathcal{CT}(\Pi_1, \circ\Delta \to A)$, $\Delta = \{A_1, \ldots, A_n\} \neq \emptyset$.

$$\Pi^T = \begin{array}{ll}
\multicolumn{2}{l}{\circ\mathbf{H}\Delta \to \mathbf{H}A} \\ \hline
\circ\mathbf{H}\Delta \to \mathbf{H}\circ\Delta & \mathbf{H}\circ\Delta \to \mathbf{H}A \\ \hline
\bullet((\circ\Delta)^*) \to (\circ\mathbf{H}\Delta)^* & \bullet(A^*) \to (\mathbf{H}\circ\Delta)^* \\ \hline
\vdots & \vdots \\
\circ\mathbf{H}\Delta \to (\bullet((\circ\Delta)^*))^* & \mathbf{H}\circ\Delta \to (\bullet(A^*))^* \\ \cline{2-2}
\vdots & \Pi_1 \\
\mathbf{H}A_1 \to (\bullet((\circ\Delta)^*))^* & \\ \cline{1-1}
A_1 \to \circ\Delta & \\ \cline{1-1}
\vdots & \\
A_1 \circ (A_n^* \circ \ldots \circ A_2^*) \to A_1 & \\ \cline{1-1}
A_1 \to A_1 &
\end{array}$$

Suppose $\mathcal{CT}(\Pi_1, \mathbf{I} \to A)$. Then

$$\Pi^T = \begin{array}{c}
\mathbf{I} \to \mathbf{H}A \\ \hline
\bullet(A^*) \to \mathbf{I}^* \\ \hline
\Pi_1 .
\end{array}$$

$F16$: Suppose $\mathcal{CT}(\Pi_1, \circ\Delta_1 \to B)$ and $\mathcal{CT}(\Pi_2, \circ\Delta_2 \to \neg B)$. Then there is a tableau Π_3 such that $\mathcal{CT}(\Pi_3, \circ\Delta_1 \cup \Delta_2 \to \mathbf{0})$. Let

$$\Pi_4 = \begin{array}{ll}
\multicolumn{2}{l}{\mathbf{G}\neg A \supset \mathbf{0} \to \neg\mathbf{G}\neg A} \\ \hline
\multicolumn{2}{l}{\mathbf{G}\neg A \supset \mathbf{0} \to (\mathbf{G}\neg A)^*} \\ \hline
\multicolumn{2}{l}{(\mathbf{G}\neg A \supset \mathbf{0}) \circ \mathbf{I} \to (\mathbf{G}\neg A)^*} \\ \hline
\multicolumn{2}{l}{\mathbf{G}\neg A \supset \mathbf{0} \to (\mathbf{G}\neg A)^* \circ \mathbf{I}^*} \\ \hline
\mathbf{G}\neg A \to \mathbf{G}\neg A & \mathbf{0} \to \mathbf{I}^* \\ \hline
& \mathbf{0} \to \mathbf{I}
\end{array}$$

and let

$$\Pi_5 = \begin{array}{l}
\neg\mathbf{G}\neg A \to \mathbf{F}A \\ \hline
(\mathbf{G}\neg A)^* \to \mathbf{F}A \\ \hline
(\mathbf{F}A)^* \to \mathbf{G}\neg A \\ \hline
\bullet((\mathbf{F}A)^*) \to \neg A \\ \hline
\bullet((\mathbf{F}A)^*) \to A^* \\ \hline
(\mathbf{F}A)^* \to \bullet(A^*) \\ \hline
(\bullet(A^*))^* \to \mathbf{F}A \\ \hline
A \to A.
\end{array}$$

Then

$$\Pi^T = \begin{array}{l} \dfrac{\dfrac{\circ(\Delta \cup \Delta_2) \setminus \{\mathbf{G}\neg A\} \to \mathbf{F}A}{\circ(\Delta_1 \cup \Delta_2) \setminus \{\mathbf{G}\neg A\} \to \neg\mathbf{G}\neg A} \quad \Pi_5}{\circ(\Delta_1 \cup \Delta_2) \setminus \{\mathbf{G}\neg A\} \to \mathbf{G}\neg A \supset \mathbf{0} \quad \Pi_4} \\ \vdots \\ \Pi_3 \end{array}$$

$F17$: Analogous to the previous case.

$F18$: Suppose $\mathcal{CT}(\Pi_1, \circ\Delta_1 \to B)$ and $\mathcal{CT}(\Pi_2, \circ\Delta_2 \to \neg B)$. Then there is a tableau Π_3 such that $\mathcal{CT}(\Pi_3, \circ\Delta_1 \cup \Delta_2 \to \mathbf{0})$. Let

$$\Pi_4 = \begin{array}{l} \dfrac{\neg\mathbf{F}A \supset \mathbf{0} \to \neg\mathbf{G}\neg A}{\neg\mathbf{F}A \supset \mathbf{0} \to (\mathbf{G}\neg A)^*} \\ \hline \vdots \\ \dfrac{\neg\mathbf{F}A \to \mathbf{G}\neg A}{\bullet(\neg\mathbf{F}A) \to \neg A} \qquad \dfrac{\mathbf{0} \to \mathbf{I}^*}{\mathbf{0} \to \mathbf{I}} \\ \hline \bullet(\neg\mathbf{F}A) \to A^* \\ \hline \neg\mathbf{F}A \to \bullet(A^*) \\ \hline (\mathbf{F}A)^* \to \bullet(A^*) \\ \hline (\bullet(A^*))^* \to \mathbf{F}A \\ \hline A \to A. \end{array}$$

Then

$$\Pi^T = \begin{array}{l} \dfrac{\circ(\Delta_1 \cup \Delta_2) \setminus \{\neg\mathbf{F}A\} \to \neg\mathbf{G}\neg A}{\circ(\Delta_1 \cup \Delta_2) \setminus \{\neg\mathbf{F}A\} \to \neg\mathbf{F}A \supset \mathbf{0} \quad \Pi_4} \\ \vdots \\ \Pi_3 \end{array}$$

$F19$: Analogous to the previous case.

$F20$: Suppose $\mathcal{CT}(\Pi_1, \circ\Delta \to A)$.

$$\Pi^T = \begin{array}{c} \dfrac{\circ\Delta \to \mathbf{HF}A}{\bullet((\mathbf{F}A)^*) \to (\circ\Delta)^*} \\ \hline (\mathbf{F}A)^* \to \bullet((\circ\Delta)^*) \\ \hline (\bullet((\circ\Delta)^*))^* \to \mathbf{F}A \\ \hline \Pi_1 \end{array}$$

$F21$: Analogous to the previous case.

$\langle F1 \rangle$: Suppose $\mathcal{CT}(\Pi_1, \circ\Delta \to A)$, $\Delta = \{A_1, \ldots, A_n\} \neq \emptyset$. Then

$$\Pi^T = \begin{array}{ll}
\underline{\circ\mathbf{G}\Delta \to \mathbf{GG}A} & \\
\underline{\bullet\circ\mathbf{G}\Delta \to \mathbf{G}A} & \\
\underline{\bullet\bullet\circ\mathbf{G}\Delta \to A} & \\
\underline{\bullet\circ\mathbf{G}\Delta \to \bullet A} & \\
\underline{\circ\mathbf{G}\Delta \to \bullet\bullet A} & \\
\multicolumn{2}{l}{\underline{\circ\mathbf{G}\Delta \to \bullet A \qquad\qquad\qquad\qquad\quad}} \\
\underline{\circ\mathbf{G}\Delta \to \mathbf{G}\circ\Delta} & \underline{\mathbf{G}\circ\Delta \to \bullet A} \\
\vdots & \qquad\qquad \Pi_1 \\
A_1 \to A_1 &
\end{array}$$

If $\Delta = \emptyset$, then we get

$$\Pi^T = \begin{array}{l}
\underline{\mathbf{I} \to \mathbf{GG}A} \\
\underline{\bullet\mathbf{I} \to \mathbf{G}A} \\
\underline{\bullet\bullet\mathbf{I} \to A} \\
\underline{\bullet\mathbf{I} \to \bullet A} \\
\underline{\mathbf{I} \to \bullet\bullet A} \\
\underline{\mathbf{I} \to \bullet A} \\
\underline{\bullet\mathbf{I} \to A} \\
\mathbf{I} \to A.
\end{array}$$

$\langle F2 \rangle$: Similar to the previous case.

$\langle F3 \rangle$: Suppose $\mathcal{CT}(\Pi_1, \circ\Delta \to A)$. Then

$$\Pi^T = \begin{array}{l}
\underline{\circ\Delta \to \mathbf{FP}A} \\
\underline{(\bullet((\bullet\circ\Delta)^*))^* \to \mathbf{FP}A} \\
\underline{\bullet\circ\Delta \to \mathbf{P}A} \\
\quad\Pi_1
\end{array}$$

$\langle F4 \rangle$: Similar to the previous case.

$\langle F5 \rangle$: Suppose $\mathcal{CT}(\Pi_1, \circ\Delta \to \mathbf{F}A)$. Let

$$\Pi_2 = \begin{array}{l}
\underline{(\bullet(A^*))^* \to (\mathbf{F}A \circ A) \circ \mathbf{P}A} \\
\underline{(\bullet(A^*))^* \circ (\mathbf{P}A)^* \to \mathbf{F}A \circ A} \\
\underline{(\bullet(A^*))^* \to \mathbf{F}A \circ A} \\
\underline{(\bullet(A^*))^* \circ A^* \to \mathbf{F}A} \\
\underline{(\bullet(A^*))^* \to \mathbf{F}A} \\
A \to A
\end{array}$$

and let

$$\Pi_3 = \begin{array}{l} A \to (\mathbf{F}A \circ A) \circ \mathbf{P}A \\ \hline A \circ (\mathbf{P}A)^* \to \mathbf{F}A \circ A \\ \hline A \to \mathbf{F}A \circ A \\ \hline (\mathbf{F}A)^* \circ A \to A \\ \hline A \to A. \end{array}$$

Let

$$\Pi_4 = \begin{array}{l} \bullet A \to (\mathbf{F}A \circ A) \circ \mathbf{P}A \\ \hline (\mathbf{F}A \circ A)^* \circ \bullet A \to \mathbf{P}A \\ \hline \bullet A \to \mathbf{P}A \\ \hline A \to A \end{array}$$

and let

$$\Pi_5 = \begin{array}{l} \mathbf{F}A \to \mathbf{G}(\mathbf{F}A \vee A \vee \mathbf{P}A) \\ \hline (\bullet(A^*))^* \to \mathbf{G}(\mathbf{F}A \vee A \vee \mathbf{P}A) \\ \hline \bullet((\bullet(A^*))^*) \to \mathbf{F}A \vee A \vee \mathbf{P}A \\ \hline \bullet((\bullet(A^*))^*) \to (\mathbf{F}A \vee A) \circ \mathbf{P}A \\ \hline \bullet((\bullet(A^*))^*) \circ (\mathbf{P}A)^* \to \mathbf{F}A \vee A \\ \hline \bullet((\bullet(A^*))^*) \circ (\mathbf{P}A)^* \to \mathbf{F}A \circ A \\ \hline \bullet((\bullet(A^*))^*) \to (\mathbf{F}A \circ A) \circ \mathbf{P}A \\ \hline \Pi_2 \quad \Pi_3 \quad \Pi_4. \end{array}$$

Then

$$\Pi^T = \begin{array}{l} \circ\Delta \to \mathbf{G}(\mathbf{F}A \vee A \vee \mathbf{P}A) \\ \hline \circ\Delta \to \mathbf{F}A \qquad \Pi_5 \\ \hline \quad \Pi_1 \end{array} .$$

$\langle F6 \rangle$: Similar to the previous case.

From this definition it is clear that

LEMMA 5.2 $\mathcal{P}_N(\Pi, A, \{A_1, \ldots, A_n\})$ *implies* $\mathcal{CT}(\Pi^T, A_1 \circ \ldots \circ A_n \to A)$.

5.3 *From Closed Tableaux to Sequent Calculus*

The mapping $(\cdot)^S$ from closed tableaux into sequent calculus proofs is as simple as could be: just turn the closed tableaux and their sequents upside down. Obviously,

LEMMA 5.3 $\mathcal{CT}(\Pi, A_1 \circ \ldots \circ A_n \to A)$ *implies* $\mathcal{P}_S(\Pi^S, A_1 \circ \ldots \circ A_n \to A)$.

5.4 From Sequent Calculus to Axioms

In order to inductively define the mapping $(\cdot)^H$ from sequent calculus proofs into axiomatic proofs, we use a translation τ of sequents into wffs:

$$\tau(X \to Y) = \tau_1(X) \supset \tau_2(Y),$$

where τ_i $(i = 1, 2)$ is defined as follows:

$$\begin{aligned}
\tau_i(A) &= A \\
\tau_1(\mathbf{I}) &= \mathbf{1} \\
\tau_2(\mathbf{I}) &= \mathbf{0} \\
\tau_1(X^*) &= \neg\tau_2(X) \\
\tau_2(X^*) &= \neg\tau_1(X) \\
\tau_1(X \circ Y) &= \tau_1(X) \wedge \tau_1(Y) \\
\tau_2(X \circ Y) &= \tau_2(X) \vee \tau_2(Y) \\
\tau_1(\bullet X) &= \mathbf{P}\tau_1(X) \\
\tau_2(\bullet X) &= \mathbf{G}\tau_2(X).
\end{aligned}$$

The axiomatic sequents are thus translated into instantiations of the axiom scheme $A \supset A$. Suppose $\mathcal{P}_S(\Pi, X \to Y)$ and $\mathcal{P}_S(\frac{\Pi}{X' \to Y'}, X_1 \circ \ldots \circ X_k \to Y')$. By the induction hypothesis there is an axiomatic proof Π^H of $\tau(X \to Y)$ from $\emptyset$.

$$\left(\frac{\Pi}{X' \to Y'}\right)^H = \begin{array}{l} \tau_1(X_1) \\ \vdots \\ \tau_1(X_k) \\ \Pi^H \\ \Pi', \end{array}$$

where Π' is a certain axiomatic proof of $\tau(X' \to Y')$ from $\tau(X \to Y)$. Sequent rules with two or three premise sequents are dealt with similarly. We shall not specify Π' here; the existence of Π' is clear from the possible worlds semantics. We have

LEMMA 5.4 *$\mathcal{P}_S(\Pi, A_1 \circ \ldots \circ A_n \to A)$ implies $\mathcal{P}_H(\Pi^H, A, \{A_1, \ldots, A_n\})$.*

6 THE THEOREM

THEOREM 6.1 *The full-circle theorem:*

1. *$\mathcal{P}_H(\Pi, A, \{A_1, \ldots, A_n\})$ implies*
2. *$\mathcal{P}_N(\Pi^N, A, \{A_1, \ldots, A_n\})$ implies*

3. $\mathcal{CT}((\Pi^N)^T, A_1 \circ \ldots \circ A_n \rightarrow A)$ *implies*

4. $\mathcal{P}_S(((\Pi^N)^T)^S, A_1 \circ \ldots \circ A_n \rightarrow A)$ *implies*

5. $\mathcal{P}_H((((\Pi^N)^T)^S)^H, A, \{A_1, \ldots, A_n\})$

Proof. By Lemmas 5.1–5.4. ⊣

COROLLARY 6.2 *The presented tableau calculi are strongly complete wrt their corresponding possible worlds semantics:* $A_1, \ldots, A_n \models A$ *iff there exists a closed display tableau for* $A_1 \circ \ldots \circ A_n \rightarrow A$.

7 QUESTIONS

Display logic offers a most natural and methodologically attractive proof-theoretic framework for simple tense logic. An obvious question is: How can the class of displayable tense logics be characterized? This question has recently been answered by [Kracht, 1996], who was able to show that a normal propositional tense logic can be presented as a display calculus admitting of cut-elimination iff it is axiomatizable by formulas of the form $A \supset B$ where both A and B contain only propositional variables, $\mathbf{1}$, $\wedge$, $\vee$, $\mathbf{F}$, and $\mathbf{P}$, and A contains each propositional variable at most once. Another question is whether this approach also works nicely for modal and tense logics based on non-classical, say, constructive systems. And finally, how does display logic relate to other developments in the direction of a general proof theory of modal logic?

Institute of Logic and Philosophy of Science
University of Leipzig

Acknowledgements. This paper was written at the Graduiertenkolleg Kognitionswissenschaft of the University of Hamburg.

I wish to thank Tijn Borghuis and Maarten de Rijke for their helpful and constructive remarks on an earlier version of this paper.

REFERENCES

[Barth and Krabbe, 1982] E.M. Barth and E.C.W. Krabbe. *From Axiom to Dialogue*. De Gruyter, Berlin, 1982.

[Belnap, 1982] N. Belnap. Display logic. *Journal of Philosophical Logic*, 11:14–25, 1982.

[Belnap, 1990] N. Belnap. Linear logic displayed. *Notre Dame Journal of Formal Logic*, 31:14–25, 1990.

[Blamey and Humberstone, 1991] S. Blamey and I. Humberstone. A perspective on modal sequent logic. *Publ. RIMS Kyoto University*, 27:763–782, 1991.

[Borghuis, 1993] T. Borghuis. Interpreting modal natural deduction in type theory. In M. de Rijke, editor, *Diamonds and Defaults*, pages 41–66, Kluwer, Dordrecht, 1993.

[Bull and Segerberg, 1984] R.A. Bull and K. Segerberg. Basic modal logic. In D. Gabbay and F. Guenthner, editors, *Handbook of Philosophical Logic. Vol. 2*, pages 1–88. Reidel, Dordrecht, 1984.

[Došen, 1985] K. Došen. Sequent-systems for modal logic. *Journal of Symbolic Logic*, 50:149–168, 1985.

[Kracht, 1996] M. Kracht. Power and weakness of the modal display calculus. In H. Wansing, editor, *Proof Theory of Modal Logic*, pages 95–122. Kluwer Academic Publishers, Dordrecht, 1996.

[Masini, 1992] A. Masini. 2-Sequent calculus: a proof theory of modalities. *Annals of Pure and Applied Logic*, 58:229–246, 1992.

[Wansing, 1994] H. Wansing. Sequent calculi for normal propositional logics. *Journal of Logic and Computation*, 4:125–142, 1994.

MICHAEL ZAKHARYASCHEV

CANONICAL FORMULAS FOR MODAL AND SUPERINTUITIONISTIC LOGICS: A SHORT OUTLINE

Every time when I am asked what canonical formulas are, I have some difficulty. Strictly speaking, they are modal or intuitionistic formulas of a rather special form associated with finite frames. But I have never used them as *formulas*. As *formulas* they seem to be useless. All one needs to know about them is that they are.

Actually, the canonical formulas are just a more or less convenient language for conveying the fact that the 'geometry and topology' of transitive general frames refuting a given modal or intuitionistic formula φ can be reconstructed in a certain canonical way starting from some finite pattern frames which are effectively determined by φ. For instance, the equality

$$\mathbf{K4.1} = \mathbf{K4} \oplus \alpha(\bullet, \emptyset, \bot) \oplus \alpha((\circ\,\circ), \emptyset, \bot)$$

(where $\alpha(\bullet, \emptyset, \bot)$ and $\alpha((\circ\,\circ), \emptyset, \bot)$ are the modal canonical formulas associated with the frames $\bullet$ and $(\circ\,\circ)$) means that a transitive general frame $\mathfrak{F}$ refutes the McKinsey axiom $\Box\Diamond p \to \Diamond\Box p$ if and only if there is a generated subframe of $\mathfrak{F}$ containing a cofinal (not necessarily generated) subframe which can be p-morphically mapped either onto the irreflexive point $\bullet$ or onto the two point cluster $(\circ\,\circ)$.

The most significant consequence of this fact is that every modal logic containing **K4** and every superintuitionistic (alias intermediate) logic can be represented like **K4.1** above, i.e. by its canonical axioms displaying explicitly the structure of their refutation frames and so the structure of frames for the logic. And as a result we obtain a universal language which makes it possible to speak about modal and superintuitionistic logics in purely frame-theoretic terms such as subframe, cofinal subframe, p-morphism, etc., without any reference to formulas and models.

I came across the idea of canonical formulas in 1981, when I discovered that refutation frames for every intuitionistic implicative formula can be reconstructed from some finite pattern frames with the help of only one notion, namely, that of partial p-morphism. The purpose of this essay is to collect together the results (including unpublished ones) which I have succeeded to obtain using the apparatus of the canonical formulas and present them as an integrated picture without proofs, confining only to motivations, definitions, examples and formulations of theorems. This picture is far from being completed. On the contrary, it is rather a starting

M. de Rijke (ed.), Advances in Intensional Logic, 195–248.

point of an extensive research program. So far the canonical formulas worked well enough for superintuitionistic logics and monomodal logics with transitive frames, and I believe they will still help us to solve new difficult problems in this field (some of them are mentioned below). But I hope also that this approach can be extended to the polymodal intransitive case, though for that purpose the conventional frame-theoretic language should be enriched with some new notions.

1 PRELIMINARY DEFINITIONS AND NOTATION

This section contains the basic definitions and notation which concern modal and superintuitionistic logics and are relevant to what follows. The meanings of some symbols are defined in Table 1. The missing definitions can be found, for instance, in [Segerberg, 1971], [Bull and Segerberg, 1984], and [van Dalen, 1986].

1.1 Logics

In this paper we deal with modal propositional system **K4**, intuitionistic propositional logic **Int** and their extensions. So by a *logic* we mean here a set of modal or intuitionistic formulas containing **K4** or, respectively, **Int** and closed under modus ponens and substitution.

The smallest logic containing a logic L and a set of formulas $\{\varphi_i : i \in I\}$ is denoted by $L + \{\varphi_i : i \in I\}$. Logics containing **Int** are known as *superintuitionistic* or *intermediate logics*. In the modal case we distinguish two kinds of extensions of **K4**: those defined above, which sometimes are called *quasi-normal logics*, and *normal* ones which presuppose taking the closure under necessitation $(\varphi/\Box\varphi)$ as well. The smallest normal logic containing L and $\{\varphi_i : i \in I\}$ is denoted by $L \oplus \{\varphi_i : i \in I\}$. Ext$L$ and NExt$L$ are the classes of all extensions and normal extensions of the logic L, respectively. By **For** we denote the set of all modal (or intuitionistic) formulas.

Tables 2 and 3 contain collections of 'standard' normal modal and superintuitionistic logics[1]. Almost all these logics are well-known from the literature (consult [Lemmon, 1977; Segerberg, 1971; Bull and Segerberg, 1984; Hosoi and Ono, 1973; Gabbay, 1981; van Dalen, 1986]). Only $\mathbf{A}^*$, $\mathbf{D4G}_1$ and $\mathbf{ND}_k$ are comparatively new. $\mathbf{A}^*$ is due to Artemov; it was considered by [Shavrukov, 1990]. The axiom of $\mathbf{D4G}_1$ is a modal analog of a topological axiom of Kuratowski [1922]; this logic was studied by [Shehtman, 1990]. The logics $\mathbf{ND}_k$ were introduced by [Maksimova, 1986].

[1] We assume $\wedge$ and $\vee$ to connect formulas stronger than $\rightarrow$

Table 1. Meanings of Principal Symbols

Symbol	Meaning
$\bullet$	an irreflexive point
$\circ$	a reflexive point
$*$	any point
$(\circ\,\circ)$	a two point cluster
$\mathfrak{F}, \mathfrak{G}, \mathfrak{H}$	Kripke frames, general frames
card$\mathfrak{F}$	cardinality of the set of points in $\mathfrak{F}$
$\mathfrak{F} \subseteq_{\rightarrow} \mathfrak{G}$	$\mathfrak{F}$ is a generated subframe of $\mathfrak{G}$
$C(x)$	cluster generated by a point x
C	a cluster
$\rho\mathfrak{F}$	skeleton of $\mathfrak{F}$
$\mathfrak{d}, \mathfrak{e}$	closed domains (alias antichains)
$\mathfrak{D}, \mathfrak{E}$	sets of closed domains
$\mathfrak{D}^\sharp$	set of all antichains (in $\mathfrak{F}$)
dom$f$	domain of the partial map f
p, q, r	propositional variables
$\bot$	falsehood
φ, ψ, χ	modal and intuitionistic formulas
$\alpha(\mathfrak{F}, \mathfrak{D}, \bot)$	modal canonical formula for $\mathfrak{F}$ and $\mathfrak{D}$
$\alpha(\mathfrak{F}, \mathfrak{D})$	modal negation free canonical formula for $\mathfrak{F}$ and $\mathfrak{D}$
$\beta(\mathfrak{F}, \mathfrak{D}, \bot)$	intuitionistic canonical formula for $\mathfrak{F}$ and $\mathfrak{D}$
$\beta(\mathfrak{F}, \mathfrak{D})$	intuitionistic negation free canonical formula for $\mathfrak{F}$ and $\mathfrak{D}$
Γ, Δ, Σ	sets of formulas
$\Box^+\varphi$	$\varphi \wedge \Box\varphi$
$\Box^n\varphi$	$\underbrace{\Box \ldots \Box}_{n} \varphi$
L, M	logics
$L + \{\varphi_i : i \in I\}$	closure of $L \cup \{\varphi_i : i \in I\}$ under modus ponens and substitution
$L \oplus \{\varphi_i : i \in I\}$	closure of $L \cup \{\varphi_i : i \in I\}$ under modus ponens, substitution and necessitation
Ext$L$	class of all logics containing L
NExt$L$	class of all normal logics in Ext$L$

1.2 Frames

A *modal general frame* $\mathfrak{F} = \langle W, R, P\rangle$ consists of a nonempty set (of worlds or points) W, a binary transitive (accessibility) relation R on W and a set (of possible values) P of subsets of W containing $\emptyset$ and closed under the operations $\cap$, $\cup$, $-$ and $\downarrow$. (For the definition of $\downarrow$ see Section 1.3.)

An *intuitionistic general frame* is a triple $\mathfrak{F} = \langle W, R, P\rangle$ where $W \neq \emptyset$, R is a partial order on W and P is a collection of upward closed (i.e. closed under the operation $\uparrow$ defined in Section 1.3) subsets of W containing $\emptyset$ and W and closed under $\cap$, $\cup$ and $\rightarrow$. (For the definition of $\rightarrow$ see Section 1.3.)

A *modal Kripke frame* is a general frame of the form $\mathfrak{F} = \langle W, R, 2^W\rangle$; for brevity we denote it by $\mathfrak{F} = \langle W, R\rangle$. An *intuitionistic Kripke frame* is an intuitionistic general frame of the form $\mathfrak{F} = \langle W, R, \mathrm{Up}W\rangle$ where $\mathrm{Up}W$ is the set of all upward closed subsets of W; it is also denoted by $\mathfrak{F} = \langle W, R\rangle$.

Without loss of generality we may (and will) assume every finite general frame to be a Kripke frame.

A nonempty set $X \subseteq W$ is an *antichain* in a frame $\mathfrak{F} = \langle W, R, P\rangle$ if, for all $x, y \in X$, xRy only if $x = y$. A point y is an *upper bound* for a set $X \subseteq W$ in $\mathfrak{F}$ if xRy for all $x \in X$.

1.3 Arrows

For a frame $\mathfrak{F} = \langle W, R, P\rangle$, $X, Y \subseteq W$ and $z \in W$, we put:

$$X\uparrow = \{y : \exists x \in X\; xRy\}, \quad X\underline{\uparrow} = X\uparrow \cup X,$$

$$X\downarrow = \{y : \exists x \in X\; yRx\}, \quad X\overline{\downarrow} = X\downarrow \cup X,$$

$$z\uparrow = \{z\}\uparrow, \quad z\underline{\uparrow} = \{z\}\underline{\uparrow}, \quad \text{etc.},$$

$$X \rightarrow Y = -((X \cap -Y)\downarrow).$$

$\mathfrak{F}$ is *rooted* if there is a point $x \in W$, the *root* of $\mathfrak{F}$, such that $W = W\underline{\uparrow}$.

1.4 Subframes and Subreductions

A Kripke frame $\mathfrak{G} = \langle V, S\rangle$ is a *subframe* of a Kripke frame $\mathfrak{F} = \langle W, R\rangle$ if $V \subseteq W$ and S is the restriction of R to V. A frame $\mathfrak{G} = \langle V, S, Q\rangle$ is a *generated subframe* of $\mathfrak{F} = \langle W, R, P\rangle$ (notation: $\mathfrak{G} \subseteq \mathfrak{F}$) if $\langle V, S\rangle$ is a subframe of $\langle W, R\rangle$ with $V = V\underline{\uparrow}$ in $\mathfrak{F}$ and $Q = \{X \cap V : X \in P\}$.

A partial (i.e. not completely defined, in general) map f from V onto W is called a *subreduction* (or a *partial p-morphism*) *of* $\mathfrak{G} = \langle V, S, Q\rangle$ *to* $\mathfrak{F} = \langle W, R, P\rangle$ if the following conditions are satisfied for all $x, y \in \mathrm{dom} f$, $z \in W$ and $X \subseteq W$

$$xSy \Rightarrow f(x)Rf(y);$$

Table 2. A list of 'standard' normal modal logics containing **K4**

D4	=	$\mathbf{K4} \oplus \Diamond\top$
S4	=	$\mathbf{K4} \oplus \Box p \to p$
GL	=	$\mathbf{K4} \oplus \Box(\Box p \to p) \to \Box p$
For	=	$\mathbf{K4} \oplus p$
Grz	=	$\mathbf{K4} \oplus \Box(\Box(p \to \Box p) \to p) \to p$
S4.1	=	$\mathbf{S4} \oplus \Box\Diamond p \to \Diamond\Box p$
K4.1	=	$\mathbf{K4} \oplus \Box\Diamond p \to \Diamond\Box p$
Triv	=	$\mathbf{K4} \oplus \Box p \leftrightarrow p$
Verum	=	$\mathbf{K4} \oplus \Box p$
S5	=	$\mathbf{S4} \oplus p \to \Box\Diamond p$
K4B	=	$\mathbf{K4} \oplus p \to \Box\Diamond p$
S4.2	=	$\mathbf{S4} \oplus \Diamond\Box p \to \Box\Diamond p$
S4.3	=	$\mathbf{S4} \oplus \Box(\Box p \to q) \vee \Box(\Box q \to p)$
$\mathbf{A}^*$	=	$\mathbf{GL} \oplus \Box\Box p \to \Box(\Box^+ p \to q) \vee \Box(\Box^+ q \to p)$
K4.2	=	$\mathbf{K4} \oplus \Diamond(p \wedge \Box q) \to \Box(p \vee \Diamond q)$
K4.3	=	$\mathbf{K4} \oplus \Box(\Box^+ p \to q) \vee \Box(\Box^+ q \to p)$
K4Z	=	$\mathbf{K4} \oplus \Box(\Box p \to p) \to (\Diamond\Box p \to \Box p)$
Dum	=	$\mathbf{S4} \oplus \Box(\Box(p \to \Box p) \to p) \to (\Diamond\Box p \to p)$
$\mathbf{D4G}_1$	=	$\mathbf{D4} \oplus \Box(\Box^+ p \vee \Box^+ \neg p) \to \Box p \vee \Box\neg p$
K4H	=	$\mathbf{K4} \oplus p \to \Box(\Diamond p \to p)$
$\mathbf{K4Alt}_n$	=	$\mathbf{K4} \oplus \Box p_1 \vee \Box(p_1 \to p_2) \vee \ldots \vee \Box(p_1 \wedge \ldots \wedge p_n \to p_{n+1})$
$\mathbf{K4I}_n$	=	$\mathbf{K4} \oplus \bigwedge_{i=0}^{n} \Diamond p_i \to \bigvee_{0 \le i \ne j \le n} \Diamond(p_i \wedge (p_j \vee \Diamond p_j))$
$\mathbf{K4B}_n$	=	$\mathbf{K4} \oplus \beta_n$, with $\beta_1 = \Diamond\Box p_1 \to p_1$, $\beta_{m+1} = \Diamond(\Box p_{m+1} \wedge \neg\beta_m) \to p_{m+1}$
$\mathbf{K4}_{n,m}$	=	$\mathbf{K4} \oplus \Box^n p \to \Box^m p$, for $1 \le m < n$

$$f(x)Rz \Rightarrow \exists y \in f^{-1}(z)\; xSy;$$

$$X \in P \Rightarrow \begin{cases} f^{-1}(X) \in Q & \text{if } \mathfrak{G} \text{ and } \mathfrak{F} \text{ are modal frames} \\ -(f^{-1}(-X){\downarrow}) \in Q & \text{if } \mathfrak{G} \text{ and } \mathfrak{F} \text{ are intuitionistic.} \end{cases}$$

In this case $\mathfrak{G}$ is said to be *subreducible to* $\mathfrak{F}$. If $\mathrm{dom} f = V$ then f is called a *reduction* (or a *p-morphism*) *of* $\mathfrak{G}$ *to* $\mathfrak{F}$, and $\mathfrak{G}$ is *reducible to* $\mathfrak{F}$.

1.5 Clusters

A set $C \subseteq W$ is called a *cluster generated by* $x \in W$ in a frame $\mathfrak{F} = \langle W, R, P\rangle$ if $C = x{\uparrow} \cap x{\downarrow}$; the cluster generated by x is denoted by $C(x)$. A cluster is *degenerate* if it has only one irreflexive point, *simple* if it consists of one reflexive point and *proper* if it contains at least two points.

The quotient frame of a Kripke frame $\mathfrak{F} = \langle W, R\rangle$ with respect to clusters is denoted by $\rho\mathfrak{F} = \langle \rho W, \rho R\rangle$ and called the *skeleton* of $\mathfrak{F}$. That is, $\rho W = \{C(x) : x \in W\}$ and $C(x)\rho RC(y)$ iff xRy. The skeleton of a quasi-ordered modal frame is an intuitionistic one. The skeleton of a quasi-ordered general frame $\mathfrak{F} = \langle W, R, P\rangle$ is the intuitionistic general frame $\rho\mathfrak{F} = \langle \rho W, \rho R, \rho P\rangle$ in which ρP is the set of skeletons of the upward closed sets in P.

1.6 Frames and Logics

A logic L is *characterized* (or *determined*) by a class of frames $\mathcal{C}$ if $L = \{\varphi \in \mathbf{For} : \forall \mathfrak{F} \in \mathcal{C}\; \mathfrak{F} \models \varphi\}$. L is said to be *Kripke complete* if it is characterized by a class of Kripke frames, and L has the *finite model property* (FMP) if it is characterized by a class of finite frames. $\mathfrak{F}$ is called a *frame for* L if all the formulas in L are valid in $\mathfrak{F}$.

2 THE CHARACTERIZATION PROBLEM

As I already mentioned in the preamble, canonical formulas reflect some universal principles of the construction of refutation frames for modal and intuitionistic formulas. This section discusses these principles in the context of the general characterization problem.

2.1 The Problem

If you happen to investigate some modal or superintuitionistic logic L, you will most likely try to inquire first what the frames for L are, even if you are a steadfast adherent of purely syntactical methods. And in order to do this you will have to find out the structure of frames for the additional axioms of L. That is, you will be solving the following

Table 3. A list of 'standard' superintuitionistic logics

For	=	$\mathbf{Int} + p$
Cl	=	$\mathbf{Int} + p \vee \neg p$
KC	=	$\mathbf{Int} + \neg p \vee \neg\neg p$
LC	=	$\mathbf{Int} + (p \to q) \vee (q \to p)$
SL	=	$\mathbf{Int} + ((\neg\neg p \to p) \to \neg p \vee p) \to \neg p \vee \neg\neg p$
KP	=	$\mathbf{Int} + (\neg p \to q \vee r) \to (\neg p \to q) \vee (\neg p \to r)$
WKP	=	$\mathbf{Int} + (\neg p \to \neg q \vee \neg r) \to (\neg p \to \neg q) \vee (\neg p \to \neg r)$
$\mathbf{ND}_k$	=	$\mathbf{Int} + (\neg p \to \neg q_1 \vee \ldots \vee \neg q_k) \to$ $\to (\neg p \to \neg q_1) \vee \ldots \vee (\neg p \to \neg q_k),\ k \geq 2$
$\mathbf{BH}_n$	=	$\mathbf{Int} + \chi_n$, where $\chi_1 = p_1 \vee \neg p_1,\ \chi_{m+1} = p_{m+1} \vee (p_{m+1} \to \chi_m)$
$\mathbf{BW}_n$	=	$\mathbf{Int} + \bigvee_{i=0}^{n} (p_i \to \bigvee_{j \neq i} p_j)$
$\mathbf{BTW}_n$	=	$\mathbf{Int} + \bigwedge_{0 \leq i < j \leq n} \neg(\neg p_i \wedge \neg p_j) \to \bigvee_{i=0}^{n} (\neg p_i \to \bigvee_{j \neq i} \neg p_j)$
$\mathbf{T}_n$	=	$\mathbf{Int} + \bigwedge_{i=0}^{n} ((p_i \to \bigvee_{i \neq j} p_j) \to \bigvee_{i \neq j} p_j) \to \bigvee_{i=0}^{n} p_i$
$\mathbf{B}_n$	=	$\mathbf{Int} + \bigwedge_{i=0}^{n} (\neg p_i \leftrightarrow \bigvee_{i \neq j} p_j) \to \bigvee_{i=0}^{n} p_i$

CHARACTERIZATION PROBLEM: *Given a (modal or intuitionistic) formula φ, to characterize in one way or another the constitution of Kripke and/or general frames validating φ.*

2.2 A Trivial Solution

Like any other problem of that kind, the characterization one has a trivial solution, which uses directly the definition of validity or a reformulation of it in some other language. For example, every modal formula φ with propositional variables $p_1, p_2, \ldots$ can be translated into a second-order formula $ST(\varphi)$, often called the *standard translation* of φ, which has a free individual variable x, set variables $X_1, X_2, \ldots$ and which is defined by induction on the construction of φ in the following way:

$$ST(p_i) = x \in X_i, \quad ST(\bot) = \bot,$$

$$ST(\psi \odot \chi) = ST(\psi) \odot ST(\chi), \quad \text{for } \odot \in \{\to, \wedge, \vee\},$$

$$ST(\Box\psi) = \forall y(xRy \to ST(\psi)\{y/x\});$$

in the intuitionistic case the equality for $\to$ should be replaced with

$$ST(\psi \to \chi) = \forall y(xRy \to (ST(\psi) \to ST(\chi))\{y/x\}).$$

And then a general frame $\mathfrak{F} = \langle W, R, P \rangle$ will validate φ if and only if $\mathfrak{F}$ is a classical model for the second-order formula $\forall x\, \forall X_1 \in P\, \forall X_2 \in P \ldots ST(\varphi)$.

Of course this solution can hardly be regarded as a satisfactory one. However, starting from it, one can sometimes succeed in eliminating all the second-order variables $X_1, X_2, \ldots$ and obtain a first-order sentence with only two binary predicates R and $=$ characterizing Kripke frames or even a certain class of general frames for φ.

2.3 *First-Order Characterization*

It was the very earliest papers of Kripke [1963] and Dummett and Lemmon [1959] on possible worlds semantics that discovered in fact that the axioms of some standard modal and superintuitionistic logics express on Kripke frames such familiar first-order properties of relational structures as reflexivity, transitivity, convergence, etc. Further results of [Lemmon, 1977; Segerberg, 1971] and others, culminated in [Sahlqvist, 1975], which characterized Kripke frames for a good many modal formulas by first-order sentences, gave rise to correspondence theory studying connections between non-classical propositional formulas and classical predicate ones (see [van Benthem, 1983, 1984]).

As an illustration I show here first-order characterizations of (transitive) Kripke frames for three formulas which will be used in subsequent examples: for every frame $\mathfrak{F} = \langle W, R \rangle$,

(i) $\mathfrak{F}$ validates the Density Axiom $\Box\Box p \rightarrow \Box p$ iff $\mathfrak{F}$ is *dense*, i.e.

$$\forall x, z(xRz \rightarrow \exists y\, xRyRz);$$

(ii) $\mathfrak{F}$ validates the Mckinsey Axiom $\Box\Diamond p \rightarrow \Diamond\Box p$ iff $\mathfrak{F}$ is *atomic*, i.e.

$$\forall x \exists y(xRy \wedge \forall z(yRz \rightarrow y = z));$$

(iii) (intuitionistic) $\mathfrak{F}$ validates the so-called Weak Kreisel-Putnam Axiom

$$(\neg p \rightarrow \neg q \vee \neg r) \rightarrow (\neg p \rightarrow \neg q) \vee (\neg p \rightarrow \neg r)$$

iff

$$\forall x \forall y_1, y_2 \in x{\uparrow}(y_1{\uparrow} \cap y_2{\uparrow} = \emptyset \rightarrow$$
$$\exists u \in x{\uparrow} \cap y_1{\uparrow}{\downarrow} \cap y_2{\uparrow}{\downarrow} \forall z \in u{\uparrow}\ (z{\uparrow} \cap y_1{\uparrow} \neq \emptyset \vee z{\uparrow} \cap y_2{\uparrow} \neq \emptyset)).$$

(It is not hard to rewrite the last sentence using R instead of arrows, but then it will be twice as long).

However, this first-order approach to the characterization problem comes across two obstacles which cannot be overcome. The first one is the phenomenon of

Kripke incompleteness discovered by Fine [1974] and Thomason [1974] (for a discussion of a liaison between Kripke completeness and first-order definability, see [Kracht, 1993]). But even among Kripke complete logics there are those whose frames cannot be characterized by any first-order sentences; examples include such respectable modal systems as the Gödel-Löb provability logic **GL**, the Grzegorczyk logic **Grz** and the superintuitionistic Scott logic **SL** (for proofs and other examples consult [van Benthem, 1983, 1984; Rodenburg, 1986]). Moreover, as was discovered by Chagrova [1991], the sets of modal and intuitionistic formulas having first-order equivalents turn out to be algorithmically undecidable.

2.4 *Prelude to the Frame-Theoretic Characterization*

Another approach to the characterization problem is based upon the observation that every refutation frame $\mathfrak{G}$ for a given formula φ contains a subframe $\mathfrak{H}$ which is reducible to a finite refutation frame for φ of size bounded by a constant N_φ depending only on φ. This result was obtained independently by Fine [1985] and Zakharyaschev [1983, 1984, 1984a, 1992]. However, its special cases for particular formulas were known long before that.

For some φ the frame $\mathfrak{H}$ turns out to be (or can be extended to) a generated subframe of $\mathfrak{G}$. And then, by the Generation Theorem, we arrive at a nice refutability criterion for such φ: for every general frame $\mathfrak{G}$, $\mathfrak{G} \not\models \varphi$ iff there is $\mathfrak{H} \underset{\to}{\subseteq} \mathfrak{G}$ which is reducible to a refutation frame $\mathfrak{F}$ for φ such that card$\mathfrak{F} \leq N_\varphi$. For example,

- $\mathfrak{G} \not\models \Box\Diamond p \to \Diamond\Box p$ iff $\exists\mathfrak{H} \underset{\to}{\subseteq} \mathfrak{G} : \mathfrak{H}$ is reducible either to $\bullet$ or to ;

- $\mathfrak{G} \not\models ((\neg\neg p \to p) \to \neg p \vee p) \to \neg p \vee \neg\neg p$ iff $\exists\mathfrak{H} \underset{\to}{\subseteq} \mathfrak{G} : \mathfrak{H}$ is reducible to

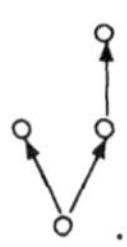 .

(A lot of examples of that kind can be found in [Rautenberg, 1979].)

For many other formulas φ, $\mathfrak{H}$ is just a subframe of $\mathfrak{G}$ (and so the Generation Theorem does not work), but the class of frames for φ turns out to be closed under forming subframes, and we again obtain a nice criterion: $\mathfrak{G} \not\models \varphi$ iff $\mathfrak{G}$ is subreducible to a frame $\mathfrak{F}$ refuting φ and containing at most N_φ points. For example, for every general frame $\mathfrak{G}$,

- $\mathfrak{G} \not\models \Box(\Box p \to p) \to \Box p$ iff $\mathfrak{G}$ is subreducible to $\circ$;

- $\mathfrak{G} \not\models \Box(\Box(p \to \Box p) \to p) \to p$ iff $\mathfrak{G}$ is subreducible either to $\bullet$ or to ;

- $\mathfrak{G} \not\models (p \to q) \vee (q \to p)$ iff $\mathfrak{G}$ is subreducible to

As was observed by Zakharyaschev [1981, 1983], refutation frames for every intuitionistic implicative formula can be described in such a manner.

Both these kinds of characterizations can be represented in a more compact way using frame and subframe formulas defined by Kit Fine. With each finite rooted frame $\mathfrak{F}$ Fine [1974a] associated a modal formula $A_{\mathfrak{F}}$, called the *frame formula* for $\mathfrak{F}$, such that, for every $\mathfrak{G}$,

$$\mathfrak{G} \not\models A_{\mathfrak{F}} \text{ iff } \exists \mathfrak{H} \subsetneq \mathfrak{G} : \mathfrak{H} \text{ is reducible to } \mathfrak{F}.$$

An intuitionistic analog of the frame formulas was introduced ten years earlier by Jankov [1963] under the name of *characteristic formulas* (although they were associated with finite pseudo-Boolean algebras). That is why $A_{\mathfrak{F}}$'s are also known as the *Jankov-Fine formulas*; another nickname — *splittings* — was given to them for some lattice-theoretic reasons (see [Kracht, 1990]). Thus we have:

$$\mathbf{K4.1} = \mathbf{K4} \oplus A_{\bullet} \oplus A_{(\circ\,\circ)}$$

$$\mathbf{SL} = \mathbf{Int} + A_{\mathfrak{F}}, \text{ where } \mathfrak{F} \text{ is}$$

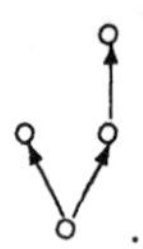

Afterwards, Fine [1985] associated with each finite rooted $\mathfrak{F}$ another modal formula $B_{\mathfrak{F}}$, called the *subframe formula for* $\mathfrak{F}$, such that, for every general frame $\mathfrak{G}$,

$$\mathfrak{G} \not\models B_{\mathfrak{F}} \text{ iff } \mathfrak{G} \text{ is subreducible to } \mathfrak{F}.$$

An intuitionistic version of subframe formulas was defined by Zakharyaschev [1981, 1983]. So we can write:

$$\mathbf{GL} = \mathbf{K4} \oplus B_{\circ}$$

$$\mathbf{Grz} = \mathbf{K4} \oplus B_{\bullet} \oplus B_{(\circ\,\circ)}$$

$$\mathbf{LC} = \mathbf{Int} + B_{\mathfrak{F}}, \text{ where } \mathfrak{F} \text{ is}$$

However, far from all modal and intuitionistic formulas have such characterizations of their frames, witness the Density Axiom, Kreisel-Putnam and Weak Kreisel-Putnam Axioms, etc.

2.5 *Cofinal Subreduction*

The necessary refutability condition, formulated at the beginning of Section 2.4, can be made also a sufficient one using a pair of new notions. One of them is suggested by the condition itself: it is the well known notion of cofinality.

Recall that a set of points $X \subseteq V$ is said to be *cofinal* in a frame $\mathfrak{G} = \langle V, S, Q \rangle$ if $X\overline{\downarrow} = V$. And in that condition the set of worlds in $\mathfrak{H}$ turns out to be cofinal in the subframe of $\mathfrak{G}$ generated by $\mathfrak{H}$. This observation leads to the following definition.

A subreduction f of $\mathfrak{G}$ to $\mathfrak{F}$ is called *cofinal* if $\mathrm{dom}f$ is cofinal in the subframe of $\mathfrak{G}$ generated by $\mathrm{dom}f$ or, in symbols, if

$$\mathrm{dom}f{\uparrow} \subseteq \mathrm{dom}f\overline{\downarrow}.$$

For example,

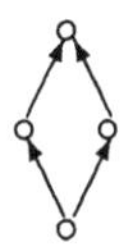 is subreducible to  but not cofinally,

while

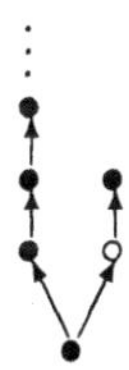 is cofinally subreducible (but not reducible) to .

Using the notion of cofinal subreduction we can characterize refutation frames for the majority of the standard axioms like this:

- $\mathfrak{G} \not\models \Box\Diamond p \to \Diamond\Box p$ iff $\mathfrak{G}$ is cofinally subreducible either to $\bullet$ or to ;
- $\mathfrak{G} \not\models \Diamond(p \wedge \Box q) \to \Box(p \vee \Diamond q)$ iff $\mathfrak{G}$ is cofinally subreducible to one of the following eight frames (each $*$ is to be replaced with $\circ$ or $\bullet$):

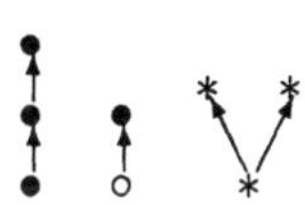

- $\mathfrak{G} \not\models \neg p \vee \neg\neg p$ iff $\mathfrak{G}$ is cofinally subreducible to

.

As was observed by Zakharyaschev [1981, 1983], refutation frames for every disjunction free intuitionistic formula can be characterized in this way.

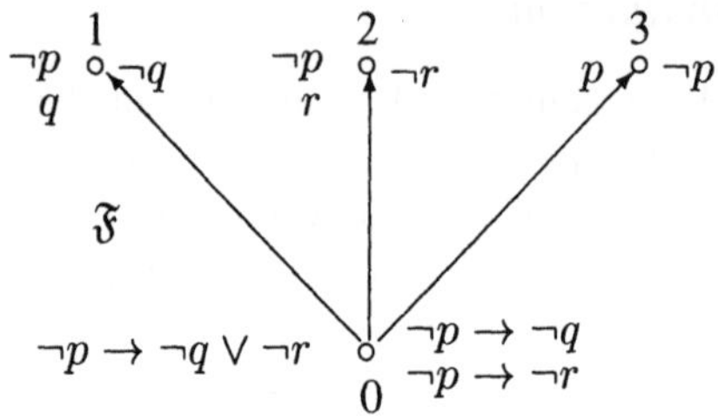

Figure 1.

2.6 Closed Domain Condition

The second notion we need is a bit more complicated. To motivate it, let us try to characterize refutation frames for the Weak Kreisel-Putnam Axiom $(\neg p \to \neg q \vee \neg r) \to (\neg p \to \neg q) \vee (\neg p \to \neg r)$. Using, say, the semantic tableau technique, we can first construct its simplest countermodel which looks like it is shown in Figure 1 (a formula on the left-hand (right-hand) side of a point means that it is true (respectively, not true) at the point).

Then we observe that every general frame $\mathfrak{G}$ refuting our formula is cofinally subreducible to $\mathfrak{F}$, for instance, by the partial map f defined by

$$f(x) = \begin{cases} 0 & \text{if } x \models \neg p \to \neg q \vee \neg r,\ \ x \not\models (\neg p \to \neg q) \vee (\neg p \to \neg r) \\ 1 & \text{if } x \models \neg p \to \neg q \vee \neg r,\ \ x \models \neg p \text{ and } x \models q \\ 2 & \text{if } x \models \neg p \to \neg q \vee \neg r,\ \ x \models \neg p \text{ and } x \models r \\ 3 & \text{if } x \models p \text{ or } x \models \neg p \wedge \neg q \wedge \neg r \\ \text{undefined} & \text{otherwise.} \end{cases}$$

However, this is only a necessary refutability condition, for the frame shown in Figure 2 is cofinally subreducible to $\mathfrak{F}$ but does not refute the Weak Kreisel-Putnam Axiom. The reason for this is that if we insert in the countermodel in Figure 1 a new point x just below 1 and 2 (and above 0) then it is impossible to extend the valuation to x without changing the truth-value of $\neg p \to \neg q \vee \neg r$ and so of the full Axiom at the root of $\mathfrak{F}$. This means in fact that the cofinal subreduction f of $\mathfrak{G}$ to $\mathfrak{F}$ defined above satisfies the following condition:

(1) $\neg\exists x \in \text{dom} f\!\uparrow\ f(x\!\uparrow) = \{1,2\}$

And this condition turns out to be a sufficient one!

Figuratively speaking, every refutation frame for the Weak Kreisel-Putnam Axiom can be constructed starting from $\mathfrak{F}$ in two steps. First we take a frame $\mathfrak{H}$ which is reducible to $\mathfrak{F}$ and then insert into it new points. The cofinality condition requires these points not to be placed above $\mathfrak{H}$. And (1) says that no point should be inserted in such a place from which it could see only inverse images of

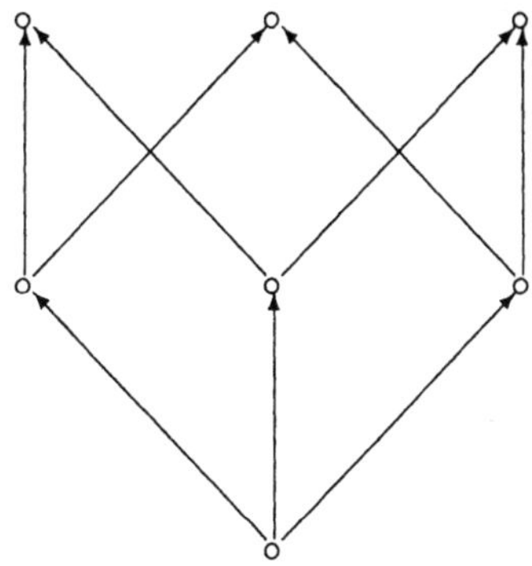

Figure 2.

1 and 2. The place or domain just below 1 and 2 in $\mathfrak{F}$ is *closed for inserting new points*, while all other domains (say, below 1 or below 1 and 3) are open. These observations motivate the following definition.

Let $\mathfrak{F}$ be a frame and $\mathfrak{D}$ a (possibly empty) set of antichains in $\mathfrak{F}$. We say a subreduction f of $\mathfrak{G}$ to $\mathfrak{F}$ satisfies the *closed domain condition for* $\mathfrak{D}$ if

(CDC) *there is no point* $x \in \mathrm{dom}f\uparrow$ *such that* $x \notin \mathrm{dom}f$ *and* $f(x\uparrow) = \mathfrak{d}\underline{\uparrow}$ *for some* $\mathfrak{d} \in \mathfrak{D}$,

or, which is equivalent, if

(CDC) $x \in \mathrm{dom}f\uparrow$ *and* $f(x\uparrow) = \mathfrak{d}\underline{\uparrow}$, *for some* $\mathfrak{d} \in \mathfrak{D}$, *imply* $x \in \mathrm{dom}f$.

Note that, by the definition, every subreduction satisfies (CDC) for $\mathfrak{D} = \emptyset$. Also, without loss of generality we may assume every reflexive singleton to be always open.

Using this notion, we can say now that

- $\mathfrak{G} \not\models (\neg p \to \neg q \vee \neg r) \to (\neg p \to \neg q) \vee (\neg p \to \neg r)$ iff there is a cofinal subreduction of $\mathfrak{G}$ to $\mathfrak{F}$ in Figure 1 satisfying (CDC) for $\mathfrak{D} = \{\{1, 2\}\}$;

- $\mathfrak{G} \not\models \Box\Box p \to \Box p$ iff there is a subreduction of $\mathfrak{G}$ to

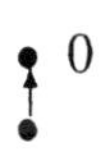

satisfying (CDC) for $\mathfrak{D} = \{\{0\}\}$.

Other examples can be found in Tables 4 and 5 below.

2.7 The Characterization

We are in a position now to formulate the fundamental result obtained by Zakharyaschev [1983, 1984, 1992]:

THEOREM 2.1 (Characterization Theorem) *For every modal or intuitionistic formula φ there is a constant N_φ such that a general frame $\mathfrak{G}$ refutes φ iff there exists a cofinal subreduction of $\mathfrak{G}$ to a refutation frame $\mathfrak{F}$ for φ with $card\,\mathfrak{F} \leq N_\varphi$, satisfying (CDC) for a set $\mathfrak{D}$ of antichains in $\mathfrak{F}$ effectively determined by φ. If φ is negation free (i.e. contains no occurrences of $\neg$, $\bot$, $\Diamond$), then instead of cofinal subreductions one can take plain subreductions. And if, in addition, φ is an intuitionistic disjunction free formula then $\mathfrak{D} = \emptyset$.*

This theorem means in particular that the role of implication in the axioms of superintuitionistic logics can be characterized at the frame-theoretic level only in terms of subreduction, the role of $\bot$ in terms of the cofinality condition and that of $\vee$ can be characterized in terms of the closed domain condition. In modal axioms $\Box$ interacts with other connectives in a more sophisticated manner.

3 THE CANONICAL FORMULAS

'Canonizing' this characterization, one can say that refutation frames for a given formula φ, and so frames for the logic axiomatized by φ, are completely determined by a finite number of finite rooted frames $\mathfrak{F}_1, \ldots, \mathfrak{F}_n$ and sets $\mathfrak{D}_1, \ldots, \mathfrak{D}_n$ of closed domains in them. And now, to close the circle, one needs to show that for every finite rooted $\mathfrak{F}$ and every set $\mathfrak{D}$ of antichains in $\mathfrak{F}$ there exists a formula whose refutation frames are determined by $\mathfrak{F}$ and $\mathfrak{D}$. If this is so then these formulas can be used as bricks from which we can construct any modal and superintuitionistic logic. It is they that are distinguished as the canonical formulas.

3.1 The Modal Case

There are infinitely many ways of defining the canonical formulas. Here is one of them. Suppose $\mathfrak{F} = \langle W, R\rangle$ is a finite rooted frame with points $a_0, \ldots, a_n$ and root a_0 and $\mathfrak{D}$ is a set of antichains in $\mathfrak{F}$. $\mathfrak{D}$ is almost arbitrary. In particular, it may be empty. The only restriction is that it contains no reflexive singletons.

Now, with $\mathfrak{F}$ and $\mathfrak{D}$ we associate a formula $\alpha(\mathfrak{F}, \mathfrak{D}, \bot)$, called the *(normal) modal canonical formula for $\mathfrak{F}$ and $\mathfrak{D}$*, which is defined as follows:

$$\alpha(\mathfrak{F}, \mathfrak{D}, \bot) = \bigwedge_{a_i R a_j} \alpha_{ij} \wedge \bigwedge_{i=0}^{n} \alpha_i \wedge \bigwedge_{\mathfrak{d} \in \mathfrak{D}} \alpha_{\mathfrak{d}} \wedge \alpha_\bot \to p_0$$

where

$$\begin{aligned}
\alpha_{ij} &= \Box^+(\Box p_j \to p_i) \\
\alpha_i &= \Box^+((\bigwedge_{\neg a_i R a_k} \Box p_k \wedge \bigwedge_{j \neq i} p_j \to p_i) \to p_i) \\
\alpha_{\mathfrak{d}} &= \Box^+(\bigwedge_{a_i \in W - \mathfrak{d}\underline{\uparrow}} \Box p_i \wedge \bigwedge_{i=0}^{n} p_i \to \bigvee_{a_j \in \mathfrak{d}} \Box p_j) \\
\alpha_\bot &= \Box^+(\bigwedge_{i=0}^{n} \Box^+ p_i \to \bot).
\end{aligned}$$

If we delete from $\alpha(\mathfrak{F}, \mathfrak{D}, \bot)$ the conjunct $\alpha_\bot$ then the resultant formula will contain neither $\bot$ nor $\neg$, even hidden in $\Diamond$. We denote this formula by $\alpha(\mathfrak{F}, \mathfrak{D})$ and call it the *(normal) modal negation free canonical formula for $\mathfrak{F}$ and $\mathfrak{D}$.*

The canonical formulas were constructed in such a way that the following condition is satisfied:

THEOREM 3.1 (Refutability Criterion for Modal Canonical Formulas) *For every general frame $\mathfrak{G}$,*

(i) $\mathfrak{G} \not\models \alpha(\mathfrak{F}, \mathfrak{D}, \bot)$ *iff there is a cofinal subreduction of $\mathfrak{G}$ to $\mathfrak{F}$ satisfying the closed domain condition (CDC) for $\mathfrak{D}$;*

(ii) $\mathfrak{G} \not\models \alpha(\mathfrak{F}, \mathfrak{D})$ *iff there is a (not necessarily cofinal) subreduction of $\mathfrak{G}$ to $\mathfrak{F}$ satisfying (CDC) for $\mathfrak{D}$.*

Another variant of the canonical formulas, formulated in the spirit of Fine's frame and subframe formulas, can be found in [Zakharyaschev, 1992].

With every rooted $\mathfrak{F}$ one can associate several (at least two) canonical formulas. In general they are connected by the following inclusions. Let $\mathfrak{D}$ and $\mathfrak{C}$ be any sets of antichains in $\mathfrak{F}$ such that $\mathfrak{D} \subseteq \mathfrak{C}$ and $\mathfrak{D}^\sharp$ the set of all antichains (save reflexive singletons of course). Then, according to Refutability Criterion, we have:

$$\begin{array}{ccccccc}
\mathbf{K4} & \oplus & \alpha(\mathfrak{F}, \emptyset, \bot) & \subseteq & \mathbf{K4} & \oplus & \alpha(\mathfrak{F}, \emptyset) \\
& \cup\!\!| & & & & \cup\!\!| & \\
\mathbf{K4} & \oplus & \alpha(\mathfrak{F}, \mathfrak{D}, \bot) & \subseteq & \mathbf{K4} & \oplus & \alpha(\mathfrak{F}, \mathfrak{D}) \\
& \cup\!\!| & & & & \cup\!\!| & \\
\mathbf{K4} & \oplus & \alpha(\mathfrak{F}, \mathfrak{C}, \bot) & \subseteq & \mathbf{K4} & \oplus & \alpha(\mathfrak{F}, \mathfrak{C}) \\
& \cup\!\!| & & & & \cup\!\!| & \\
\mathbf{K4} & \oplus & \alpha(\mathfrak{F}, \mathfrak{D}^\sharp, \bot) & \subseteq & \mathbf{K4} & \oplus & \alpha(\mathfrak{F}, \mathfrak{D}^\sharp)
\end{array}$$

Fine's subframe and frame formulas for $\mathfrak{F}$ take the first and the last places in this hierarchy, respectively. For it follows from Refutability Criterion that

$$\begin{aligned}
\mathbf{K4} \oplus \alpha(\mathfrak{F}, \mathfrak{D}^\sharp, \bot) &= \mathbf{K4} \oplus A_{\mathfrak{F}} \\
\mathbf{K4} \oplus \alpha(\mathfrak{F}, \emptyset) &= \mathbf{K4} \oplus B_{\mathfrak{F}}.
\end{aligned}$$

Now, combining the Characterization Theorem and Refutability Criterion, we immediately obtain the following

THEOREM 3.2 (Completeness Theorem for Modal Canonical Formulas)

(i) *There is an algorithm which, given a modal formula* φ, *returns canonical formulas* $\alpha(\mathfrak{F}_1, \mathfrak{D}_1, \bot), \ldots, \alpha(\mathfrak{F}_n, \mathfrak{D}_n, \bot)$ *such that*

$$\mathbf{K4} \oplus \varphi = \mathbf{K4} \oplus \alpha(\mathfrak{F}_1, \mathfrak{D}_1, \bot) \oplus \ldots \oplus \alpha(\mathfrak{F}_n, \mathfrak{D}_n, \bot).$$

(ii) *There is an algorithm which, given a modal negation free formula* φ *(containing only* $\to$, $\wedge$, $\vee$ *and* $\Box$*), returns negation free canonical formulas* $\alpha(\mathfrak{F}_1, \mathfrak{D}_1), \ldots, \alpha(\mathfrak{F}_n, \mathfrak{D}_n)$ *such that*

$$\mathbf{K4} \oplus \varphi = \mathbf{K4} \oplus \alpha(\mathfrak{F}_1, \mathfrak{D}_1) \oplus \ldots \oplus \alpha(\mathfrak{F}_n, \mathfrak{D}_n).$$

The term 'completeness' in the title of this theorem means that every normal modal logic L containing **K4** can be axiomatized by canonical formulas, i.e. represented in the form

$$L = \mathbf{K4} \oplus \{\alpha(\mathfrak{F}_i, \mathfrak{D}_i, \bot) : i \in I\};$$

and if L is axiomatized by some negation free formulas then there is also another representation:

$$L = \mathbf{K4} \oplus \{\alpha(\mathfrak{F}_i, \mathfrak{D}_i) : i \in I\}.$$

Table 4 contains canonical axiomatizations of all the standard logics in Table 2.

I must warn the reader that the algorithm mentioned in Theorem 3.2 constructs first all the rooted countermodels for a given formula φ with at most N_φ points and then, according to them, writes down the corresponding canonical axioms. The known upper bound for N_φ is extremely high, and so the algorithm is too burdensome to be used in practice. However, usually the following heuristic (non-mechanical) procedure works well enough. First you construct one or two elementary countermodels for φ and, inserting into these countermodels new points, check which domains in them are closed and whether it is possible to do without the cofinality condition. As a result, you will have one or two canonical formulas which probably axiomatize $\mathbf{K4} \oplus \varphi$. If you cannot prove this then construct a more complicated countermodel and another canonical axiom, and so on. Usually one or two iterations are enough to find the desired canonical axiomatization.

3.2 *Quasi-Normal Modal Logics*

The Characterization Theorem above pays no attention to what particular point in a frame refutes a given formula, i.e. it can be used for characterizing frames only for *normal* modal logics. As a result, the set of normal modal canonical formulas

Table 4. Canonical axioms of the standard modal logics

D4	$= \mathbf{K4} \oplus \alpha(\bullet, \emptyset, \bot)$
S4	$= \mathbf{K4} \oplus \alpha(\bullet, \emptyset)$
GL	$= \mathbf{K4} \oplus \alpha(\circ, \emptyset)$
For	$= \mathbf{K4} \oplus \alpha(*, \emptyset)$ (2 axioms)
Grz	$= \mathbf{K4} \oplus \alpha(\bullet, \emptyset) \oplus \alpha($$, \emptyset)$
S4.1	$= \mathbf{S4} \oplus \alpha($$, \emptyset, \bot)$
K4.1	$= \mathbf{K4} \oplus \alpha(\bullet, \emptyset, \bot) \oplus \alpha($$, \emptyset, \bot)$
Triv	$= \mathbf{K4} \oplus \alpha(\bullet, \emptyset) \oplus \alpha($$, \emptyset) \oplus \alpha($$, \emptyset)$
Verum	$= \mathbf{K4} \oplus \alpha(\circ, \emptyset) \oplus \alpha($$, \emptyset)$
S5	$= \mathbf{S4} \oplus \alpha($$, \emptyset)$
K4B	$= \mathbf{K4} \oplus \alpha($$, \emptyset)$ (4 axioms)
S4.2	$= \mathbf{S4} \oplus \alpha($$, \emptyset, \bot)$
S4.3	$= \mathbf{S4} \oplus \alpha($$, \emptyset)$
$\mathbf{A}^*$	$= \mathbf{GL} \oplus \alpha($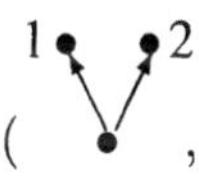$, \{\{1\}, \{1, 2\}\})$
K4.2	$= \mathbf{K4} \oplus \alpha($$, \emptyset, \bot) \oplus \alpha($$, \emptyset, \bot) \oplus \alpha($$, \emptyset, \bot)$ (8 axioms)
K4.3	$= \mathbf{K4} \oplus \alpha($$, \emptyset)$ (6 axioms)
K4Z	$= \mathbf{K4} \oplus \alpha($$, \emptyset) \oplus \alpha($$, \emptyset) \oplus \alpha($$, \emptyset) \oplus \alpha($$, \emptyset)$
Dum	$= \mathbf{S4} \oplus \alpha($$, \emptyset) \oplus \alpha($$, \emptyset)$
$\mathbf{D4G}_1$	$= \mathbf{D4} \oplus \alpha($$, \mathfrak{D}^\sharp, \bot)$

Table 4. Canonical axioms of the standard modal logics (cont'd)

K4H $= \mathbf{K4} \oplus \alpha($  $, \emptyset) \oplus \alpha($ 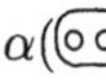 $, \emptyset)$ (9 axioms)

K4Alt$_n$ $= \mathbf{K4} \oplus \{\alpha(\mathfrak{F}, \emptyset)$: exactly $n+1$ points are seen from the root of $\mathfrak{F}\}$

K4I$_n$ $= \mathbf{K4} \oplus \alpha($ 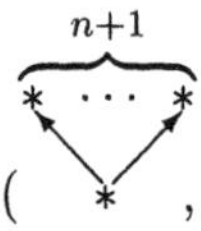 $, \emptyset)$ $(2n+4$ axioms)

K4B$_n$ $= \mathbf{K4} \oplus \alpha($  $0, \emptyset)$ $(2^{n+1}$ axioms)

K4$_{n,m}$ $= \mathbf{K4} \oplus \alpha($  $0, \mathfrak{D}^\sharp)$

is incomplete for the class of quasi-normal extensions of **K4**. However, modifying appropriately the notion of subreduction, it is not hard to generalize the theorem to frames with distinguished points and define accordingly quasi-normal canonical formulas which can axiomatize all logics in Ext**K4**. We will not deal with the whole class of quasi-normal logics here (it is still a rather unexplored territory) and for details refer the reader to [Zakharyaschev, 1992, 1996].

It is worth noting however that the set of normal canonical formulas built upon *reflexive* frames turns out to be complete for the class of quasi-normal extensions of **S4**. Namely, the following result holds:

THEOREM 3.3 (Completeness Theorem for Canonical Formulas in Ext**S4**)

(i) *There is an algorithm which, given a modal formula φ, returns canonical formulas $\alpha(\mathfrak{F}_1, \mathfrak{D}_1, \bot), \ldots, \alpha(\mathfrak{F}_n, \mathfrak{D}_n, \bot)$, associated with reflexive frames $\mathfrak{F}_1, \ldots, \mathfrak{F}_n$, such that*

$$\mathbf{S4} + \varphi = \mathbf{S4} + \alpha(\mathfrak{F}_1, \mathfrak{D}_1, \bot) + \ldots + \alpha(\mathfrak{F}_n, \mathfrak{D}_n, \bot).$$

(ii) *There is an algorithm which, given a modal negation free formula φ, returns negation free canonical formulas $\alpha(\mathfrak{F}_1, \mathfrak{D}_1), \ldots, \alpha(\mathfrak{F}_n, \mathfrak{D}_n)$, built upon reflexive frames, such that*

$$\mathbf{S4} + \varphi = \mathbf{S4} + \alpha(\mathfrak{F}_1, \mathfrak{D}_1) + \ldots + \alpha(\mathfrak{F}_n, \mathfrak{D}_n).$$

3.3 *The Intuitionistic Case*

The intuitionistic canonical formulas are defined similarly to the modal ones. Let $\mathfrak{F} = \langle W, R \rangle$ be a finite rooted partially ordered frame with points $a_0, \dots, a_n$ and root a_0 and let $\mathfrak{D}$ be a set of antichains in $\mathfrak{F}$ (containing no singletons).

The *intuitionistic canonical formula* $\beta(\mathfrak{F}, \mathfrak{D}, \bot)$ for $\mathfrak{F}$ and $\mathfrak{D}$ may look like this:

$$\beta(\mathfrak{F}, \mathfrak{D}, \bot) = \bigwedge_{a_i R a_j} \beta_{ij} \wedge \bigwedge_{\mathfrak{d} \in \mathfrak{D}} \beta_{\mathfrak{d}} \wedge \beta_{\bot} \to p_0$$

where

$$\begin{aligned} \beta_{ij} &= (\bigwedge_{\neg a_j R a_k} p_k \to p_j) \to p_i \\ \beta_{\mathfrak{d}} &= \bigwedge_{a_i \in W - \mathfrak{d}\uparrow} (\bigwedge_{\neg a_i R a_k} p_k \to p_i) \to \bigvee_{a_j \in \mathfrak{d}} p_j \\ \beta_{\bot} &= \bigwedge_{i=0}^{n} (\bigwedge_{\neg a_i R a_k} p_k \to p_i) \to \bot. \end{aligned}$$

$\beta(\mathfrak{F}, \mathfrak{D})$, the *intuitionistic negation free canonical formula* for $\mathfrak{F}$ and $\mathfrak{D}$, is obtained from $\beta(\mathfrak{F}, \mathfrak{D}, \bot)$ by deleting the conjunct $\beta_{\bot}$, which is the only one containing $\bot$.

THEOREM 3.4 (Refutability Criterion for Intuitionistic Canonical Formulas) *For every intuitionistic general frame* $\mathfrak{G}$,

(i) $\mathfrak{G} \not\models \beta(\mathfrak{F}, \mathfrak{D}, \bot)$ *iff there is a cofinal subreduction of* $\mathfrak{G}$ *to* $\mathfrak{F}$ *satisfying the closed domain condition (CDC) for* $\mathfrak{D}$;

(ii) $\mathfrak{G} \not\models \beta(\mathfrak{F}, \mathfrak{D})$ *iff there is a (not necessarily cofinal) subreduction of* $\mathfrak{G}$ *to* $\mathfrak{F}$ *satisfying (CDC) for* $\mathfrak{D}$.

The inclusions similar to those in Section 3.1 hold for the intuitionistic canonical formulas as well. It is also worth noting that $\beta(\mathfrak{F}, \emptyset)$ is (equivalent to) a purely implicative formula and $\beta(\mathfrak{F}, \emptyset, \bot)$ is disjunction free. We call them the *implicative* and *disjunction free canonical formulas* for $\mathfrak{F}$, respectively.

THEOREM 3.5 (Completeness Theorem for Intuitionistic Canonical Formulas)

(i) *There is an algorithm which, given an intuitionistic formula* φ, *returns canonical formulas* $\beta(\mathfrak{F}_1, \mathfrak{D}_1, \bot), \dots, \beta(\mathfrak{F}_n, \mathfrak{D}_n, \bot)$ *such that*

$$\mathbf{Int} + \varphi = \mathbf{Int} + \beta(\mathfrak{F}_1, \mathfrak{D}_1, \bot) + \ldots + \beta(\mathfrak{F}_n, \mathfrak{D}_n, \bot).$$

(ii) *There is an algorithm which, given a negation free formula* φ, *returns negation free canonical formulas* $\beta(\mathfrak{F}_1, \mathfrak{D}_1), \dots, \beta(\mathfrak{F}_n, \mathfrak{D}_n)$ *such that*

$$\mathbf{Int} + \varphi = \mathbf{Int} + \beta(\mathfrak{F}_1, \mathfrak{D}_1) + \ldots + \beta(\mathfrak{F}_n, \mathfrak{D}_n).$$

(iii) *Given a disjunction free formula φ, one can effectively construct disjunction free canonical formulas $\beta(\mathfrak{F}_1, \emptyset, \bot), \ldots, \beta(\mathfrak{F}_n, \emptyset, \bot)$ such that*

$$\mathbf{Int} + \varphi = \mathbf{Int} + \beta(\mathfrak{F}_1, \emptyset, \bot) + \ldots + \beta(\mathfrak{F}_n, \emptyset, \bot).$$

(iv) *Given an implicative formula φ, one can effectively construct implicative canonical formulas $\beta(\mathfrak{F}_1, \emptyset), \ldots, \beta(\mathfrak{F}_n, \emptyset)$ such that*

$$\mathbf{Int} + \varphi = \mathbf{Int} + \beta(\mathfrak{F}_1, \emptyset) + \ldots + \beta(\mathfrak{F}_n, \emptyset).$$

It follows from this theorem that every superintuitionistic logic L can be represented as

$$L = \mathbf{Int} + \{\beta(\mathfrak{F}_i, \mathfrak{D}_i, \bot) : i \in I\}$$

and if L is axiomatized by negation free formulas then it can be also represented as

$$L = \mathbf{Int} + \{\beta(\mathfrak{F}_i, \mathfrak{D}_i) : i \in I\}.$$

Besides, if the additional axioms of L are disjunction free then in both these cases $\mathfrak{D}_i = \emptyset$ for all $i \in I$.

Table 5 shows canonical representations of the standard superintuitionistic logics.

3.4 Independence

Having established the axiomatic completeness of the set of the canonical formulas, it is natural to ask if it is *independent* in the sense that it contains no complete proper subsets and, if the answer is 'no', to try to find out such a complete and independent set of formulas. In general we say a set of formulas Γ is an *axiomatic basis* for a class of logics $\mathcal{L} \subseteq \text{Ext}L_0$ if

$$\mathcal{L} = \{L_0 + \Delta : \Delta \subseteq \Gamma\},$$

but no proper subset of Γ meets this property. (If we deal only with normal logics then + in this definition should be replaced with $\oplus$.)

The notion of independence turns out to be closely related to the following notion of prime logic and formula (which will be given here only for normal logics; in the quasi-normal case $\oplus$ should be replaced with +).

Suppose $\mathcal{L}$ is a class of (normal) logics in NExtL_0. A logic $L \in \mathcal{L}$ is called *prime* in $\mathcal{L}$ if, for every $L_1, L_2 \in \mathcal{L}$, $L = L_1 \oplus L_2$ implies $L = L_1$ or $L = L_2$. And a formula φ is *prime* if the logic $L_0 \oplus \varphi$ is prime.

Now, it is not hard to see that a set of formulas Γ, which is complete for $\mathcal{L}$ and contains no distinct deductively equal formulas, is a basis for $\mathcal{L}$ iff every formula in Γ is prime. Thus, all we need is to find prime modal and intuitionistic canonical

Table 5. Canonical axioms of the standard superintuitionistic logics

For	=	$\mathbf{Int} + \beta(\circ, \emptyset)$
Cl	=	$\mathbf{Int} + \beta($  $, \emptyset)$
KC	=	$\mathbf{Int} + \beta($ 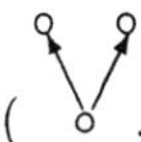 $, \emptyset, \bot)$
LC	=	$\mathbf{Int} + \beta($ 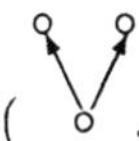 $, \emptyset)$
SL	=	$\mathbf{Int} + \beta($ 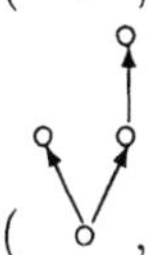 $, \mathfrak{D}^\sharp, \bot)$
KP	=	$\mathbf{Int} + \beta($ 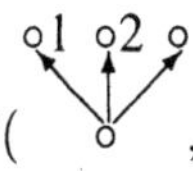 $, \{\{1,2\}\}, \bot) + \beta($ 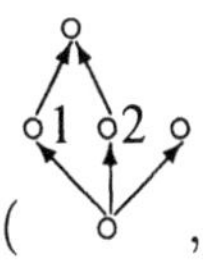 $, \{\{1,2\}\}, \bot)$
WKP	=	$\mathbf{Int} + \beta($ 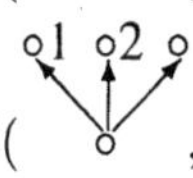 $, \{\{1,2\}\}, \bot)$
$\mathbf{ND}_k$	=	$\mathbf{Int} + \beta($ 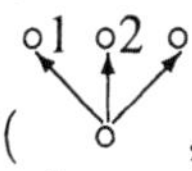 $, \{\{1,2\}\}, \bot) + \ldots$
		$+ \beta($ 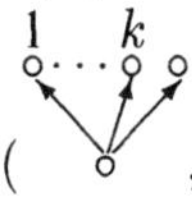 $, \{\{1, \ldots, k\}\}, \bot)$
$\mathbf{BH}_n$	=	$\mathbf{Int} + \beta($ 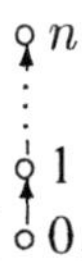 $, \emptyset)$
$\mathbf{BW}_n$	=	$\mathbf{Int} + \beta($ 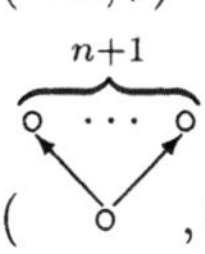 $, \emptyset)$
$\mathbf{BTW}_n$	=	$\mathbf{Int} + \beta($ 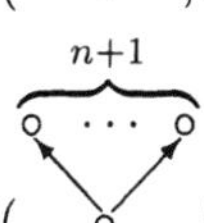 $, \emptyset, \bot)$
$\mathbf{T}_n$	=	$\mathbf{Int} + \beta($ 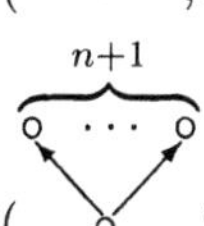 $, \mathfrak{D}^\sharp)$

Table 5. Canonical axioms of the standard superintuitionistic logics (cont'd)

$\mathbf{B}_n$	$= \mathbf{Int} + \beta($ 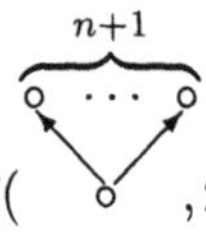 $, \mathfrak{D}^\sharp, \bot)$

formulas and check whether the set of them is axiomatically complete. Clearly, the first candidates to this role are the frame formulas, i.e. the canonical formulas of the form $\alpha(\mathfrak{F}, \mathfrak{D}^\sharp, \bot)$ and $\beta(\mathfrak{F}, \mathfrak{D}^\sharp, \bot)$. For we have the following result.

LEMMA 3.6 (i) *$\alpha(\mathfrak{F}, \mathfrak{D}^\sharp, \bot) \in \mathbf{K4} \oplus \{\alpha(\mathfrak{G}_i, \mathfrak{C}_i, \bot) : i \in I\}$ iff for some $i \in I$ $\mathfrak{F} \not\models \alpha(\mathfrak{G}_i, \mathfrak{C}_i, \bot)$ iff for some $i \in I$ $\alpha(\mathfrak{F}, \mathfrak{D}^\sharp, \bot) \in \mathbf{K4} \oplus \alpha(\mathfrak{G}_i, \mathfrak{C}_i, \bot)$.*

(ii) *$\beta(\mathfrak{F}, \mathfrak{D}^\sharp, \bot) \in \mathbf{Int} + \{\beta(\mathfrak{G}_i, \mathfrak{C}_i, \bot) : i \in I\}$ iff $\mathfrak{F} \not\models \beta(\mathfrak{G}_i, \mathfrak{C}_i, \bot)$ for some $i \in I$ iff $\beta(\mathfrak{F}, \mathfrak{D}^\sharp, \bot) \in \mathbf{Int} + \beta(\mathfrak{G}_i, \mathfrak{C}_i, \bot)$ for some $i \in I$.*

This lemma means in fact that all Jankov-Fine's formulas are prime in NExt**K4** and Ext**Int**. And the next lemma shows that no other (modulo the deductive equality) prime formulas exist.

LEMMA 3.7 (Decomposition Lemma)

(i) *For every canonical formula $\alpha(\mathfrak{F}, \mathfrak{D}, \bot)$ one can effectively construct canonical formulas*

$$\alpha(\mathfrak{F}_1, \mathfrak{D}_1, \bot), \dots, \alpha(\mathfrak{F}_n, \mathfrak{D}_n, \bot)$$

with card $\mathfrak{F}_i >$ card$\mathfrak{F}$, for all $i = 1, \dots, n$, and

$$\mathbf{K4} \oplus \alpha(\mathfrak{F}, \mathfrak{D}, \bot) = \mathbf{K4} \oplus \alpha(\mathfrak{F}, \mathfrak{D}^\sharp, \bot) \oplus \alpha(\mathfrak{F}_1, \mathfrak{D}_1, \bot) \oplus \dots \oplus \alpha(\mathfrak{F}_n, \mathfrak{D}_n, \bot).$$

(ii) *For every canonical formula $\beta(\mathfrak{F}, \mathfrak{D}, \bot)$ one can effectively construct canonical formulas*

$$\beta(\mathfrak{F}_1, \mathfrak{D}_1, \bot), \dots, \beta(\mathfrak{F}_n, \mathfrak{D}_n, \bot)$$

such that card $\mathfrak{F}_i >$ card$\mathfrak{F}$, for all $i = 1, \dots, n$, and

$$\mathbf{Int} + \beta(\mathfrak{F}, \mathfrak{D}, \bot) = \mathbf{Int} + \beta(\mathfrak{F}, \mathfrak{D}^\sharp, \bot) + \beta(\mathfrak{F}_1, \mathfrak{D}_1, \bot) + \dots + \beta(\mathfrak{F}_n, \mathfrak{D}_n, \bot).$$

Thus we have

THEOREM 3.8

(i) *A modal formula* φ *is prime in* NExt**K4** *iff, for some* $\mathfrak{F}$, $\mathbf{K4} \oplus \varphi = \mathbf{K4} \oplus \alpha(\mathfrak{F}, \mathfrak{D}^\sharp, \bot)$.

(ii) *An intuitionistic formula* φ *is prime in* Ext**Int** *iff, for some* $\mathfrak{F}$, $\mathbf{Int} + \varphi = \mathbf{Int} + \beta(\mathfrak{F}, \mathfrak{D}^\sharp, \bot)$.

For the intuitionistic case this result was first obtained by Jankov [1969].

However, the class of prime formulas turns out to be too small to axiomatize all logics in NExt**K4** and Ext**Int**. For example, even the logics

$$\mathbf{GL} = \mathbf{K4} \oplus \alpha(\circ, \emptyset) \text{ and } \mathbf{Int} + \beta($$ 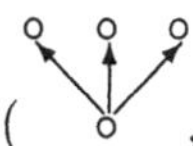 $$, \emptyset, \bot)$$

cannot be axiomatized by prime formulas. As a result we arrive at the following

THEOREM 3.9 NExt**K4** *and* Ext**Int** *have no axiomatic bases at all.*

For exactly the same reason NExt**GL**, NExt**Grz**, NExt**S4** and many other natural classes of logics have no axiomatic bases, which is one more evidence of their extremely complicated structures. In Section 5.1 we shall see examples of non-trivial classes of modal and superintuitionistic logics with axiomatic bases.

I conclude this section with an open problem which seems to have already appeared in the literature but nevertheless deserves to be repeated once again.

PROBLEM 3.10 *Does every modal and superintuitionistic logic have an independent axiomatization?*[2]

4 MODAL COMPANIONS OF SUPERINTUITIONISTIC LOGICS

The fact that refutation frames for both modal and intuitionistic formulas can be characterized in the same terms is far from being accidental. Actually, it is another manifestation of the connection between superintuitionistic and modal logics that was discovered by Gödel [1933] and then investigated by McKinsey and Tarski [1948], Dummett and Lemmon [1959], Maksimova and Rybakov [1974], Blok [1976] and Esakia [1979]. The canonical formulas make this connection quite clear and visual. Moreover, they open a way for obtaining new results in this direction.

[2]Chagrov and Zakharyaschev [1995] gave a negative solution to this problem. In fact they constructed concrete logics in NExt**K4**, NExt**GL**, NExt**Grz** and Ext**Int** having no independent sets of axioms.

4.1 *The Modal Companion Theorem*

A modal logic M is said to be a *modal companion* of a superintuitionistic logic L if, for every intuitionistic formula φ,

$$\varphi \in L \text{ iff } T\varphi \in M$$

where T is the *Gödel translation* prefixing $\Box$ to every subformula of φ. If M is a modal companion of L then we write $\rho M = L$ and say that L is the *superintuitionistic fragment* of M.

Gödel [1933] predicted and McKinsey and Tarski [1948] proved that **S4** is a modal companion of **Int**. Then Dummett and Lemmon [1959] showed that every superintuitionistic logic $\mathbf{Int} + \{\varphi_i : i \in I\}$ has $\mathbf{S4} \oplus \{T\varphi_i : i \in I\}$ as its modal companion. Grzegorczyk [1967] found another companion for **Int**, namely **Grz**. Then Maksimova and Rybakov [1974] established that ρ is a lattice homomorphism from NExt**S4** onto Ext**Int** and that the set $\rho^{-1}L$ of all modal companions of every consistent $L \in$ Ext**Int** forms the infinite segment of logics

$$\{M : \tau L \subseteq M \subseteq \sigma L\}$$

where τL, the *smallest companion* of L, is $\mathbf{S4} \oplus \{T\varphi : \varphi \in L\}$. Finally, Blok [1976] and Esakia [1979] discovered that σL, the *greatest companion* of L, is $\tau L \oplus \mathbf{Grz}$ and that σ is a lattice isomorphism from Ext**Int** onto NExt**Grz**.

All these results turn out to be practically immediate consequences of the following general characterization of the map ρ in terms of the canonical formulas.

THEOREM 4.1 (Modal Companion Theorem) *A logic $M \in$ NExt***S4** *is a modal companion of a superintuitionistic logic*

$$L = \mathbf{Int} + \{\beta(\mathfrak{F}_i, \mathfrak{D}_i, \bot) : i \in I\}$$

iff M can be represented in the form

$$M = \mathbf{S4} \oplus \{\alpha(\mathfrak{F}_i, \mathfrak{D}_i, \bot) : i \in I\} \oplus \{\alpha(\mathfrak{G}_j, \mathfrak{C}_j, \bot) : j \in J\}$$

where each $\mathfrak{G}_j$, for $j \in J$, contains at least one proper cluster.

For example, $\rho\mathbf{S4.1} = \rho\mathbf{Dum} = \mathbf{Int}$, $\rho\mathbf{S4.2} = \mathbf{KC}$, $\rho\mathbf{S4.3} = \mathbf{LC}$, $\rho\mathbf{S5} = \rho\mathbf{Triv} = \mathbf{Cl}$ (see Tables 4 and 5).

As a direct consequence of Theorem 4.1 we obtain that for every superintuitionistic logic $L = \mathbf{Int} + \{\beta(\mathfrak{F}_i, \mathfrak{D}_i, \bot) : i \in I\}$,

$$\tau L = \mathbf{S4} \oplus \{\alpha(\mathfrak{F}_i, \mathfrak{D}_i, \bot) : i \in I\}$$

and

$\sigma L = \tau L \oplus \alpha($  $, \emptyset).$

In particular,

$\sigma\mathbf{Int} = \mathbf{S4} \oplus \alpha($$, \emptyset) = \mathbf{Grz}.$

The maps τ and σ can be characterized by the canonical formulas in the following way (for the definition of the skeleton operator ρ see Section 1.5).

THEOREM 4.2 *For every logic* $L \in \mathrm{Ext}\mathbf{Int}$ *and every modal canonical formula* $\alpha(\mathfrak{F}, \mathfrak{D}, \bot)$,

(i) $\alpha(\mathfrak{F}, \mathfrak{D}, \bot) \in \tau L$ *iff* $\beta(\rho\mathfrak{F}, \mathfrak{D}, \bot) \in L$;

(ii) $\alpha(\mathfrak{F}, \mathfrak{D}, \bot) \in \sigma L$ *iff either* $\mathfrak{F}$ *contains a proper cluster or* $\mathfrak{F}$ *is partially ordered and* $\beta(\mathfrak{F}, \mathfrak{D}, \bot) \in L$.

It follows from this result and Completeness Theorems of Sections 3.1 and 3.3 that L is decidable iff τL is decidable iff σL is decidable. In fact, from the technical point of view the main role of the maps ρ, τ and σ is that they preserve many properties of logics, which makes it possible to transfer results from the field of modal logics to superintuitionistic ones and backwards.

Another interesting consequence of Modal Companion and Completeness Theorems is the following result establishing effectiveness of the transition from a (finitely axiomatizable) modal logic to its superintuitionistic fragment (try to obtain it in some other way!).

THEOREM 4.3 *Given a modal formula* φ, *one can effectively construct an intuitionistic formula* ψ *such that*

$$\rho(\mathbf{S4} \oplus \varphi) = \mathbf{Int} + \psi.$$

4.2 *Semantic Characterization of* ρ, τ *and* σ

It follows from McKinsey and Tarski [1946] that, for every intuitionistic formula φ and every modal general frame $\mathfrak{F}$, $\rho\mathfrak{F} \models \varphi$ iff $\mathfrak{F} \models T\varphi$, and so

$$\mathfrak{F} \models M \text{ iff } \rho\mathfrak{F} \models \rho M$$

for every $M \in \mathrm{NExt}\mathbf{S4}$.

It should be clear from this result that all important properties of logics (Kripke completeness, FMP, the disjunction property, etc.) are preserved while passing from a modal logic to its superintuitionistic fragment.[3]

Suppose now that $\mathfrak{F} = \langle W, R, P \rangle$ is an intuitionistic frame and let $\sigma\mathfrak{F} = \langle W, R, \sigma P \rangle$ be the modal frame in which σP is the Boolean closure of P in the space W. Then, as it follows from [Maksimova and Rybakov, 1974],

$$\mathfrak{F} \models L \text{ iff } \sigma\mathfrak{F} \models \sigma L$$

[3] I know of only one property which is not preserved under ρ. That is independent axiomatizability (see [Chagrov and Zakharyaschev, 1995]).

for every $L \in \mathbf{ExtInt}$. So L is characterized by a class of frames $\mathcal{C}$ iff σL is characterized by $\sigma\mathcal{C} = \{\sigma\mathfrak{F} : \mathfrak{F} \in \mathcal{C}\}$. This characterization means, in particular, that σ preserves the tabularity, pretabularity, finite model and disjunction property. However, Kripke completeness is not in general preserved under σ, as was shown by Shehtman [1980].

It is much harder to characterize semantically the transition from a superintuitionistic logic to its smallest companion. Given an intuitionistic frame $\mathfrak{F} = \langle W, R, P\rangle$ and an ordinal ξ such that $0 < \xi \leq \omega$, by $\tau_\xi\mathfrak{F}$ we denote a modal frame $\langle \xi W, \xi R, \xi P\rangle$ in which $\langle \xi W, \xi R\rangle$ is the direct product of $\langle W, R\rangle$ and the cluster with ξ points (in other words, $\langle \xi W, \xi R\rangle$ is the result of replacing each point in W with a ξ-point cluster) and ξP is a subset of $2^{\xi W}$ such that $\rho\tau_\xi\mathfrak{F} = \mathfrak{F}$ and $\{\zeta\} \times X \in \xi P$ for each $\zeta \in \xi$ and each $X \in \sigma P$. For instance, one can take ξP to be the Boolean closure of the set $\{X \times Y : X \subseteq \xi,\ Y \in \sigma P\}$. If $P = \mathrm{Up}W$, i.e. $\mathfrak{F}$ is a Kripke frame, then clearly we can put $\xi P = 2^{\xi W}$, and so $\tau_\xi\mathfrak{F} = \langle \xi W, \xi R\rangle$.

THEOREM 4.4 *The following conditions are equivalent for every superintuitionistic logic L:*

(i) *L is characterized by a class of frames $\mathcal{C}$;*

(ii) *τL is characterized by the class $\bigcup_{0<\xi<\omega} \tau_\xi\mathcal{C}$, where $\tau_\xi\mathcal{C} = \{\tau_\xi\mathfrak{F} : \mathfrak{F} \in \mathcal{C}\}$;*

(iii) *τL is characterized by the class $\tau_\omega\mathcal{C}$.*

As a consequence we obtain a positive solution to the conjecture of Dummett and Lemmon [1959]:

THEOREM 4.5 *If a superintuitionistic logic L is characterized by an intuitionistic Kripke frame $\mathfrak{F} = \langle W, R\rangle$ then τL is characterized by the modal Kripke frame $\tau_\omega\mathfrak{F} = \langle \omega W, \omega R\rangle$.*

Another consequence:

THEOREM 4.6 (Preservation Theorem) *The map τ preserves Kripke completeness, and the finite model and disjunction property.*

PROBLEM 4.7 *Does τ preserve canonicity and compactness?*[4]

(For the definition of these notions see Section 5.2. The results of that section show that in general σ does not preserve these properties.)

[4] That τ and ρ preserve canonicity is shown in [Zakharyaschev, 1996]. Using the algebraic characterization of compactness found by Wolter [1993], one can show that τ and ρ preserve this property as well.

4.3 *Quasi-normal companions of* **Int**

As was already mentioned in Section 4.1, **Grz** is the greatest modal companion of **Int** among the normal extensions of **S4**. However, if we give up the requirement of normality then it turns out (see [Chagrov, 1985]) that there are proper extensions of **Grz** in which **Int** is embeddable by T. And all of them can also be characterized through their canonical axioms.

Call a point x a *focus* of an antichain X in a frame $\mathfrak{F}$ if $x\uparrow = \{x\} \cup X\uparrow$.

THEOREM 4.8 *The following conditions are equivalent:*

(i) *A logic* $M \in$ Ext**S4** *is a modal companion of* **Int**;

(ii) *M can be represented in the form*

$$M = \mathbf{S4} + \{\alpha(\mathfrak{F}_i, \mathfrak{D}_i, \bot) : i \in I\} + \{\alpha(\mathfrak{G}_j, \mathfrak{C}_j, \bot) : j \in J\}$$

where each $\mathfrak{F}_i$, for $i \in I$, contains a proper cluster, each $\mathfrak{G}_j$, for $j \in J$, is partially ordered and at least one antichain in $\mathfrak{C}_j$ has no focus in $\mathfrak{G}_j$;

(iii) $\alpha(\mathfrak{G}, \mathfrak{C}, \bot) \notin M$ *for any partially ordered $\mathfrak{G}$ and any set $\mathfrak{C}$ of antichains having focuses in $\mathfrak{G}$.*

For example, the quasi-normal logic

Grz $+ \alpha($ 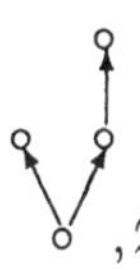 $, \mathfrak{D}^\sharp, \bot)$

is a modal companion of **Int** properly extending **Grz**. In fact, it is an easy exercise now to show that there are a continuum of modal companions of **Int** containing **Grz** (see Figure 3 and Lemma 5.9).

Two more consequences of Theorems 4.8 and 3.3 also deserve mentioning.

THEOREM 4.9 *There is an algorithm which is capable of deciding, given a modal formula φ, whether* **Int** *is embeddable in* **S4** $+ \varphi$ *by the Gödel translation T.*

THEOREM 4.10 *Let Δ be the set of all canonical formulas $\alpha(\mathfrak{F}, \mathfrak{D}, \bot)$ such that $\mathfrak{F}$ is partially ordered and $\mathfrak{D}$ contains an antichain having no focus in $\mathfrak{F}$. Then the logic* $\mathbf{M}^* = \mathbf{Grz} + \Delta$ *is the greatest modal companion of* **Int** *in* Ext**S4**.

PROBLEM 4.11 *Is* $\mathbf{M}^*$ *finitely axiomatizable?*

To characterize $\mathbf{M}^*$ semantically, we define by induction a special Kripke frame $\mathfrak{F}_\omega = \langle W_\omega, R_\omega \rangle$ with a distinguished (actual) world w. Let $\langle W_0, R_0 \rangle$ be the disjoint union of all finite rooted partially ordered frames. For $i \geq 1$ we define $\langle W_i, R_i \rangle$ by taking

$$W_i = W_{i-1} \cup \{x_{\mathfrak{e}} : \mathfrak{e} \text{ is a finite antichain in } \langle W_{i-1}, R_{i-1} \rangle\}$$

and R_i to be the reflexive and transitive closure of the relation

$$R_{i-1} \cup \{\langle x_{\mathfrak{e}}, x \rangle : x \in \mathfrak{e}\}.$$

Finally, we put

$$W_\omega = \bigcup_{i<\omega} W_i \cup \{w\}, \quad R_\omega = \bigcup_{i<\omega} R_i \cup \{\langle w, x \rangle : x \in W_\omega\}.$$

THEOREM 4.12 (Properties of $\mathbf{M}^*$)

(i) $\mathbf{M}^* = \{\varphi : (\mathfrak{F}_\omega, w) \models \varphi\}$;

(ii) $\mathbf{M}^*$ *is decidable;*

(iii) $\mathbf{M}^*$ *does not have FMP;*

(iv) $\mathbf{M}^*$ *is Halldén complete and has the disjunction property.*

One can show however that there is a superintuitionistic logic which cannot be embedded by T in any extension of $\mathbf{M}^*$. So, the Blok-Esakia Theorem does not hold in the quasi-normal case.

For a survey of results concerning embeddings of superintuitionistic logics into modal ones consult [Chagrov and Zakharyaschev, 1992]. Proofs of the theorems presented in Section 4, except those in Section 4.3, can be found in [Zakharyaschev, 1989, 1989a].

5 SUBFRAME AND COFINAL SUBFRAME LOGICS

Canonical axioms of almost all the standard modal logics in Table 4 (save $\mathbf{A}^*$, $\mathbf{D4G}_1$ and $\mathbf{K4}_{n,m}$) contain no closed domains at all. This section considers the class of all modal and superintuitionistic logics having axioms of that kind. Unlike NExt**K4** and Ext**Int**, this smaller class turns out to be quite manageable and at the same time by no means trivial. First we shall deal only with modal logics and then transfer the relevant results to superintuitionistic ones. For proofs see [Zakharyaschev, 1996].

5.1 Decidability and FMP

Subframe and cofinal subframe logics can be defined both syntactically and semantically. We shall take the syntactical definition as the basic one and formulate the latter as a theorem.

A logic $L \in$ NExt**K4** is said to be a *subframe logic* if it can be represented in the form

$$L = \mathbf{K4} \oplus \{\alpha(\mathfrak{F}_i, \emptyset) : i \in I\}.$$

Subframe logics were introduced and studied by Fine [1985]. L is called a *cofinal subframe logic* if it is representable as

$$L = \mathbf{K4} \oplus \{\alpha(\mathfrak{F}_i, \emptyset, \bot) : i \in I\}.$$

The classes of all subframe and cofinal subframe logics will be denoted by $\mathcal{SF}$ and $\mathcal{CSF}$, respectively. The names of these logics are justified by the following

THEOREM 5.1 (i) (Fine [1985]) *L is a subframe logic iff it is characterized by a class of frames that is closed under forming (not necessarily generated) subframes.*

(ii) *L is a cofinal subframe logic iff it is characterized by a class of frames that is closed under forming cofinal subframes.*

It follows in particular that $\mathcal{SF} \subseteq \mathcal{CSF}$. Actually, this inclusion is a proper one, since, for example, the classes of frames for **K4.1** and **K4.2** are closed under cofinal subframes, but not under plain subframes. Moreover, one can show that there are a continuum of logics in $\mathcal{CSF} - \mathcal{SF}$.

It follows also that $\mathcal{CSF}$ is a complete sublattice of NExt**K4** and $\mathcal{SF}$ a complete sublattice of $\mathcal{CSF}$.

Our next theorem characterizes the canonical formulas which are provable in subframe and cofinal subframe logics. It is easily obtained with the help of the fact that a composition of (cofinal) subreductions is again a (cofinal) subreduction.

THEOREM 5.2

(i) *For every subframe logic $L = \mathbf{K4} \oplus \{\alpha(\mathfrak{F}_i, \emptyset) : i \in I\}$ and every canonical formula $\alpha(\mathfrak{F}, \mathfrak{D}, \bot)$ the following conditions are equivalent:*

- $\alpha(\mathfrak{F}, \mathfrak{D}, \bot) \in L$;
- $\exists i \in I \; \mathfrak{F} \not\models \alpha(\mathfrak{F}_i, \emptyset)$;
- *$\mathfrak{F}$ is subreducible to $\mathfrak{F}_i$ for some $i \in I$;*
- *$\alpha(\mathfrak{F}, \mathfrak{D}, \bot) \in \mathbf{K4} \oplus \alpha(\mathfrak{F}_i, \emptyset)$ for some $i \in I$;*
- $\alpha(\mathfrak{F}, \emptyset) \in L$.

(ii) *For every cofinal subframe logic* $L = \mathbf{K4} \oplus \{\alpha(\mathfrak{F}_i, \emptyset, \bot) : i \in I\}$ *and every canonical formula* $\alpha(\mathfrak{F}, \mathfrak{D}, \bot)$ *the following conditions are equivalent:*

- $\alpha(\mathfrak{F}, \mathfrak{D}, \bot) \in L$;
- $\exists i \in I\ \mathfrak{F} \not\models \alpha(\mathfrak{F}_i, \emptyset, \bot)$;
- $\mathfrak{F}$ *is cofinally subreducible to* $\mathfrak{F}_i$ *for some* $i \in I$;
- $\alpha(\mathfrak{F}, \mathfrak{D}, \bot) \in \mathbf{K4} \oplus \alpha(\mathfrak{F}_i, \emptyset, \bot)$ *for some* $i \in I$;
- $\alpha(\mathfrak{F}, \emptyset, \bot) \in L$.

This result has a number of consequences. First, we observe that

(i) $L \in \mathrm{NExt}\mathbf{K4}$ is a subframe logic iff we have, for every canonical formula $\alpha(\mathfrak{F}, \mathfrak{D}, \bot)$,

$$\alpha(\mathfrak{F}, \mathfrak{D}, \bot) \in L \Rightarrow \alpha(\mathfrak{F}, \emptyset) \in L.$$

and that

(ii) L is a cofinal subframe logic iff, for every $\alpha(\mathfrak{F}, \mathfrak{D}, \bot)$,

$$\alpha(\mathfrak{F}, \mathfrak{D}, \bot) \in L \Rightarrow \alpha(\mathfrak{F}, \emptyset, \bot) \in L.$$

Second, the sets of canonical formulas of the form $\alpha(\mathfrak{F}, \emptyset)$ and $\alpha(\mathfrak{F}, \emptyset, \bot)$ turn out to be axiomatic bases for the classes $\mathcal{SF}$ and $\mathcal{CSF}$, respectively. More exactly, the following theorem holds:

THEOREM 5.3 *Every subframe (cofinal subframe) logic can be axiomatized by an independent set of canonical axioms of the form* $\alpha(\mathfrak{F}, \emptyset)$ *(or* $\alpha(\mathfrak{F}, \emptyset, \bot)$*, respectively), and such an axiomatization is unique.*

Our third consequence is the most important one.

THEOREM 5.4 *Every cofinal subframe logic has FMP and so is decidable if finitely axiomatizable.*

FMP of subframe logics was first established by Fine [1985].

Notice by the way that this result immediately yields the FMP of all extensions of **S4.3** which was first proved by Bull [1966] and Fine [1971]. Indeed, every logic $L \in \mathrm{NExt}\mathbf{S4.3}$ ($= \mathrm{Ext}\mathbf{S4.3}$) can be constructed by adding to **S4.3** a (finite) number of canonical axioms whose frames are chains of clusters and so contain no closed domains, which means that $L \in \mathcal{CSF}$. One more example may be also of some interest. Using the semantic characterization of cofinal subframe logics it is easy to show that if φ is a Boolean combination of modalities (i.e. φ is constructed from formulas of the form $\Box^k \Diamond^l \dots \Box^m \Diamond^n p$ with the help of $\wedge$, $\vee$, $\rightarrow$ and $\neg$) then $\mathbf{S4} \oplus \varphi$ belongs to $\mathcal{CSF}$ and so has FMP.

Theorem 5.4 can be strengthened in two directions. First, one can give an upper bound for the size of minimal refutation frames for the logics in $\mathcal{CSF}$. Namely, we have the following

THEOREM 5.5 *For every cofinal subframe logic L and every formula $\varphi \notin L$ there is a frame for L refuting φ and containing at most $2^{3length(\varphi)}$ points.*

This theorem gives approximately the same upper bound for the size of minimal refutation frames as the usual filtration. But the standard filtration method cannot be applied to all cofinal subframe logics.

For some logics in $\mathcal{CSF}$, like **S4.3** and its extensions, the exponential upper bound can be reduced to a polynomial one (cf. [Ono and Nakamura, 1980] and [Chagrov, 1983]). However it is impossible to do this for all cofinal subframe logics. For the following result holds:

THEOREM 5.6 *There is a sequence of modal formulas φ_n, $n = 1, 2, \ldots$, such that $\varphi_n \notin$ **Grz** (or **S4** or **K4**) and, for some constant $c > 0$, every frame for **Grz** (or **S4** or **K4**) refuting φ_n contains at least $2^{c\,length(\varphi_n)}$ points.*

Moreover, Chagrov [1983] extended this result to all logics between **S4** and **S4.2Grz**. By the way, Chagrov [1985a] proved that there exists no function which is a common upper bound for the size of minimal refutation frames for all (modal or superintuitionistic) logics having FMP. However the following question remains open:

PROBLEM 5.7 *Is there such a common upper bound for finitely axiomatizable logics with FMP?*

The second way of strengthening the theorem on the FMP and decidability of cofinal subframe logics is to consider recursive sets of axioms.

THEOREM 5.8 *Every cofinal subframe logic which is recursively axiomatizable by canonical formulas is decidable.*

In fact this result follows directly from Theorem 5.2. However it does not mean that every recursively axiomatizable logic in $\mathcal{CSF}$ is decidable.

Let $\mathfrak{F}_n$, for $n = 3, 4, \ldots$, be the sequence of (reflexive) frames shown in Figure 3.

LEMMA 5.9 *For no $n \neq m$, $\mathfrak{F}_n$ is subreducible to $\mathfrak{F}_m$.*

As a consequence of Lemma 5.9 and Theorem 5.2, one can readily prove the following

THEOREM 5.10

(i) *There are subframe logics with infinite independent sets of canonical axioms.*

(ii) *The cardinality of both $\mathcal{SF}$ and $\mathcal{CSF}$ is that of continuum.*

(iii) *There are a continuum of undecidable subframe logics and infinitely many of them are recursively axiomatizable (but not by canonical formulas).*

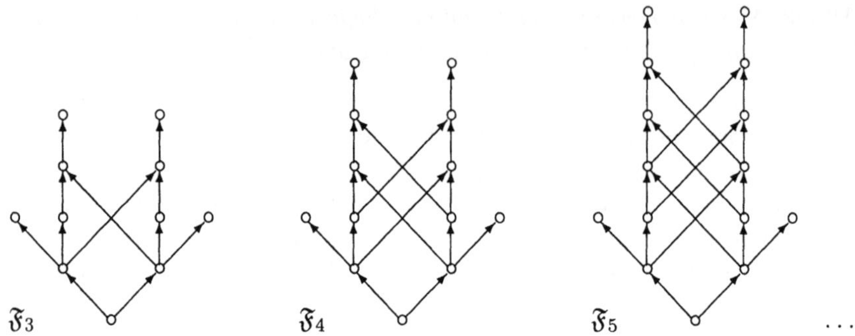

Figure 3.

5.2 *Elementarity, Canonicity and Compactness*

Let us first recall the definitions of these properties. A modal logic L is *elementary* if there is a set Σ of first-order sentences with = and R as their only predicates such that, for every Kripke frame $\mathfrak{F} = \langle W, R\rangle$

$$\mathfrak{F} \models L \text{ iff } \mathfrak{F} \models \Sigma$$

(where the former occurrence of $\models$ is modal, while the latter one is classical).

L is *canonical* if the underlying Kripke frame of every descriptive frame for L also validates L. (Canonicity is also known as *d-persistence*; see [Goldblatt, 1976].)

Finally, L is *compact* if every L-consistent set of formulas is satisfiable in a model whose underlying Kripke frame is a frame for L. (Don't confuse this notion of compactness with Thomason's [1972] compactness! One can encounter in the literature with such a synonym of compactness as *strong Kripke completeness*.)

Fine [1985] gave a nice frame-theoretic characterization of these properties for a subframe logic L: each of them turns out to be equivalent to the finite embedding property of the class of Kripke frames for L (i.e. $\mathfrak{F} \models L$ whenever every finite subframe of $\mathfrak{F}$ validates L). Moreover, all these properties are decidable for finitely axiomatizable subframe logics.

To extend Fine's characterization to the class $\mathcal{CST}$ of cofinal subframe logics we require three more definitions.

Given a frame $\mathfrak{F}_C = \langle W_C, R_C\rangle$ with a distinguished cluster C and an ordinal ξ, $0 < \xi \leq \omega$, by $\mathfrak{F}_\xi^{ir}$ we denote the frame which is obtained from $\mathfrak{F}_C$ by replacing C with an ascending chain of ξ irreflexive points. And if we replace C with an ascending chain containing ξ reflexive points then the resulting frame is denoted by $\mathfrak{F}_\xi^r$. Figure 4 illustrates these constructions.

A finite frame $\mathfrak{F}$ is called (*cofinally*) *quasi-embeddable* in a frame $\mathfrak{G}$ if there is a (cofinal) subreduction f of $\mathfrak{G}$ to $\mathfrak{F}$ such that $f^{-1}(x)$ is a singleton for every

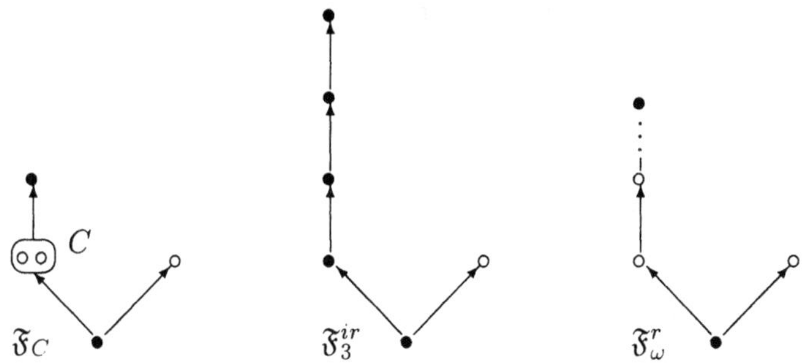

Figure 4.

point x whose cluster $C(x)$ is not final in $\mathfrak{F}$. E.g., the frame $\mathfrak{F}$ in Figure 5 is quasi-embeddable in $\mathfrak{G}$ and cofinally quasi-embeddable in $\mathfrak{H}$.

Finally, we say a logic L has the *finite cofinal quasi-embedding property* if a Kripke frame $\mathfrak{F}$ validates L whenever every finite frame, which is cofinally quasi-embeddable in $\mathfrak{F}$, validates L.

THEOREM 5.11 *For every $L \in \mathcal{CSF}$ the following conditions are equivalent:*

(1) *L is elementary;*

(2) *L is canonical;*

(3) *L is compact;*

(4) *for every finite rooted frame $\mathfrak{F}_C$ with a non-degenerate non-final cluster C,*

$$\mathfrak{F}_C \not\models L \text{ only if } \mathfrak{F}^{ir}_\xi \not\models L \text{ and } \mathfrak{F}^{r}_\zeta \not\models L \text{ for some } \xi, \zeta < \omega;$$

(5) *L has the finite cofinal quasi-embedding property.*

It should be clear that for a subframe logic L the requirement of non-finality of C in (4) is redundant, and (5) becomes just the finite embedding property.

Thus, elementarity, canonicity and compactness of logics in $\mathcal{CSF}$ are completely determined by non-degenerate clusters in the frames of their canonical axioms.

EXAMPLES 5.12 (i) **Grz** has none of these properties, since every finite linearly ordered reflexive frame validates **Grz**, while $(\circ\,\circ)$ fails to do this. By a similar argument **GL**, **K4Z** and **Dum** are not elementary, canonical or compact.

(ii) All the other cofinal subframe logics in Table 4 enjoy these properties. For instance, a finite frame $\mathfrak{F}$ is not a frame for **K4.1** iff it contains either a final $\bullet$ or a final $(\circ\,\circ)$, and so $\mathfrak{F}$ will retain this status after any replacement of its non-final clusters.

(iii) All subframe and cofinal subframe logics whose (negation free) canonical axioms are built on irreflexive frames are elementary, and the cardinality of this class is clearly that of continuum. It is easy to see also that a logic L in the continual class $\mathcal{CSF} \cap \mathbf{NExtGrz}$ is elementary iff $\mathbf{K4B}_n \subseteq L$ for some $n < \omega$. On the other hand, the continual classes

$$\{L \in \mathcal{SF} : \mathbf{K4} \subset L \subseteq \mathbf{GL}\}$$

and

$$\{L \in \mathcal{SF} : \mathbf{S4} \subset L \subseteq \mathbf{Grz}\}$$

contain only non-elementary logics.

As to elementarity, Theorem 5.11 simply expresses the fact that the ascending (non-cofinal) chain condition is not first-order definable. Much more curious is that the elementarity of cofinal subframe logics turns out to be equivalent to their canonicity and compactness. Of course, by Fine's [1975] and van Benthem's [1980] theorem, canonicity follows from completeness with respect to some elementary class of Kripke frames. Moreover, all known canonical logics are characterized by elementary classes of Kripke frames. But I cannot believe that these properties are equivalent for all logics in NExt**K4**. To conclude this discussion, I just repeat the well known

PROBLEM 5.13 *Is it true that for every logic $L \in$ NExt**K4**, L is compact only if L is canonical only if L is characterized by some elementary class of Kripke frames?*[5]

THEOREM 5.14 *There is an algorithm which, given a formula φ such that $\mathbf{K4} \oplus \varphi \in \mathcal{CSF}$, decides whether $\mathbf{K4} \oplus \varphi$ is elementary and, if it is, constructs a first-order equivalent for φ.*

As was shown by Fine [1985], all elementary subframe logics are universal, i.e. their classes of Kripke frames can be described by universal first-order sentences. However for φ's axiomatizing cofinal subframe logics this algorithm constructs in general only Π^0_4-first-order equivalents.

Theorem 5.14 contrasts in a sense with Chagrova's [1991] result according to which the elementarity problem for an arbitrary logic $\mathbf{K4} \oplus \varphi$ is undecidable. But the condition of Theorem 5.14 is very strong. For the following was proved in [Chagrov and Zakharyaschev, 1993]:

[5] F. Wolter has recently constructed a compact logic above **K** which is not canonical; for details, see [Wolter, 1996]. Above **K4**, as follows from [Fine 1974b], such logics also exist, if we restrict them to a finite language.

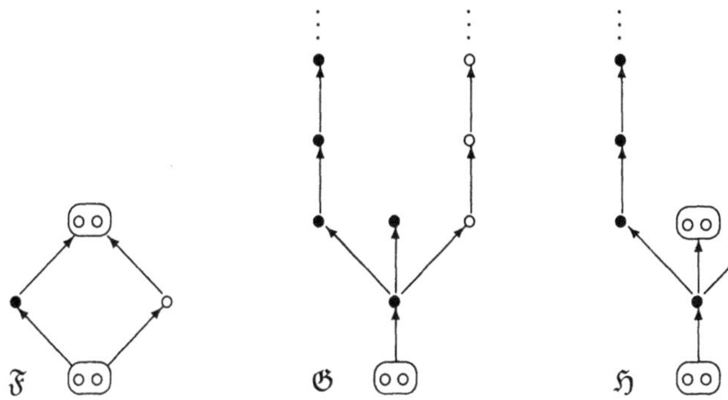

Figure 5.

THEOREM 5.15

(i) *There is no algorithm which is capable of deciding, given a formula* φ, *whether* $\mathbf{K4} \oplus \varphi \in \mathcal{SF}$.

(ii) *There is no algorithm which is capable of deciding, given a formula* φ, *whether* $\mathbf{K4} \oplus \varphi \in \mathcal{CSF}$.

The result below generalizes Fine's [1985] characterization of modal classes of frames closed under subframes. Recall that a class $\mathcal{C}$ of Kripke frames is called *modal* if it is the class of all Kripke frames for some modal logic. Say that $\mathcal{C}$ has the *finite (cofinal) subreduction property* if $\mathfrak{F} \in \mathcal{C}$ whenever every finite rooted (cofinal) subreduct of $\mathfrak{F}$ is in $\mathcal{C}$.

THEOREM 5.16 *Suppose that a class* $\mathcal{C}$ *of transitive Kripke frames is closed under (cofinal) subframes. Then* $\mathcal{C}$ *is modal iff it is closed under reducibility and has the finite (cofinal) subreduction property.*

Thus $\mathcal{CSF}$ is a very nice class of logics. And it will be even nicer if we can solve the following

PROBLEM 5.17 *What are ordinary (not necessarily canonical) modal formulas axiomatizing (cofinal) subframe logics?*

For an extensive study of the class of subframe logics above $\mathbf{K}$ (in the sense of Theorem 5.1) see [Wolter, 1993].

5.3 Superintuitionistic Logics with Implicative and Disjunction Free Axioms

Superintuitionistic subframe and cofinal subframe logics, i.e., the logics in Ext**Int** which are axiomatizable by intuitionistic canonical formulas of the form $\beta(\mathfrak{F}, \emptyset)$

and $\beta(\mathfrak{F}, \emptyset, \bot)$, are, by Theorem 3.5, nothing but extensions of **Int** with purely implicative and disjunction free axioms, respectively. The results of Sections 5.1 and 5.2 can readily be reformulated for these two kinds of superintuitionistic logics. I will show here only some of them.

First of all, we have a nice frame-theoretic characterization of these logics:

THEOREM 5.18

(i) *A logic $L \in$* Ext**Int** *is axiomatized by purely implicative formulas iff it is characterized by a class of frames that is closed under subframes.*

(ii) *$L \in$* Ext**Int** *is axiomatizable by disjunction free formulas iff it is characterized by a class of frames that is closed under cofinal subframes.*

Intuitionistic implicative and disjunction free canonical formulas, as it follows from Theorem 5.2, turn out to be prime in the classes of logics under consideration, and so we obtain the following

THEOREM 5.19 *Every superintuitionistic logic axiomatizable by implicative (disjunction free) formulas can be uniquely axiomatized by an independent set of canonical formulas of the form $\beta(\mathfrak{F}, \emptyset)$ (respectively, $\beta(\mathfrak{F}, \emptyset, \bot)$).*

The intuitionistic version of Theorem 5.4 is simply McKay's [1968] theorem on FMP of superintuitionistic logics with disjunction free axioms. Moreover, by Theorem 5.5, every such logic has the exponential finite model property. And this upper bound cannot be made smaller, for instance, for **Int** (cf. [Zakharyaschev and Popov, 1980]) and even for all logics between **Int** and **KC** [Chagrov, 1983]).

Since all the frames $\mathfrak{F}_n$ in Figure 3 are reflexive and contain no proper clusters, Lemma 5.9 provides us with

THEOREM 5.20 *There are a continuum of superintuitionistic logics axiomatizable by purely implicative formulas; a continuum of them are undecidable, although some are recursively axiomatizable (but not by canonical formulas).*

Intuitionistic frames are homogeneous in the sense that all clusters in them are simple. So the criterion 5.11 becomes the following statement:

THEOREM 5.21 *Every superintuitionistic logic axiomatizable by disjunction free formulas is elementary, canonical and compact.*[6]

[6] A superintuitionistic logic L is called *compact* (or *strongly Kripke complete*) if, for every pair (Γ, Δ) of sets of formulas such that $L \vdash \bigwedge_{\psi \in \Gamma'} \psi \to \bigvee_{\chi \in \Delta'} \chi$ for no finite $\Gamma' \subseteq \Gamma$ and $\Delta' \subseteq \Delta$, all formulas in Γ are simultaneously true and all formulas in Δ are not true at some point in a model whose underlying Kripke frame validates L.

First-order definability ($\forall\exists$-definability, to be more exact) of intuitionistic disjunction free formulas was first established by Chagrova [1986] and Rodenburg [1986]. With the help of this result and McKay's FMP theorem, one can derive canonicity (and so compactness) simply by applying Fine's [1975] theorem, which can be easily extended to superintuitionistic logics using the fact that the map τ (defined in Section 4.1) preserves Kripke completeness. Shimura [1993], having taken advantage of modified canonical formulas, found a direct way of proving the canonicity of superintuitionistic (propositional and certain predicate) logics with disjunction free axioms.

Of the logics in Table 3 which are not cofinal subframe, **KP**, **WKP** and $\mathbf{ND}_k$ are elementary, Kripke complete and so canonical and compact. All $\mathbf{T}_n$'s and $\mathbf{B}_n$'s, for $n \geq 2$, are neither elementary nor canonical nor compact (although have FMP). For, according to recent Shimura's [1995] result, no logic L in the continual interval

$$\mathbf{Int} \subset L \subseteq T_2$$

is compact. As to the Scott system **SL**, it was proved by Rodenburg [1986] that it is not elementary. However as far as I know the following question remains open still:

PROBLEM 5.22 *Is* **SL** *canonical?*[7]

If the intuitionistic variant of Problem 5.13 has a negative solution then **SL** may be regarded as a possible candidate to separate elementarity from canonicity (the McKinsey system **KM** held the same status until [Goldblatt, 1991] proved that it is not canonical).

6 BEYOND CSF

Let us now leave the class of cofinal subframe logics. Almost immediately we find ourselves surrounded by a wild jungle of unknownness with only few paths cut by Fine [1974b], Sahlqvist [1975], Sobolev [1977] and some others. Chagrovs' undecidability results (see [Chagrov, 1990; Chagrova, 1991; Chagrov and Zakharyaschev, 1993]) showed that no effective criteria exist for establishing borders of almost all important properties of modal and superintuitionistic logics. However, partial and yet quite general sufficient or necessary conditions can still be found. And the language of the canonical formulas opens new perspectives here. In this section we consider several results (of both positive and negative character) concerning the finite model property of logics lying outside of $\mathcal{CSF}$.

[7] Shimura [1995] gave a negative solution to this problem.

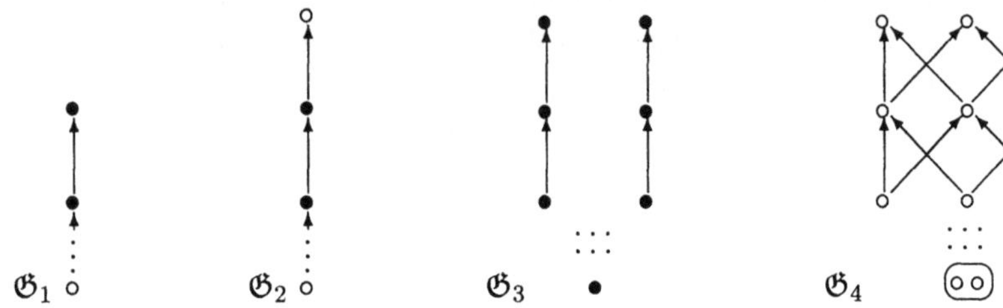

Figure 6.

6.1 *Without FMP*

First I illustrate advantages of using canonical formulas for finding logics lacking the finite model property (compare the logics below with standard examples, say, in [Fine, 1972; Bull and Segerberg, 1984] or [Gabbay, 1981]).

EXAMPLE 6.1 The logic

$L_1 = \mathbf{K4} \oplus \alpha($  $, \mathfrak{D}^\sharp)$

does not have FMP. For the formula

$\alpha($  $, \emptyset)$

does not belong to L_1, since the frame $\mathfrak{G}_1$ in Figure 6 refutes it and validates L_1 but every finite refutation frame for this formula refutes L_1's axiom as well, because at least one reflexive point in it has an irreflexive immediate successor.

In fact L_1 turns out to be the 'simplest' normal extension of **K4** without FMP in the sense that all normal logics whose canonical axioms are built upon frames with only one point enjoy this property (however this is not so in the quasi-normal case where the Solovay logic $\mathbf{S} = \mathbf{GL} + \alpha(\circ, \emptyset)$ not only lacks FMP, but has no Kripke frames at all).

EXAMPLE 6.2 The logic

$L_2 = \mathbf{K4} \oplus \alpha($  $, \mathfrak{D}^\sharp, \bot)$

does not have FMP either. For the frame $\mathfrak{G}_2$ in Figure 6 separates

$\alpha($  $, \emptyset, \bot)$

from L_2 but no finite frame can do this. Moreover, the same argument remains valid for

$$\mathbf{K4.3} \oplus \alpha($$  $$,\mathfrak{D}^\sharp,\bot).$$

It is worth noting that L_2 is the first example of a logic without FMP having a single axiom which is a frame formula.

PROBLEM 6.3 *What about FMP of logics of the form* $\mathbf{S4} \oplus \alpha(\mathfrak{F},\mathfrak{D}^\sharp,\bot)$ *and* $\mathbf{Int} + \beta(\mathfrak{F},\mathfrak{D}^\sharp,\bot)$*?*[8]

EXAMPLE 6.4 The logic

$$L_3 = \mathbf{GL} \oplus \alpha($$  $$,\mathfrak{D}^\sharp)$$

is the 'simplest' normal extension of **GL** without FMP. Indeed, the frame $\mathfrak{G}_3$ in Figure 6 separates

$$\alpha($$  $$,\emptyset)$$

from L_3, but every irreflexive finite refutation frame for this formula contains a point with two incomparable immediate successors and so also refutes the axiom of L_3.

EXAMPLE 6.5 The 'simplest' extension of **S4** without FMP is

$$L_4 = \mathbf{S4} \oplus \alpha($$ 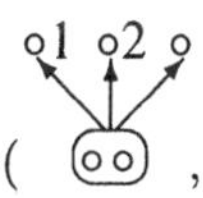 $$,\{\{1,2\}\}).$$

For $\mathfrak{G}_4$ in Figure 6 separates

$$\alpha($$ 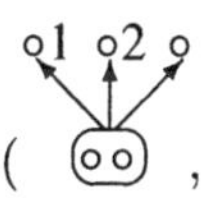 $$,\emptyset)$$

from L_4, but no finite frame is in a position to do this. Observe by the way that $\rho L_4 = \mathbf{Int}$.

EXAMPLE 6.6 Here is a superintuitionistic logic without FMP:

$$L_5 = \mathbf{Int} + \beta($$ 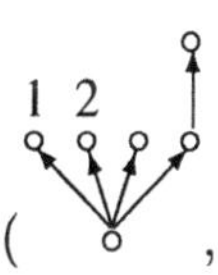 $$,\{\{1,2\}\}).$$

[8] That $\mathbf{GL} \oplus \alpha(\mathfrak{F},\mathfrak{D}^\sharp,\bot)$ has FMP was shown by Kracht [1993a].

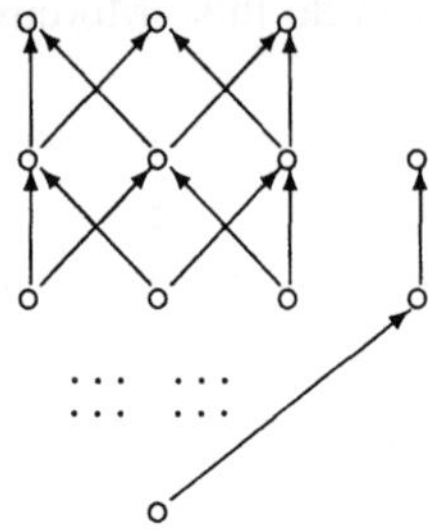

$\mathfrak{G}_5$

Figure 7.

Indeed, the frame $\mathfrak{G}_5$ in Figure 7 separates

$\beta($ 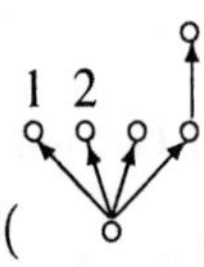

$, \emptyset)$

from L_5, and one can prove that every finite frame for L_5 validates this formula. However I do not know whether L_5 is the 'simplest' superintuitionistic logic lacking FMP.

EXAMPLE 6.7 Clearly, the smallest modal companion

$\tau(\mathbf{BW}_4 + L_5) = \mathbf{S4I}_4 \oplus \alpha($ 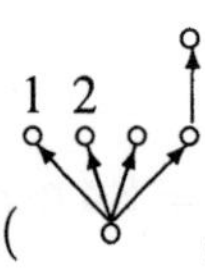

$, \{\{1,2\}\})$

of $\mathbf{BW}_4 + L_5$ does not have FMP either. What is not so evident is that $\tau(\mathbf{BW}_4 + L_5)$ turns out to be a Sahlqvist logic, since it can be axiomatized by adding to $\mathbf{S4I}_4$ the formula

$$\phi = \Diamond\phi_1 \wedge \Diamond\phi_2 \wedge \Diamond\phi_3 \wedge \Diamond\phi_4 \to \Diamond\psi$$

where

$$\begin{aligned}
\phi_1 &= p_1 \wedge \Box\neg p_2 \wedge \Box p_3 \wedge \Box q \\
\phi_2 &= \Box\neg p_1 \wedge p_2 \wedge \Box p_3 \wedge \Box q \\
\phi_3 &= \Box\neg p_1 \wedge \Box\neg p_2 \wedge \neg p_3 \wedge \Box q \\
\phi_4 &= \Box\neg p_1 \wedge \Box\neg p_2 \wedge \Box p_3 \wedge q \wedge r \wedge \Diamond\phi_5 \\
\phi_5 &= \Box\neg p_1 \wedge \Box\neg p_2 \wedge \Box p_3 \wedge \neg q \wedge \Box\neg r \\
\psi &= \Diamond p_1 \wedge \Diamond p_2 \wedge \Box((p_3 \wedge q) \vee \Diamond p_1 \vee \Diamond p_2).
\end{aligned}$$

Thus, by Sahlqvist's [1975] theorem, $\tau(\mathbf{BW}_4 + L_5)$ and so $\mathbf{BW}_4 + L_5$ itself are elementary and canonical, but lack FMP. The first logic of that kind was constructed by Hughes and Cresswell [1984]; however it lies in the field of **K** (not **K4**) where 'everything is possible'.

It is also worth noting (see [Chagrov and Zakharyaschev, 1995a]) that there is no algorithm which, given a modal formula φ, could recognize whether $\mathbf{S4} \oplus \varphi$ is a Sahlqvist logic or not, and that the class of logics above **S4** axiomatizable by Sahlqvist formulas is continual.

These results show that the language of Sahlqvist's formulas has rather strong expressive capacities.

PROBLEM 6.8 *Are finitely axiomatizable Sahlqvist logics above* **K**, **K4** *and* **S4** *decidable?*

Of course it is not difficult now to produce a continuum of modal and superintuitionistic logics without FMP. It would be more useful however to find a solution to the following

PROBLEM 6.9 *Give a general sufficient (syntactical or in terms of canonical formulas) conditions for a logic to lack FMP (completeness or decidability).*

6.2 *With FMP*

Now I show a pair of rather general sufficient conditions for FMP of logics whose canonical axioms may contain closed domains.

The first one is based on the following lemma which is of some interest by itself.

LEMMA 6.10 *Suppose that both* $\alpha(\mathfrak{F}, \mathfrak{D}, \bot)$ *and* $\alpha(\mathfrak{G}, \mathfrak{C}, \bot)$ *(or both* $\alpha(\mathfrak{F}, \mathfrak{D})$ *and* $\alpha(\mathfrak{G}, \mathfrak{C})$*) are two canonical formulas such that*

- *there is a cofinal subreduction (respectively, a plain subreduction) f of $\mathfrak{G}$ to $\mathfrak{F}$ satisfying (CDC) for $\mathfrak{D}$ and*
- *an antichain $\mathfrak{e} \subseteq dom f\uparrow$ is in $\mathfrak{C}$ whenever $f(\mathfrak{e}\underline{\uparrow}) = \mathfrak{d}\underline{\uparrow}$ for some $\mathfrak{d} \in \mathfrak{D}$.*

Then $\alpha(\mathfrak{G}, \mathfrak{C}, \bot) \in \mathbf{K4} \oplus \alpha(\mathfrak{F}, \mathfrak{D}, \bot)$ $(\alpha(\mathfrak{G}, \mathfrak{C}) \in \mathbf{K4} \oplus \alpha(\mathfrak{F}, \mathfrak{D})$*, respectively).*

The same result holds of course for intuitionistic canonical formulas.

THEOREM 6.11 *Suppose that*

$$L = \mathbf{K4} \oplus \{\alpha(\mathfrak{F}_i, \mathfrak{D}_i, \bot) : i \in I\} \oplus \{\alpha(\mathfrak{F}_j, \mathfrak{D}_j) : j \in J\}$$

and either

- *for every $i \in I \cup J$, all points in $\mathfrak{F}_i$ are reflexive, or*
- *for every $i \in I \cup J$, all points in $\mathfrak{F}_i$ are irreflexive.*

Then L has FMP.

Usually to establish FMP of a logic L we remove some points from the canonical model for L. On the contrary, Theorem 6.11 can be proved by inserting new points. Suppose that all $\mathfrak{F}_i$'s in the axioms of L above are irreflexive and $\alpha(\mathfrak{G}, \mathfrak{C}, \bot) \notin L$. Then with the help of Lemma 6.10 one can show that by inserting new reflexive points below every open domain in $\mathfrak{G}$ we can construct a finite frame separating $\alpha(\mathfrak{G}, \mathfrak{C}, \bot)$ from L.

It follows from this theorem that all logics

$$\mathbf{K4}_{n,m} = \mathbf{K4} \oplus \Box^n p \to \Box^m p, \text{ for } 1 \leq m < n$$

have FMP and so are decidable. However, as far as I know the following questions still remain open:

PROBLEM 6.12 *Do the logics*

$$\mathbf{K}_{n,m} = \mathbf{K} \oplus \Box^n p \to \Box^m p$$

have FMP? More generally, what about FMP of extensions of **K** *or* **K4** *with modal reduction principles (for the definition, consult [van Benthem, 1984])?*[9]

Theorem 6.11 provides us also with another consequence:

THEOREM 6.13 *There are finitely axiomatizable logics L_1, $L_2 \in$ NExt***K4** *such that*

- *both L_1 and L_2 have FMP, but*
- *$L_1 \oplus L_2$ lacks the property.*

Indeed, it suffices to take

$$L_1 = \mathbf{CL} = \mathbf{K4} \oplus \alpha(\circ, \emptyset)$$

and

$$L_2 = \mathbf{K4} \oplus \alpha($$  $$, \mathfrak{D}^\sharp)$$

and use Theorem 6.11 and Example 6.4.

[9]Theorem 6.11 can be generalized to show that all extensions of **K4** with modal reduction principles do have FMP and are decidable, see [Zakharyaschev, 1996a].

The first example of that sort was presented by Blok [1976]: he defined algebraically two superintuitionistic logics with FMP such that their sum does not have this property. Recently Kracht [1992] gave a syntactical variant of Blok's result. It may be worth noting that Kripke completeness is not preserved under sums either. For example, Shehtman [1980]'s incomplete logic $L \in$ NExt**Grz** with the complete superintuitionistic fragment ρL can be represented as the sum $L = \mathbf{Grz} \oplus \tau\rho L$ of complete (by Theorem 4.6) logics.

Since new logics are often constructed by joining known ones, in such a situation the following problem may be of interest:

PROBLEM 6.14 *Give sufficient conditions for preserving FMP (completeness, decidability, etc.) under sums of logics.*

For FMP a number of preservation results concerning frame axioms were obtained by Kracht [1993]. As to the other properties, someone may still hope to solve positively the next

PROBLEM 6.15 *Are decidability and compactness preserved under sums of (finitely axiomatizable) logics?*

In fact, the proof of Theorem 6.11 uses the difference between reflexive and irreflexive points. For logics above **S4** one can take advantage of the difference between simple and proper clusters and obtain the following

THEOREM 6.16 *Every logic $L \in$* NExt**S4** *axiomatizable by canonical and negation free canonical formulas, built upon frames in which all clusters are proper, has FMP.*

The second sufficient condition for FMP can be applied to both modal and superintuitionistic logics. It is proved using more or less conventional process of dropping points; see [Zakharyaschev, 1993].

Given a finite frame $\mathfrak{G}$, say that each point of depth 1 is 1-*stable* in $\mathfrak{G}$, and a point x is $n+1$-*stable* if it is not m-stable, for any $m \leq n$, and at least n-stable point is not accessible from x. An antichain $\mathfrak{d}$ in $\mathfrak{G}$ is said to be n-*stable* if it contains an n-stable point *in the subframe of $\mathfrak{G}$ generated by* $\mathfrak{d}$ and no $n+1$-stable point in that subframe.

EXAMPLE 6.17 (a) Suppose $\mathfrak{G}$ is a finite rooted generated subframe of one of the frames shown in Figure 8. Then each point in $\mathfrak{G}$ different from its root is n-stable where n is the number located near the point. Every antichain $\mathfrak{d}$ in $\mathfrak{G}$, containing at least two points, is also n-stable, with n being the maximal degree of stability of points in $\mathfrak{d}$.

(b) If $\mathfrak{G}$ is a finite tree of clusters then every antichain in $\mathfrak{G}$, different from a non-final singleton, is either 1- or 2-stable in $\mathfrak{G}$. More generally, if no point in $\mathfrak{d}$

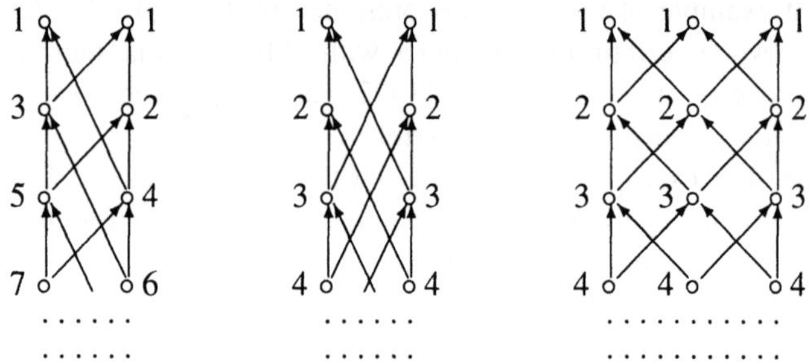

Figure 8.

sees all the points of depth 1 in $\mathfrak{d}\underline{\uparrow}$, in particular, if $\mathfrak{d}$ has no upper bound in $\mathfrak{G}$, then $\mathfrak{d}$ is also either 1- or 2-stable.

(c) The antichains $\{a\}$ in

and $\{1,2\}$ in

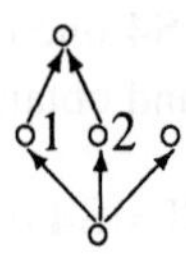

are not n-stable for any n; neither is any antichain $\mathfrak{d}$ such that $\mathfrak{d}\uparrow$ contains only one final cluster.

THEOREM 6.18 *Suppose that, for some $d > 0$, a modal (or superintuitionistic) logic L can be axiomatized by a (finite or infinite) set of canonical axioms $\alpha(\mathfrak{F},\mathfrak{D},\bot)$ (respectively, $\beta(\mathfrak{F},\mathfrak{D},\bot)$) in which every closed domain $\mathfrak{d} \in \mathfrak{D}$ is n-stable in $\mathfrak{F}$ for some $n \leq d$. Then L has FMP.*

EXAMPLE 6.19 It follows from [Anderson, 1972] that every superintuitionistic logic with additional axioms of one variable can be also axiomatized by one or two intuitionistic canonical formulas of the form $\beta(\mathfrak{F},\mathfrak{D}^\sharp,\bot)$ where $\mathfrak{F}$ is a finite rooted generated subframe of the first frame depicted in Figure 8 and $\mathfrak{D}^\sharp$ is the set of all antichains in $\mathfrak{F}$. Therefore, by Theorem 6.18, every such logic has FMP, and so is decidable. This result was first proved by Sobolev [1977] who used a completely different technique.

Above **K4**, there are Kripke incomplete logics axiomatizable by formulas in one variable, and variable free formulas can axiomatize logics without FMP. Above **S4** and **Int**, two variables are enough to produce a logic without FMP. However, I hope to find a positive solution to the following

PROBLEM 6.20 *Is it true that every logic in* NExt**S4** *axiomatizable by formulas in one variable has FMP?*[10]

COROLLARY 6.21 *If a logic $L \in$* NExt**K4** *(or $L \in$* Ext**Int***) can be axiomatized by a (finite or infinite) set of canonical formulas $\alpha(\mathfrak{F}, \mathfrak{D}, \bot)$ (or, respectively, $\beta(\mathfrak{F}, \mathfrak{D}, \bot)$) such that in every $\mathfrak{d} \in \mathfrak{D}$ at least one point does not see all the final points in $\mathfrak{d}\underline{\uparrow}$, then L has FMP. In particular, L has FMP if $\mathfrak{F}$ is a tree of clusters and $\mathfrak{D}$ contains no non-final irreflexive singletons (for logics in* NExt**S4** *and* Ext**Int** *the last condition is always satisfied).*

Of the standard modal and superintuitionistic logics in Tables 2 and 3 which are not cofinal subframe, Corollary 6.21 is applicable to $\mathbf{D4G}_1$, **SL**, **WKP**, $\mathbf{ND}_k$ and $\mathbf{B}_n$, but not to $\mathbf{A}^*$ whose canonical axiom is negation free (it was shown by Shavrukov [1990] that $\mathbf{A}^*$ does not have FMP). Moreover, it gives a positive solution to the conjecture of [Shehtman, 1990] on FMP of the logics

$$\mathbf{D4G}_n = \mathbf{D4} \oplus \Box \bigvee_{j=0}^{n} \Box^+ (p_j \wedge \bigwedge_{i \neq j} \neg p_i) \to \bigvee_{j=0}^{n} \Box \neg (p_j \wedge \bigwedge_{i \neq j} \neg p_i).$$

Indeed, it is not hard to see that

$$\mathbf{D4G}_n = \mathbf{K4} \oplus \alpha(\bullet, \emptyset, \bot) \oplus \alpha(\quad$$ 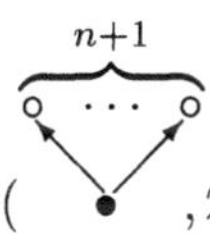 $$, \mathfrak{D}^\sharp, \bot),$$

and so, by Corollary 6.21, all $\mathbf{D4G}_n$'s have FMP.

6.3 *Decidability without FMP*

Proving FMP still remains the main method for establishing decidability of modal and superintuitionistic logics. However, actually the *finiteness* of models is not so important. As far as decidability is concerned, FMP of a finitely axiomatizable logic L is but one possible sufficient condition for the existence of a recursively enumerable class $\mathcal{C}$ of (not necessarily finite) frames such that

(1) every formula $\varphi \in L$ is separated from L by a frame in $\mathcal{C}$ and

(2) one can effectively check whether a frame in $\mathcal{C}$ validates a given formula,

[10] It is true for finitely axiomatizable logics; see [Zakharyaschev, 1996a].

which in turn implies decidability.

Here is a good example showing how the canonical formulas can be used for proving decidability of a big class of logics which in general do not possess FMP, namely the finitely axiomatizable normal extensions of **K4.3** (see Example 6.2). Proofs can be found in [Zakharyaschev and Alekseev, 1995].

Observe first that each finitely axiomatizable logic $L \in \mathrm{NExt}\mathbf{K4.3}$ can be represented in the form

$$L = \mathbf{K4.3} \oplus \{\alpha(\mathfrak{F}_i, \mathfrak{D}_i, \bot) : i \in I\}$$

with chains of clusters $\mathfrak{F}_i$ and a finite I. The decidability problem reduces then to finding an algorithm which, given a formula $\alpha(\mathfrak{F}, \mathfrak{D}, \bot)$ built on a chain of clusters $\mathfrak{F}$, could decide whether $\alpha(\mathfrak{F}, \mathfrak{D}, \bot) \in L$. Recall also that, according to Fine [1974b], L is Kripke complete. So we may consider only Kripke frames of width 1.

Let $C_0, \ldots, C_n$ be all the distinct clusters in $\mathfrak{F}$ ordered in such a way that $C_0 \subset C_1\overline{\downarrow} \subset \ldots \subset C_n\overline{\downarrow}$. A *type for* $\alpha(\mathfrak{F}, \mathfrak{D}, \bot)$ is any n–tuple $t = \langle \xi_1, \ldots, \xi_n \rangle$ such that, for $i \in \{1, \ldots, n\}$, either $\xi_i = m$ or $\xi_i = m+$, for some $m < \omega$, or $\xi_i = \omega$, with $\xi_i = 0$ if $C_i \in \mathfrak{D}$. Given a type $t = \langle \xi_1, \ldots, \xi_n \rangle$ for $\alpha(\mathfrak{F}, \mathfrak{D}, \bot)$, define a *t–extension* of $\mathfrak{F}$ as the frame $\mathfrak{G}$ that is obtained from $\mathfrak{F}$ by inserting between each pair C_{i-1}, C_i of $\mathfrak{F}$'s adjacent clusters either a descending chain of m irreflexive points, if $\xi_i = m < \omega$, or a descending chain of $m+1$ points of which only the last (lowest) one is reflexive, if $\xi_i = m+$, or an infinite descending chain of irreflexive points, if $\xi_i = \omega$. For instance, the frame $\mathfrak{G}_2$ in Figure 6 is the t–extension of the frame in Example 6.2, for every $t = \langle \omega, n \rangle$, $0 \leq n < \omega$.

LEMMA 6.22 *Let* $L \in \mathrm{NExt}\mathbf{K4.3}$ *and* $\alpha(\mathfrak{F}, \mathfrak{D}, \bot) \notin L$. *Then* $\alpha(\mathfrak{F}, \mathfrak{D}, \bot)$ *is separated from L by the t-extension of* $\mathfrak{F}$, *for some type t for* $\alpha(\mathfrak{F}, \mathfrak{D}, \bot)$.

Thus, the recursively enumerable class of all possible t–extensions of finite chains of clusters satisfies (1) for every $L \in \mathrm{NExt}\mathbf{K4.3}$. To establish (2) it suffices to show that there is an algorithm which, given an axiom $\alpha(\mathfrak{F}_i, \mathfrak{D}_i, \bot)$ and a type $t = \langle \xi_1, \ldots, \xi_n \rangle$ for $\alpha(\mathfrak{F}, \mathfrak{D}, \bot)$, decides whether $\alpha(\mathfrak{F}_i, \mathfrak{D}_i, \bot)$ is valid in the t-extension $\mathfrak{G}$ of $\mathfrak{F}$.

Let k be the number of irreflexive points in $\mathfrak{F}_i$. Construct a cofinal subframe $\mathfrak{G}_k$ of $\mathfrak{G}$ by 'cutting off' the infinite descending chains, inserted in $\mathfrak{F}$, (if any) just below their $(k+1)$-st points and let X be the set of all these $(k+1)$-st points.

LEMMA 6.23 $\mathfrak{G} \not\models \alpha(\mathfrak{F}_i, \mathfrak{D}_i, \bot)$ *iff there is an injective cofinal subreduction f of* $\mathfrak{G}_k$ *to* $\mathfrak{F}_i$ *satisfying (CDC) for* $\mathfrak{D}_i$ *and such that* $X \cap \mathrm{dom} f = \emptyset$.

As an immediate consequence of these two lemmas we obtain

THEOREM 6.24 *All finitely axiomatizable logics in* $\mathrm{NExt}\mathbf{K4.3}$ *are decidable.*

I believe that the method of proving Theorem 6.24 can be extended to solve positively the following

PROBLEM 6.25 *Are finitely axiomatizable logics of finite width decidable?*

Wolter [1996a] generalized Theorem 6.24 to finitely axiomatizable tense linear logics (which in general are not even Kripke complete) and Kracht to those finite width logics that are of finite tightness.

Another example of proving decidability in the absence of FMP can be found in [Zakharyaschev, 1996], where a good many quasi-normal modal logics were shown to be decidable.

7 THE DISJUNCTION PROPERTY AND HALLDÉN COMPLETENESS

One more series of results, which was obtained using the apparatus of the canonical formulas, concerns the disjunction property (DP) and Halldén completeness (HC) of superintuitionistic logics and normal modal logics containing **S4**. Those results were found mainly in the course of attacking two problems, viz., Hosoi and Ono's [1973] problem on DP of superintuitionistic logics containing disjunction free formulas that are not provable in **Int** and Maksimova's problem (formulated in [Logic Notebook, 1986]) on the decidability of DP.

The original definitions of DP and HC are purely syntactical and so are hardly suitable for applying the canonical formulas. (I remind the reader that a superintuitionistic logic L has the *disjunction property* if, for every formulas ψ and χ, $\psi \vee \chi \in L$ implies $\psi \in L$ or $\chi \in L$. A modal logic M has the (*modal*) DP if, for every ψ and χ, $\Box\psi \vee \Box\chi \in M$ implies $\Box\psi \in M$ or $\Box\chi \in M$. Finally, a (modal or superintuitionistic) logic L is *Halldén complete* if $\psi \vee \chi \in L$ implies $\psi \in L$ or $\chi \in L$ for every ψ and χ having no common variables. A brief review of results on these and related properties can be found in [Chagrov and Zakharyaschev, 1991].)

Fortunately, there are rather simple frame-theoretic equivalents of these definitions. A (modal or superintuitionistic) logic L has DP iff, for every two rooted descriptive frames $\mathfrak{F}$ and $\mathfrak{G}$ for L, there exists a rooted descriptive frame for L containing the disjoint union of $\mathfrak{F}$ and $\mathfrak{G}$ as its generated subframe. For modal logics this criterion follows from Theorem 3.4 of [Lemmon, 1977] and for superintuitionistic ones it was proved (in algebraic terms) by Maksimova [1986]. A frame-theoretic equivalent for HC of modal logics was found in 1992 independently by several people, A. Chagrov, L. Maksimova and F. Wolter among them. In fact it is a natural generalization of the sufficient condition of van Benthem and Humberstone [1983]: $L \in$ NExt**K** is HC iff, for every rooted descriptive frames $\mathfrak{F}$ and $\mathfrak{G}$ for L with roots x and y, there is a rooted descriptive frame $\mathfrak{H}$ for L with root z which is reducible to $\mathfrak{F}$ and $\mathfrak{G}$ by reductions f and g, respectively, such that $f(z) = x$ and $g(z) = y$. $L \in$ Ext**Int** is HC iff, for every two rooted intuitionistic descriptive frames $\mathfrak{F}$ and $\mathfrak{G}$ for L, there exists a rooted descriptive frame for L containing generated subframes reducible to $\mathfrak{F}$ and $\mathfrak{G}$.

The key idea, giving a solution to Hosoi and Ono's problem, is that every canonical formula $\alpha(\mathfrak{F}, \emptyset, \bot)$, built on a reflexive frame $\mathfrak{F}$ with simple first and at least one of the final clusters can be represented as a disjunction of boxed formulas which are 'stronger' than $\alpha(\mathfrak{F}, \emptyset, \bot)$. In other words we have the following theorem proved in [Zakharyaschev, 1987].

THEOREM 7.1 *Suppose that a consistent logic $M \in$ NExt**S4** contains a formula $\alpha(\mathfrak{F}, \emptyset, \bot)$ with $\mathfrak{F}$ having simple first and one of the final clusters and such that $\alpha(\mathfrak{G}, \emptyset, \bot) \notin M$ for every proper generated subframe $\mathfrak{G}$ of $\mathfrak{F}$. Then M does not have DP.*

As a consequence of Theorems 7.1 and 4.6 we immediately obtain:

THEOREM 7.2 *Every consistent superintuitionistic logic L containing a disjunction free formula which is not in* **Int** *does not have DP.*

The contraposition of this statement looks more attractive and yields a solution to the problem:

THEOREM 7.3 *Every consistent superintuitionistic logic with DP has the same disjunction free formulas as* **Int**.

Approximately at the same time (about 1984) and using a similar idea this problem was independently solved by Minari [1986]. Recently I have found another, purely frame-theoretic solution which uses the necessary condition of DP above for 'simulating' finite binary trees (which characterize the disjunction free fragment of **Int**) by general frames (see [Zakharyaschev, 1994]).

It may be of interest also to notice that the formulas φ_n, mentioned in Theorem 5.6, are deductively equal to $\alpha(\mathfrak{T}_n, \emptyset, \bot)$ where $\mathfrak{T}_n$ is the reflexive binary tree of height n. By Theorem 7.1 this means that φ_n does not belong to any modal logic with DP. As a result we obtain the following unexpected result first observed by Chagrov and Zakharyaschev [1990]:

THEOREM 7.4 *No modal or superintuitionistic logic L with DP has the polynomial finite model property; at the best L only has the exponential FMP.*

With the help of Theorem 7.1 and the frame-theoretic sufficient condition of DP it is not hard to characterize cofinal subframe logics in NExt**S4** having DP.

THEOREM 7.5 *A consistent cofinal subframe logic $M \in$ NExt**S4** different from* **S4** *has DP iff every canonical formula in its independent canonical axiomatization is built on a frame whose first cluster or each of the final clusters is proper.*

PROBLEM 7.6

(i) *Characterize the canonical axioms of cofinal subframe logics with the disjunction property in* NExt**K4**.[11]

(ii) *Characterize canonical axioms of Halldén complete cofinal subframe logics in* NExt**K4** *and* Ext**Int**.

According to Theorem 7.5, given a formula φ such that $M = \mathbf{S4} \oplus \varphi \in \mathcal{CSF}$, one can effectively determine whether M has DP or not. (The same seems to be true for $\mathbf{K4} \oplus \varphi$ and the Halldén completeness.) But beyond $\mathcal{CSF}$ the situation is quite different. For as was proved by Chagrov and Zakharyaschev [1993], no algorithm exists which is capable of deciding, given a modal (or intuitionistic) formula φ, whether $\mathbf{Grz}\oplus\varphi$ ($\mathbf{Int}+\varphi$) has DP or is HC. In order to apply Chagrov's scheme for proving the undecidability we required to construct some non- trivial logics having these properties. As a result the following three sufficient conditions for DP and HC were found.

THEOREM 7.7 *Suppose that a logic $L \in$* NExt**Grz** *or $L \in$* Ext**Int** *can be axiomatized by canonical formulas $\alpha(\mathfrak{F}, \mathfrak{D}, \bot)$ or, respectively, $\beta(\mathfrak{F}, \mathfrak{D}, \bot)$ (or negation free canonical formulas $\alpha(\mathfrak{F}, \mathfrak{D})$ or $\beta(\mathfrak{F}, \mathfrak{D})$) such that the set X of the immediate successors of the root in $\mathfrak{F}$ contains at least three points and $\mathfrak{d} \in \mathfrak{D}$ for every antichain $\mathfrak{d}$ with $\geq \mathrm{card} X/2$ points from X. Then L has DP and is HC.*

THEOREM 7.8 *Suppose that a logic $L \in$* NExt**Grz** *or $L \in$* Ext**Int** *is Kripke complete and can be axiomatized by canonical formulas built on frames whose roots have exactly one immediate successor. Then L is HC.*

THEOREM 7.9 *Suppose that a logic $L \in$* NExt**S4** *or $L \in$* Ext**Int** *can be axiomatized by canonical formulas $\alpha(\mathfrak{F}, \mathfrak{D}, \bot)$ or, respectively, $\beta(\mathfrak{F}, \mathfrak{D}, \bot)$ such that* $\mathrm{height}(\mathfrak{F}) \geq 3$ *and there is a closed domain $\mathfrak{d} \in \mathfrak{D}$ consisting of points in some final clusters and having no focus in $\mathfrak{F}$. Then L has DP.*

Each of these conditions covers a continuum of logics, and the scope of the first one can be somewhat extended by using the preservation of DP while passing from a superintuitionistic logic L to its smallest companion τL. It is worth noting however that Chagrov and Zakharyaschev [1993] constructed a continuum of Halldén complete superintuitionistic logics having no Halldén complete modal companions at all.

The sufficient condition of Theorem 7.9 can be used for clarifying the relation between DP and such properties as Kripke completeness and decidability. The only known fact in this direction was Wroński's [1973] result showing that there are undecidable infinitely axiomatizable logics with DP.

[11] Recently this problem has been solved by A. Zhvakin.

Actually, Theorem 7.9 gives a method for constructing a finitely axiomatizable logic with a combination of DP and some 'negative' properties starting from a known logic possessing these 'negative' properties and having finitely many negation free axioms.

Suppose that

$$L = \mathbf{Int} + \{\beta(\mathfrak{F}_i, \mathfrak{D}_i) : i = 1, \ldots, n\}$$

is a Kripke incomplete or undecidable superintuitionistic logic (say, one of those constructed by Shehtman [1977, 1978]). Then, by Theorem 7.9, the logic

$$\widehat{L} = \mathbf{Int} + \{\beta(\widehat{\mathfrak{F}}_i, \widehat{\mathfrak{D}}_i, \bot) : i = 1, \ldots, n\},$$

where

$$\widehat{W_i} = W_i \cup \{0, 1, 2, 3\},$$

$\widehat{R_i}$ is the reflexive and transitive closure of the relation

$$R_i \cup \{\langle x, 0\rangle : x \in W_i\} \cup \{\langle 0, j\rangle : 1 \leq j \leq 3\}$$

and

$$\widehat{\mathfrak{D}_i} = \mathfrak{D}_i \cup \{\{1, 2\}\},$$

has DP. Moreover, one can prove (using the apparatus of the canonical formulas) that it retains the 'negative' property of L. Thus we have:

THEOREM 7.10

(i) *There are finitely axiomatizable Kripke incomplete superintuitionistic and modal logics with DP.*

(ii) *There are undecidable, finitely axiomatizable superintuitionistic and modal logics with DP.*

II. Mathematisches Institut
Freie Universität Berlin

Acknowledgments. I wish to thank Alexander Chagrov for his comments and suggestions on the first draft of this paper. Also I would like to thank Marcus Kracht, Hiroakira Ono and Frank Wolter for many helpful conversations. In different periods this work was supported by grant number 93-11-16006 from the Russian Fundamental Research Foundation, by the Soros Foundation, and by the Alexander von Humboldt Foundation.

REFERENCES

[Anderson, 1972] J.G. Anderson. Superconstructive propositional calculi with extra axiom sche-mes containing one variable. *Zeitschrift für Mathematische Logik und Grundlagen der Mathematik*, 18:113–130, 1972.

[van Benthem, 1980] J. van Benthem. Some kinds of modal completeness. *Studia Logica*, 39:125–141, 1980.

[van Benthem, 1983] J. van Benthem. *Modal logic and classical logic*, Bibliopolis, Naples, 1983.

[van Benthem, 1984] J. van Benthem. Correspondence theory. In D. Gabbay and F. Guenthner, editors, *Handbook of Philosophical Logic, volume 2*, pages 167–247. Reidel, Dordrecht, 1984.

[van Benthem and Humberstone, 1983] J. van Benthem and I.L. Humberstone. Halldén completeness by gluing of Kripke frames. *Notre Dame Journal of Formal Logic*, 24:426–430, 1983.

[Blok, 1976] W.J. Blok. *Varieties of interior algebras*, PhD thesis, University of Amsterdam, 1976.

[Bull, 1966] R.A. Bull. That all normal extensions of **S4.3** have the finite model property. *Zeitschrift für Mathematische Logik und Grundlagen der Mathematik*, 12:341–344, 1966.

[Bull and Segerberg, 1984] R.A. Bull and K. Segerberg. Basic modal logic. In D. Gabbay and F. Guenthner, editors, *Handbook of Philosophical Logic, volume 2*, pages 1–88, Reidel, Dordrecht, 1984.

[Chagrov, 1983] A.V. Chagrov. On the polynomial finite model property of modal and intermediate logics. In *Mathematical Logic, Mathematical Linguistics and Algorithm Theory*, pages 75–83, Kalinin State University, Kalinin, 1983. (Russian)

[Chagrov, 1985] A.V. Chagrov. Varieties of logical matrices. *Algebra and Logic*, 24:426–489, 1985. (Russian)

[Chagrov, 1985a] A.V. Chagrov. On the complexity of propositional logics. In *Complexity Problems in Mathematical Logic*, pages 80–90, Kalinin State University, Kalinin, 1985. (Russian)

[Chagrov, 1990] A.V. Chagrov. Undecidable properties of extensions of provability logic, parts I, II. *Algebra and Logic*, 29:350–367, 613–623, 1990. (Russian)

[Chagrov and Zakharyaschev, 1989] A.V. Chagrov and M.V. Zakharyaschev. The undecidability of the disjunction property of intermediate calculi. Preprint, Institute of Applied Mathematics, the USSR Academy of Sciences, no. 57, 1989. (Russian)

[Chagrov and Zakharyaschev, 1990] A.V. Chagrov and M.V. Zakharyaschev. An essay in complexity aspects of intermediate calculi. *Proceedings of the Fourth Asian Logic Conference*, pages 26–29, CSK Education Centre, Tokyo, 1990.

[Chagrov and Zakharyaschev, 1991] A.V. Chagrov and M.V. Zakharyaschev. The disjunction property of intermediate propositional logics. *Studia Logica*, 50:189–216, 1991.

[Chagrov and Zakharyaschev, 1992] A.V. Chagrov and M.V. Zakharyaschev. Modal companions of intermediate propositional logics. *Studia Logica*, 51:49–82, 1992.

[Chagrov and Zakharyaschev, 1993] A.V. Chagrov and M.V. Zakharyaschev. The undecidability of the disjunction property of propositional logics and other related problems. *Journal of Symbolic Logic*, 58:967–1002, 1993.

[Chagrov and Zakharyaschev, 1995] A.V. Chagrov and M.V. Zakharyaschev. On the independent axiomatizability of modal and intermediate logics. *Journal of Logic and Computation*, 5:287–302, 1995.

[Chagrov and Zakharyaschev, 1995a] A.V. Chagrov and M.V. Zakharyaschev. Sahlqvist formulas are not so elementary even above **S4**. In L. Csirmaz, D.M. Gabbay and M. de Rijke, editors, *Logic Colloquium '92*, pages 61–73, Studies in Logic, Language and Information, CSLI Publications/FoLLI, Stanford, CA, 1995.

[Chagrova, 1986] L.A. Chagrova. On first-order definability of intuitionistic formulas with restrictions on occurrences of connectives. In *Logical Methods for Constructing Effective Algorithms*, pages 135–136, Kalinin State University, Kalinin, 1986. (Russian)

[Chagrova, 1991] L.A. Chagrova. An undecidable problem in correspondence theory. *Journal of Symbolic Logic*, 56:1261–1272, 1991.

[van Dalen, 1986] D. van Dalen. Intuitionistic logic. In D. Gabbay and F. Guenthner, editors, *Handbook of Philosophical Logic, volume 3*, pages 225–339, Reidel, Dordrecht, 1986.

[Dummett and Lemmon, 1959] M.A.E. Dummett and L.J. Lemmon. Modal logics between **S4** and **S5**. *Zeitschrift für Mathematische Logik und Grundlagen der Mathematik*, 5:250–264, 1959.

[Esakia, 1979] L.L. Esakia. On varieties of Grzegorczyk algebras. In *Studies in Non-Classical Logics and Set Theory*, pages 257–287, Nauka, Moscow, 1979. (Russian)

[Fine, 1979] K. Fine. The logics containing **S4.3**. *Zeitschrift für Mathematische Logik und Grundlagen der Mathematik*, 17:371–376, 1979.

[Fine, 1972] K. Fine. Logics containing **S4** without the finite model property. In W. Hodges, editor, *Conference in Mathematical Logic, London 1970*, Lecture Notes in Mathematics, vol.255, pages 88–102, Springer-Verlag, Berlin, 1972.

[Fine, 1974] K. Fine. An incomplete logic containing **S4**. *Theoria*, 40:23–29, 1974.

[Fine, 1974a] K. Fine. An ascending chain of **S4** logics. *Theoria*, 40:110–116, 1974.

[Fine, 1974b] K. Fine. Logics containing **K4**. Part I. *Journal of Symbolic Logic*, 39:31–42, 1974.

[Fine, 1975] K. Fine. Some connections between elementary and modal logic. In S. Kanger, editor, *Proceedings of the Third Scandinavian Logic Symposium*, pages 15–31, North-Holland, Amsterdam, 1975.

[Fine, 1985] K. Fine. Logics containing **K4**. Part II. *Journal of Symbolic Logic*, 50:619–651, 1985.

[Gabbay, 1981] D.M. Gabbay. *Semantical Investigations in Heyting's Intuitionistic Logic*. Reidel, Dordrecht, 1981.

[Gödel, 1933] K. Gödel. Eine Interpretation des intuitionistischen Aussagenkalküls. *Ergebnisse eines mathematischen Kolloquiums, Heft 4*, pages 39–40, 1933.

[Goldblatt, 1976] R.I. Goldblatt. Metamathematics of modal logic. *Reports on Mathematical Logic*, **6**, pp. 41–77; **7**, pp. 21–52.

[Goldblatt, 1991] R.I. Goldblatt. The McKinsey axiom is not canonical. *Journal of Symbolic Logic*, 56:554–562, 1991.

[Grzegorczyk, 1967] A. Grzegorczyk. Some relational systems and the associated topological spaces. *Fundamenta Mathematicae*, 60:223–231, 1967.

[Hosoi and Ono, 1973] T. Hosoi and H. Ono. Intermediate propositional logics (A survey). *Journal of Tsuda College*, 5:67–82, 1973.

[Hughes and Cresswell, 1984] G.E. Hughes and M.J. Cresswell *A Companion to Modal Logic*, Methuen, London, 1984.

[Jankov, 1963] V.A. Jankov. The relationship between deducibility in the intuitionistic propositional calculus and finite implicational structures. *Soviet Mathematics Doklady*, 4:1203–1204, 1963.

[Jankov, 1969] V.A. Jankov. Conjunctively indecomposable formulas in propositional calculi. *Izvestija Academii Nauk SSSR*, 33:13–38, 1969. (English translation: *Mathematics of USSR*, Izvestija, 3:17–35.)

[Kracht, 1990] M. Kracht. An almost general splitting theorem for modal logic. *Studia Logica*, 49:455–470, 1990.

[Kracht, 1992] M. Kracht. Sometimes FMP is not preserved under joins. Manuscript, 1992.

[Kracht, 1993] M. Kracht. How completeness and correspondence theory got married. In M. de Rijke, editor, *Diamonds and Defaults*, pages 175-214, Kluwer Academic Publishers, Dordrecht, 1993.

[Kracht, 1993a] M. Kracht. Splittings and the finite model property. *Journal of Symbolic Logic*, 58:139–157, 1993.

[Kripke, 1963] S. Kripke. Semantical analysis of modal logic I: Normal propositional calculi. *Zeitschrift für Mathematische Logik und Grundlagen der Mathematik*, 9:67–96, 1963.

[Kuratowksi, 1922] C. Kuratowski. Sur l'opération $\overline{A}$ de l'Analysis Situs. *Fundamenta Mathematicae*, 3:182–199, 1922.

[Lemon, 1977] E.J. Lemmon. *An Introduction to Modal Logic*, in collaboration with D. Scott, Blackwell, Oxford, 1977.

[Logic Notebook, 1986] Logic Notebook (1986), Novosibirsk, 1986. (Russian)

[Maksimova, 1986] L.L. Maksimova. On maximal intermediate logics with the disjunction property. *Studia Logica*, 45:69–75, 1986.

[Maksimova and Rybakov, 1974] L.L. Maksimova and V.V. Rybakov. On the lattice of normal modal logics. *Algebra and Logic*, 13:188–216, 1974. (Russian)

[McKay, 1968] C.G. McKay. The decidability of certain intermediate logics. *Journal of Symbolic Logic*, 33:258–264, 1968.

[McKinsey and Tarski, 1946] J.C.C. McKinsey and A. Tarski. On closed elements in closure algebras. *Annals of Mathematics*, 47:122–162, 1946.

[McKinsey and Tarski, 1948] J.C.C. McKinsey and A. Tarski. Some theorems about the sentential calculi of Lewis and Heyting. *Journal of Symbolic Logic*, 13:1–15, 1948.

[Minari, 1986] P. Minari. Intermediate logics with the same disjunctionless fragment as intuitionistic logic. *Studia Logica*, 45:207–222, 1986.

[Ono and Nakamura, 1980] H. Ono and A. Nakamura. On the size of refutation Kripke models for some linear modal and tense logics. *Studia Logica*, 39:325–333, 1980.

[Rautenberg, 1979] W. Rautenberg. *Klassische und nichtklassische Aussagenlogik*, Vieweg Verlag, Wiesbaden, 1979.

[Rodenburg, 1986] P.H. Rodenburg. *Intuitionistic Correspondence Theory*, PhD thesis, University of Amsterdam, 1986.

[Sahqlvist, 1975] H. Sahlqvist. Completeness and correspondence in the first and second order semantics for modal logic. In S. Kanger, editor, *Proceedings of the Third Scandinavian Logic Symposium*, pages 110–143, North-Holland, Amsterdam, 1975.

[Segerberg, 1971] K. Segerberg. *An Essay in Classical Modal Logic*, Philosophical Studies, vol.13, University of Uppsala, 1971.

[Shavrukov, 1990] V.Ju. Shavrukov. On two extensions of the provability logic **GL**. *Mathematical Sbornik*, 181:240–255, 1990. (English translation: *Mathematics of the USSR Sbornik*, 69:255–270.)

[Shehtman, 1977] V.B. Shehtman. On incomplete propositional logics. *Soviet Mathematics Doklady*, 235:542–545, 1977. (Russian)

[Shehtman, 1978] V.B. Shehtman. Undecidable superintuitionistic calculus. *Soviet Mathematics Doklady*, 240:549–552, 1978. (Russian)

[Shehtman, 1980] V.B. Shehtman. Topological models of propositional logics. *Semiotics and Information Science*, 15:74–98, 1980. (Russian)

[Shehtman, 1990] V.B. Shehtman. Derived sets in Euclidean spaces and modal logic. Preprint X-90-05, University of Amsterdam, 1990.

[Shimura, 1993] T. Shimura. Kripke completeness of some intermediate predicate logics with the axiom of constant domain and a variant of canonical formulas. *Studia Logica*, 52:23–40, 1993.

[Shimura, 1995] T. Shimura. On completeness of intermediate predicate logics with respect to Kripke semantics. *Bulletin of the Section of Logic*, 24:41–45, 1995.

[Sobolev, 1977] S.K. Sobolev. On the finite approximability of superintuitionistic logics. *Mathematical Sbornik*, 102:289–301, 1977. (Russian)

[Thomason, 1972] S.K. Thomason. Noncompactness in propositional modal logic. *Journal of Symbolic Logic*, 37:716–720, 1972.

[Thomason, 1974] S.K. Thomason. An incompleteness theorem in modal logic. *Theoria*, 40:30–34, 1974.

[Wolter, 1993] F. Wolter. *Lattices of Modal Logics*. PhD thesis, FU Berlin, 1993.

[Wolter, 1996] F. Wolter. Properties of tense logics. *Mathematical Logic Quarterly*, in print.

[Wolter, 1996a] F. Wolter. Tense logics without tense operators. *Mathematical Logic Quarterly*, 42:145–171, 1996.

[Wroński, 1973] A. Wroński. Intermediate logics and the disjunction property. *Reports on Mathematical Logic*, 1:39–51, 1973.

[Zakharyaschev, 1981] M.V. Zakharyaschev. Certain classes of intermediate logics. Preprint, Institute of Applied Mathematics, the USSR Academy of Sciences, no. 160, 1981. (Russian)

[Zakharyaschev, 1983] M.V. Zakharyaschev. On intermediate logics. *Soviet Mathematics Doklady*, 27:274–277, 1983.

[Zakharyaschev, 1984] M.V. Zakharyaschev. Normal modal logics containing **S4**. *Soviet Mathematics Doklady*, 29:252–255, 1984.

[Zakharyaschev, 1984a] M.V. Zakharyaschev. *Syntax and Semantics of Superintuitionistic and Modal Logics*. PhD thesis, Moscow, 1984. (Russian)

[Zakharyaschev, 1987] M.V. Zakharyaschev. On the disjunction property of intermediate and modal logics. *Mathematical Zametki*, 42:729-738, 1987. (English translation: *Mathematical Notes*, 42:901–905).

[Zakharyaschev, 1988] M.V. Zakharyaschev. Syntax and semantics of modal logics containing **S4**. *Algebra and Logic*, 27:408–428, 1988.

[Zakharyaschev, 1989] M.V. Zakharyaschev. Syntax and semantics of intermediate logics. *Algebra and Logic*, 28:262–282, 1989.

[Zakharyaschev, 1989a] M.V. Zakharyaschev. Modal companions of intermediate logics: syntax, semantics and preservation theorems. *Mathematical Sbornik*, 180:1415–1427, 1989. (English translation: *Mathematics of the USSR Sbornik*, 68:277–289.)

[Zakharyaschev, 1992] M.V. Zakharyaschev. Canonical formulas for **K4**. Part I: Basic results. *Journal of Symbolic Logic*, 57:1377–1402, 1992.

[Zakharyaschev, 1993] M.V. Zakharyaschev. A sufficient condition for the finite model property of modal logics above **K4**. *Bulletin of the IGPL*, 1:13–21, 1993.

[Zakharyaschev, 1994] M.V. Zakharyaschev. A new solution to the problem of T. Hosoi and H. Ono. *Notre Dame Journal of Formal Logic*, 35:450–457, 1994.

[Zakharyaschev, 1996] M.V. Zakharyaschev. Canonical formulas for **K4**. Part II: Cofinal subframe logics. *Journal of Symbolic Logic*, 61(2):421–449, 1996.

[Zakharyaschev, 1996a] M.V. Zakharyaschev. Canonical formulas for **K4**. Part III: The finite model property. *Journal of Symbolic Logic*, in print.

[Zakharyaschev and Alekseev, 1995] M.V. Zakharyaschev and A.V. Alekseev. All finitely axiomatizable normal extensions of **K4.3** are decidable. *Mathematical Logic Quarterly*, 41:15–23, 1995.

[Zakharyaschev and Popov, 1980] M.V. Zakharyaschev and S.V. Popov. On the complexity of countermodels for intuitionistic calculus. Preprint, Institute of Applied Mathematics, the USSR Academy of Sciences, no. 45, 1980. (Russian)

EDWARD N. ZALTA

THE MODAL OBJECT CALCULUS AND ITS INTERPRETATION

The modal object calculus is the system of logic which houses the (proper) axiomatic theory of abstract objects.[1] This calculus has some rather interesting features in and of itself, independent of the proper theory. The most sophisticated, type-theoretic incarnation of the calculus can be used to analyze the intensional contexts of natural language and so constitutes an intensional logic. However, the simpler second-order version of the calculus couches a theory of fine-grained properties, relations and propositions and serves as a framework for defining situations, possible worlds, stories, and fictional characters, among other things. In the present paper, we focus on the second-order calculus.

The second-order modal object calculus is so-called to distinguish it from the second-order modal predicate calculus. Though the differences are slight, the extra expressive power of the object calculus significantly enhances its ability to resolve logical and philosophical concepts and problems. Our primary goal in this paper is to describe a new interpretation of the modal object calculus. As a secondary and preliminary goal, we shall recast its intended interpretation in a new and interesting way, based on a reconception of logic and model theory that we take to be philosophically more perspicuous than the standard conception. But in order to accomplish these goals, we shall first need to motivate and define the calculus and then sketch some of its applications. Then we will be in a position to redescribe its intended interpretation and develop the new interpretation. In Section 1, then, we explain the ideas underlying the basic definitions that construct the calculus. In Section 2, we examine some of the applications of the calculus which do not require the proper theory of abstract objects. In Section 3, we describe the intended interpretation of the calculus in terms of the reconception of logic and model theory. In Section 4, we construct the new interpretation of the calculus by developing, in a modal setting, a suggestion due to Peter Aczel. Finally, Section 5 contains some observations about the ideas that have been presented and offers a brief outline of how to extend those ideas to produce a model of the theory of abstract objects. Readers who are familiar with the definition and applications of the object calculus may wish to skip ahead to Section 3, where the intended interpretation of the calculus is recast in a new theoretical setting.

[1] The proper theory of abstract objects is developed in [Zalta, 1983, 1988] and extended in [1987, 1993a]. This system was developed within the tradition of 'object theory', which has recently been revived by Parsons [1980], Routley [1979], Castañeda [1974], and Rapaport [1978], and which originated in the naive theories of Meinong [1904] and Mally [1912].

M. de Rijke (ed.), Advances in Intensional Logic, 249–279.

1 THE SECOND ORDER MODAL OBJECT CALCULUS

The (modal) object calculus is the logical system for asserting and proving facts about *abstract* objects. Abstract objects are to be distinguished from ordinary spatio-temporal objects such as you, me, this computer, electrons, planets, etc., and from ordinary objects that are spatio-temporally located at other possible worlds. The predicate calculus is the background logic in which we assert facts and draw inferences about ordinary objects because such objects *exemplify* the properties we discover them to have. Exemplification is the mode of predication upon which the predicate calculus is based, for the philosophical claim that objects $x_1, \ldots, x_n$ *exemplify* relation F^n is represented by ordinary atomic formulas of the form $F^n x_1 \ldots x_n$. But *abstract* objects are completely different in kind from ordinary objects. They are not the kind of thing that could have a location in spacetime; they are not contingent and they are not sparsely distributed in their domain. They are not discovered by sensory perception or postulated in scientific theories. Rather, abstract objects are *constituted* by the properties that we use to conceptualize, define, and individuate them. We conceptualize and define the number 2, for example, in terms of its number-theoretic properties; we identify the null set in terms of its set-theoretic properties; we individuate the actual world in terms of its propositional properties; we conceptualize Sherlock Holmes in terms of the properties attributed to him in the Conan Doyle novels. The groups of properties just mentioned are even more important to the identity of their respective objects than some of the properties these objects necessarily exemplify. For example, all of these objects necessarily fail to be buildings, necessarily fail to be spoons, etc., and from this it follows that they necessarily exemplify the negations of these properties. But these necessarily exemplified negations are not properties by which we conceive the objects in question.

Every abstract object is characterized by some group of constitutive properties. We need a way of predicating properties of abstract objects that is even stronger than necessary exemplification; i.e., we need to be able to predicate the property F that is constitutive of the abstract object x in a way that is stronger than the assertion that $\Box Fx$. To do this, we have introduced a second mode of predication and incorporated it into our logic. If property F is constitutive of, or intrinsic to, abstract object x, we say that *x encodes F*. We shall say that the number two encodes rather than exemplifies the property of being the successor of 1, for this is one of its constitutive number-theoretic properties. Similarly, since the property of *being such that Clinton is married* is one of the properties we use to differentiate our world from other possible worlds, this property will be one that is encoded by the actual world. And we shall suppose that Sherlock Holmes encodes rather than exemplifies the property of being a detective, for this is one of the properties attributed to him in the novels. Notice that our distinction allows us to explain the following asymmetry between Socrates and the unit set of Socrates: the property of having Socrates as an element is constitutive of the unit set of Socrates, but it is not constitutive of Socrates that he is an element of that unit set. To explain this fact, we say that the unit set of Socrates encodes the property of having Socrates as an element, but that Socrates is an ordinary object and so does not encode properties — he is not therefore 'constituted' by the property of being an element of his unit set.

We incorporate the notion of encoding into our logic by introducing an atomic formula 'xF' to express: x encodes F. We shall assume that F is a 1-place relation term and that encoding is a monadic form of predication. The second-order modal object calculus can now be described simply as a second-order modal predicate calculus that is based on both atomic exemplification and atomic encoding formulas. In what follows, we shall include complex n-place relation terms in our presentation of the calculus. With the aid of metavariables o (ranging over all object terms), ν (ranging over object variables), ρ (ranging over relation terms), α (ranging over all variables), τ (ranging over all terms), and φ, ψ (ranging over all formulas), the calculus is defined as follows:

(a) Two kinds of primitive terms: (1) object terms (constants a_1, a_2, ..., and variables x_1, x_2, ...), and (2) n-place relation terms (predicates P_1^n, P_2^n, ..., and variables F_1^n, F_2^n, ...), for $n \geq 0$.

(b) Two kinds of atomic formulas: (1) exemplification formulas of the form $\rho^n o_1 \ldots o_n$ ($n \geq 0$), and (2) encoding formulas of the form $o\rho^1$. An exemplification formula such as $F^n x_1 \ldots x_n$ should be read: objects $x_1, \ldots, x_n$ *exemplify* relation F^n. An encoding formula such as xF^1 should be read: object x *encodes* property F^1.

(c) Molecular, quantified, and modal formulas of the form $\neg\varphi$ ('it is not the case that φ'), $\varphi \rightarrow \psi$ ('if φ then ψ'), $\forall\alpha\varphi$ ('every α is such that φ'), and $\Box\varphi$ ('necessarily φ'), where α is any object or relation variable.

(d) Complex n-place relation terms ($n \geq 0$) of the form: $[\lambda\nu_1 \ldots \nu_n \varphi]$, where φ has neither encoding subformulas nor quantifiers binding relation variables, and none of the object variables ν_1, ..., ν_n in the term $[\lambda\nu_1 \ldots \nu_n\, \varphi]$ appear free in any other λ-expression occurring in φ. A λ-predicate such as $[\lambda y_1 \ldots y_n\, \varphi]$ should be read: being objects $y_1, \ldots, y_n$ such that φ.

Intuitively, the object constants denote, and object variables range over, elements of a primitive domain of objects; similarly, the predicates denote, and the predicate variables range over, elements of a primitive domain of relations. In addition to calling the 1-place relations *properties*, we call the 0-place relations *proposition*. The n-place λ-expressions ($n \geq 1$) allow us to name complex relations, and 0-place λ-expressions of the form $[\lambda\, \varphi]$ allow us to name complex propositions. Of the three restrictions on the formation of λ-expressions, only the first is really necessary—encoding formulas are banished from λ-expressions in order to avoid paradox when the comprehension schema for abstract objects is added to the system.[2] The other two restrictions are for convenience—λ-expressions may not have relation quantifiers (such as $[\lambda x\ \exists F Fx]$) or free variables in nested λ-expressions (such as $[\lambda x\ [\lambda y\ Ryx]a]$). This allows us to simplify the presentation of the interpretations of the calculus.[3] The restrictions on the formation of λ-expressions simply guarantee that only the familiar first-order definable relations and propositions will be found in the system. The main difference, then, between the languages of the object calculus and predicate calculus is the presence of a second kind of atomic formula in the former. In what follows, we employ the usual conventions for introducing formulas containing & (and), $\vee$ (or), $\equiv$ (iff), $\exists$ (some), and $\Diamond$ (possibly).

If we add to this language the primitive predicate '$E!$' to denote the property of being spatio-temporally located, we can formally define the notions of *ordinary* and *abstract* object and assert things about them:

x is an *ordinary* object ('$O!x$') $=_{df}$ $[\lambda y\ \Diamond E!y]x$

x is an *abstract* object ('$A!x$') $=_{df}$ $[\lambda y\ \neg\Diamond E!y]x$

Intuitively, these definitions tell us that ordinary objects have a location in spacetime in some possible world, whereas abstract objects are not the kind of thing that could have a location in spacetime. The usual conception of ordinary objects is captured by the following two principles: (A) ordinary objects x and y are identical iff they necessarily exemplify the same properties, and (B) ordinary objects are not the kind of thing that could encode properties:

(A) $O!x \,\&\, O!y \rightarrow (x{=}y \equiv \Box\forall F(Fx \equiv Fy))$

(B) $O!x \rightarrow \Box\neg\exists F xF$

By contrast, the two basic principles governing abstract objects are:

(C) $\exists x(A!x \,\&\, \forall F(xF \equiv \varphi))$, where x is not free in φ

(D) $A!x \,\&\, A!y \rightarrow (x{=}y \equiv \Box\forall F(xF \equiv yF))$

Principle C is a comprehension principle for abstract objects; it guarantees that for every condition φ on properties F, there is an abstract individual that *encodes* exactly the properties satisfying the condition. If we think model-theoretically for

[2] See [Zalta, 1983, pages 158–60].

[3] One can develop interpretations in which these restrictions are removed. To remove the 'no relations quantifiers' restriction, one would have to extend the algebraic semantics developed in Section 3 to higher order operations such as those developed in [Došen, 1988] or those utilized in the type-theoretic version of the calculus. To remove the 'no free nested λ-variables' restriction, one would have to incorporate into the algebraic semantics a *relativized* predication operation of the kind used in [Bealer, 1982] and [Menzel, 1986].

the moment, and allow ourselves talk about sets, then this principle correlates an abstract object with every (expressible) set of properties. Principle D is the identity principle for abstract objects; it says that abstract objects x and y are identical iff they necessarily encode the same properties. It is a simple consequence of Principles C and D that for every condition on properties φ, there is a *unique* abstract object that encodes just the properties satisfying the condition; there couldn't be two distinct abstract objects encoding exactly the properties satisfying a given condition if distinct abstract objects have to differ by at least one encoded property.

Principle C asserts the existence of a wide variety of abstract objects, some of which are complete with respect to the properties they encode, while others are incomplete in this respect. For example, one instance of Principle C asserts there exists an abstract object that encodes just the properties Gorbachev exemplifies. This object is complete because Gorbachev either exemplifies F or exemplifies the negation of F, for every property F. Another instance of comprehension asserts that there is an abstract object that encodes just the three properties: *being golden*, *being a mountain*, and *having a spatio-temporal location* ($E!$). This object is incomplete because for every *other* property F, it encodes neither F nor the negation of F.[4] But though abstract objects may be partial with respect to their encoded properties, they are all complete with respect to the properties they *exemplify*. In other words, the following principle of classical logic is preserved: for every object x and property F, either x exemplifies F or x exemplifies the negation of F. We can express this formally if we use our λ-notation to define the negation of F ('$\bar{F}$') as: $[\lambda y\, \neg Fy]$. So we preserve the following formal principle of classical logic: $Fx \vee \bar{F}x$.[5]

Since xF does not entail Fx, abstract objects may encode incompatible properties without contradiction, for incompatible properties are defined as properties that couldn't be *exemplified* by the same objects. The following, for example, are jointly consistent: x encodes roundness (xR), x encodes squareness (xS), and necessarily everything that *exemplifies* being round fails to *exemplify* being square ($\Box\forall y(Ry \rightarrow \neg Sy)$). Thus, the notorious 'round square' may simply be the abstract object that encodes just being round and being square. Since there are no restrictions on the comprehension principle, it guarantees that no matter what properties one brings to mind to conceive of a thing, there is something that encodes just the properties involved in that conception.

Given that they they necessarily fail to have a location in spacetime, abstract objects necessarily fail to exemplify the properties of ordinary objects that entail spatio-temporality. For example, they necessarily fail to have a shape, they necessarily fail to be subject to the laws of generation and decay, they necessarily fail to be buildings, people, planets, etc. Consequently, by the classical laws of complex

[4] Notice that this object is consistent with the contingent fact that nothing *exemplifies* all three properties.

[5] Note that encoding satisfies classical bivalence: $\forall F \forall x(xF \vee \neg xF)$. But the incompleteness of abstract objects is captured by the fact that the following is not in general true: $xF \vee x\bar{F}$.

properties, abstract objects necessarily exemplify the negations of these properties. But notice that the properties abstract objects encode are more important than the properties they necessarily exemplify, since the former are the ones by which we individuate them.[6] And it is important to mention that abstract objects may contingently exemplify certain relations to ordinary objects, such as being sought by y, being worshiped by z, inspiring u to action, etc.

Principles A–D are part of the proper theory of (abstract) objects. In this paper, however, we shall be more concerned with the underlying modal object calculus. In addition to being a vehicle for expressing the basic principles about ordinary and abstract objects, the calculus offers a framework for asserting the basic existence and identity principles governing n-place relations ($n \geq 0$). We shall treat these basic existence and identity principles of relations as a part of the calculus, for we conceive of these entities as logical objects. So the calculus consists of the language described above together with the principles of classical quantified modal logic, the logic of encoding, and the logic of relations. In the next section, we shall sketch some of the applications of this logical framework, for even without a commitment to Principle C, the system can be used to define a wide range of important philosophical notions and prove basic logical facts concerning them. We conclude this section, however, with a more precise specification of the logical portion of the calculus.

The logic consists of: (a) the axioms and rules of classical propositional logic, (b) the axioms and rules for classical quantification theory, (c) the axioms and rules for the classical modal logic $S5$, (d) an axiom for the logic of encoding, and (e) axioms and definitions for the logic of relations. This logical basis can be explicitly identified as follows:

(a) Classical Propositional Logic:

Axioms: Tautologies of Propositional Logic

Rule of Modus Ponens (MP): if $\vdash \varphi$ and $\vdash \varphi \rightarrow \psi$, then $\vdash \psi$

(b) Classical Quantification Theory:[7]

Axioms: $\forall\alpha\varphi \rightarrow \varphi^{\tau}_{\alpha}$, provided τ substitutable for α

Axioms: $\forall\alpha(\varphi \rightarrow \psi) \rightarrow (\varphi \rightarrow \forall\alpha\psi)$, provided α not free in φ

Rule of Generalization (Gen): if $\vdash \varphi$, then $\vdash \forall\alpha\varphi$

(c) Classical Modal Logic $S5$:

[6] In the logic that we shall describe shortly, it will be a logical axiom that $\Diamond xF \rightarrow \Box xF$. So properties that are encoded by an abstract object at some world are encoded by that object at every world. This also helps to capture the idea that encoded properties are intrinsic to abstract objects.

[7] In these axioms, τ may be any object term or relation term, and α any object variable or relation variable, with the proviso that for τ to be substitutable for α, they must both be object terms or both relation terms.

Axioms: $\Box(\varphi \rightarrow \psi) \rightarrow (\Box\varphi \rightarrow \Box\psi)$

Axioms: $\Box\varphi \rightarrow \varphi$

Axioms: $\Diamond\varphi \rightarrow \Box\Diamond\varphi$

Rule of Necessitation (RN): if $\vdash \varphi$, then $\vdash \Box\varphi$

(d) Logic of Encoding: $\Diamond xF \rightarrow \Box xF$

(e) Logic of Relations:[8]

λ-Conversion: $[\lambda y_1 \ldots y_n\, \varphi]x_1 \ldots x_n \equiv \varphi^{x_1,\ldots,x_n}_{y_1,\ldots,y_n}$, where x_i is substitutable for y_i in φ $(1 \leq i \leq n)$.

Property Identity: $F^1 = G^1 =_{df} \Box\forall x(xF \equiv xG)$

Relation Identity: $F^n = G^n =_{df}$ (for $n > 1$) $(\forall x_1)\ldots(\forall x_{n-1})$:
$[\lambda y\, F^n y x_1 \ldots x_{n-1}] = [\lambda y\, G^n y x_1 \ldots x_{n-1}]$ &
$[\lambda y\, F^n x_1 y x_2 \ldots x_{n-1}] = [\lambda y\, G^n x_1 y x_2 \ldots x_{n-1}]$ & ... &
$[\lambda y\, F^n x_1 \ldots x_{n-1} y] = [\lambda y\, G^n x_1 \ldots x_{n-1} y]$

Proposition Identity: $F^0 = G^0 =_{df} [\lambda y\, F^0] = [\lambda y\, G^0]$

λ-Identity$_1$: $[\lambda y_1 \ldots y_n\, F^n y_1 \ldots y_n] = F^n$

λ-Identity$_2$: $[\lambda y_1 \ldots y_n\, \varphi] = [\lambda y'_1 \ldots y'_n\, \varphi']$ (alphabetic variants)

This completes the definition of the calculus. With the exception of the logic of encoding and the logic of relations, the deductive apparatus should be familiar. These additional groups of axioms will be discussed in the next section.

2 APPLICATIONS OF THE CALCULUS

In previous work, we have not distinguished the results that require an appeal to Principle C from those that are derivable solely as logical theorems of the modal object calculus. In this section, we distinguish the latter for those who may be hesitant about committing themselves to a proper metaphysical theory. We discuss the following topics: relations, situations and possible worlds, modality and the Barcan formula, and the distinction between fact and fiction.

[8] To make the notation easier to read, we state the axioms and definitions using $x, x_1, \ldots$ and $y, y_1, \ldots$ as typical variables in place of the metavariables $\nu_1, \nu_2, \ldots$. The notation $\varphi^{x_1,\ldots,x_n}_{y_1,\ldots,y_n}$ stands for the result of replacing, respectively, x_i for y_i in φ, and the requirement that x_i be substitutable for y_i guarantees that x_i will not be 'captured' by a quantifier when the substitution is carried out.

Relations

From the logic of relations we may derive a precise theory of relations. The cornerstone of the theory is the comprehension principle for relations that follows from the λ-Conversion principle by applications of (Universal) Generalization, the Rule of Necessitation, and the derived rule of Existential Generalization:

> $\exists F^n \Box \forall x_1 \ldots \forall x_n (F^n x_1 \ldots x_n \equiv \varphi)$, provided φ has no free F^ns, no encoding formulas, no relation quantifiers, and no free occurrences of the x_i within λ-expressions in φ.

When $n = 1$ and $n = 0$, this becomes a comprehension principle for properties and propositions, respectively. Using variables $p, q, \ldots$ to go proxy for $F^0, G^0, \ldots$, we may formulate the comprehension schema for propositions as follows:

> $\exists p \Box (p \equiv \varphi)$, provided φ has no free ps, no encoding formulas, and no relation quantifiers.

These comprehension principles yield a wide range of complex properties, relations, and propositions, and we presume familiarity with the typical examples of such.

The definitions that stipulate when relations F and G are identical complete the theory. These definitions were stated as part of the logic of relations, and there are three cases to consider: when $n = 1$, $n > 1$, and $n = 0$. Look again at the definition of property identity ($n = 1$): properties F and G are identical iff necessarily, they are encoded by the same objects. It is useful to know at this point that in the semantics for our language, every property receives two extensions—an exemplification extension and an encoding extension. Thus, properties can be logically equivalent in one of two ways: by having the same exemplification extension at every possible world or by having the same encoding extension at every possible world. We can express this in our language as two ways in which properties F and G can be *necessarily* equivalent: $\Box \forall x (Fx \equiv Gx)$ and $\Box \forall x (xF \equiv xG)$. Property identity is equated with the latter. So properties F and G can be 'distinct' even though they are necessarily equivalent in the traditional sense. By stipulating that properties having the same encoding extension throughout a fixed domain of worlds are 'identical', we offer an extensional theory of intensional entities. In the next section, it becomes clear that whereas the exemplification extension of a property may vary from world to world, its encoding extension does not. The fact that the encoding extension of a property is fixed across possible worlds explains the Axiom of Encoding—if an object is in the encoding extension of a property at some world, it is in the encoding extension of that property at every world. Thus, if F and G have the same encoding extension at one world, they have the same encoding extension at all worlds, and so to prove $F = G$, it suffices to prove

$\forall x(xF \equiv xG)$.[9]

The definition of property identity holds the key to the definition of relation and proposition identity. Relations F^n and G^n, for $n > 1$, are defined to be identical just in case, intuitively, no matter how you plug $n - 1$ objects into F^n and G^n (provided you plug them up in the same order), the resulting properties are always identical. Again, this definition of relation identity allows us to assert that $F^n \neq G^n$ even though $\Box\forall x_1 \ldots \forall x_n(F^n x_1 \ldots x_n \equiv G^n x_1 \ldots x_n)$. Necessarily equivalent relations, therefore, may be distinct.

If we continue to use the variables p and q as substitutes for F^0 and G^0, then the definition of proposition identity tells us that propositions p and q are identical iff the properties $[\lambda y\ p]$ and $[\lambda y\ q]$ are identical. This reduces the identity of propositions to the identity of the 'propositional properties' one can construct in terms of them. And this, in turn, is a matter of encoding. Since there is no logical connection between encoding and exemplifying a property, one can assert $p \neq q$ even though $\Box(p \equiv q)$. So necessarily equivalent propositions may be distinct.

Situations and Worlds

This analysis of relations provides the foundation for the theories of situations and worlds. To define the basic notions of situation theory, we first define truth for propositions by elimination: a proposition p *is true* iff p. We say that a *propositional* property is any property F such that for some proposition p, F is the property of being such that p; i.e.,

$$\mathit{Propositional}(F) =_{df} \exists p(F = [\lambda y\ p])$$

Then, we say that a *situation* is any (abstract) object x that encodes only propositional properties; i.e., x is a *situation* iff x is abstract and for every F, if xF, then F is propositional. So each situation encodes a group of propositional properties. The propositions p encoded in a situation via $[\lambda y\ p]$ are the ones true in that situation. More specifically, where s is a variable ranging over situations, we can define the idea that p is *true in s* as: s encodes $[\lambda y\ p]$ (i.e., $s[\lambda y\ p]$). In what follows, we represent the notion that p is true in s more picturesquely as: $s \models p$. We also say that a situation s is *part-of* situation s' iff every proposition true in s is true in s'. An *actual* situation s is one such that every proposition true in s is true (simpliciter), and a situation is *possible* iff it is possible that it is actual. A situation

[9] We employ the modality in the definition of $F = G$ because, from a philosophical point of view, identity is a modal notion. If we think model-theoretically in terms of a primitive notion of identity on properties and assume Principles C and D, then we can see why properties encoded by the same objects would be the same. For suppose not, i.e., suppose F and G are properties that are encoded by the same objects, but that F and G are distinct. If F and G are distinct, there are sets of properties containing F and not G (and vice versa). But Principles C and D ensure that there is a distinct abstract object for each set of properties. So there will be lots of distinct abstract objects encoding F and not G (and vice versa), contradicting the assumption that F and G are encoded by the same objects.

s is *consistent* iff there are no propositions p and q such that (a) the conjunction of p and q is impossible, and (b) p and q are both true in s.

In [Zalta, 1993a], we have shown that from these definitions, one can derive a rather large group of basic theorems in situation theory. Though some of these theorems appeal to Principle C, most are simply logical theorems of the modal object calculus. Among the theorems that are simple consequences of the above definitions we find: that situations s and s' are identical iff the same propositions are true in both; that every part of a situation is a situation; that a situation s is a part of situation s' iff every proposition true in s is true in s'; that situations s and s' are identical iff each is part of the the other; that part-of is reflexive, antisymmetric, and transitive on the situations (should there be any); that no proposition and its negation are both true in any actual situation; that some propositions are not true in any actual situation; that if p is true in actual situation s, then s exemplifies $[\lambda y\ p]$, and that all possible situations are consistent.

The basic notions of world theory can also be defined. A *world* is a situation s such that it is possible that all and only true propositions p are true in s; i.e.,

$$World(s) =_{df} \Diamond\forall p(s \models p \equiv p)$$

Truth at a world ($w \models p$) may therefore be defined as the same notion as truth in a situation. It is provable that worlds w are maximal in the sense that for every proposition p, either $w \models p$ or $w \models \neg p$. It also follows that all worlds are possible and consistent, and that all the necessary consequences of propositions true at a world are also true at that world. By combining the definitions of 'world' and 'actual', we obtain a notion of an actual world, and though Principle C is needed to prove the existence of such an object, it is not needed to to prove that if there is one, there is a unique one. Under the assumption that there is an actual world, say w_α, it also follows that w_α is nonwellfounded in the following sense: all the facts about w_α are *true in* w_α; i.e., Fw_α iff $w_\alpha \models Fw_\alpha$, for any property F. So, in particular, if w_α exists, it exemplifies exactly the propositional properties that it encodes.

Modality and the Barcan Formula

Notice that the modal object calculus employs the simplest quantified modal logic. This is the modal logic that results from combining classical quantification theory with the axioms and rules of S5. The Barcan formula and the converse Barcan formula are both derivable in such a logic:[10]

(BF) $\forall x\Box\varphi \to \Box\forall x\varphi$

(CBF) $\Box\forall x\varphi \to \forall x\Box\varphi$

[10]The second order Barcan Formulas are also derivable, but we shall not discuss these formulas in what follows.

We know from the study of modal logic that in such a simple system, the quantifier $\forall x$ ranges over a single, fixed domain of objects. The reader will be able to verify this in the next section. But for now, it is useful to point out that by having a single fixed domain instead of the variable domains used in [Kripke, 1963], we don't have to address the question of whether an object in one domain is the same as an object in another domain. That is, the problem of trans-world identification does not arise.

We conclude this subsection with a brief account of the nature of possible objects. Recall the equivalent formulation of the Barcan formula:

(BF) $\Diamond\exists xFx \rightarrow \exists x\Diamond Fx$

To take a simple example, consider the claim that person x (who doesn't have a brother) might have had a brother. We might represent this claim as: $\Diamond\exists yByx$. From the Barcan formula it then follows that $\exists y\Diamond Byx$. So our logic, in conjunction with possibility claims, entails that there is something which possibly is x's brother. Any such object will be treated as an *ordinary* object which isn't spatio-temporally located but which might have been. That is because the relation of brotherhood is such that necessarily, any two things exemplifying this relation both exemplify $E!$. That is, our logic is consistent with the following non-logical principle:

$\Box\forall x\forall y(Bxy \rightarrow E!x \;\&\; E!y)$

Since any object that possibly is x's brother will exemplify $E!$ in any world where it is x's brother, we know that it possibly exemplifies $E!$, and so by definition, it is an ordinary object. It is important to stress that such an object is not x's brother (recall that x doesn't have a brother). We may consistently suppose that any ordinary object that might have been x's brother exemplifies many of the same properties that abstract objects exemplify, namely, the *negations* of such properties as being spatio-temporally located, having a shape, having a texture, being a building, being a person, etc. However, such ordinary objects exemplify certain modal properties that *bona fide* abstract objects lack (should there be any). An abstract object, by definition, couldn't possibly exemplify $E!$ and so given the non-logical principle displayed above, couldn't possibly be someone's brother. These consequences indicate the range of judgments we can consistently add to our calculus in connection with common sense claims such as 'x might have had a brother'.

Fact and Fiction

Finally, the distinction between fact and fiction can be analyzed in the calculus. Using a primitive 2-place relation of authorship, we may say that a *story* is any situation s authored by some ordinary object. Since stories are identified as situations, one can define 'According to the story s, p' as $s \models p$. Intuitively, the true

English sentences of the form 'According to s, p' constitute the data of fiction. We presuppose that for each story s, there is a group of propositions that satisfy the sentence 'According to s, p'. Just as one cannot drop the story operator 'According to s' and preserve truth, one cannot validly move from $s \models p$ to p. Notice that stories, unlike worlds, need not be maximal nor even consistent. Of course, two people may disagree about which propositions p are true in a given story, but that doesn't matter because both persons could nevertheless accept the identification of the story s with the abstract object that encodes all the properties of the form $[\lambda y\ p]$ such that p is a proposition true according to the story. That is, despite disagreement about exactly which propositions are true in a story, our analysis of stories is a precise philosophical characterization of what stories are *in principle*. Finally, a *character* x of story s can be analyzed as any object x such that for some F, $s \models Fx$. Not all characters of a story are fictional. We allow that real objects such as London, the Prince of Wales, the planet Jupiter, etc., can appear as characters in a story. But the fictional characters are the ones that 'originate' in the story. If the logic of fiction can be simplified by identifying such 'native characters' as abstract objects, then we have a reason for accepting Principle C. But without this principle, a definite description like 'the abstract object that encodes just the properties F that, according to story s, x exemplifies F' ('$\imath x(A!x \,\&\, \forall F(xF \equiv s \models Fx))$') is not well-defined.

This brief review of the applications of the modal object calculus should have drawn attention to one point, namely, that even if one has metaphysical scruples about accepting Principle C, the modal object calculus can still serve as a vehicle for defining interesting philosophical notions and proving basic facts about them. Though the calculus has other applications, we shall not describe these in any detail here.

3 THE INTENDED INTERPRETATION OF THE CALCULUS

Before we describe the new interpretation of the calculus (in Section 4), we must first review the intended interpretation. However, in the years since the intended interpretation of the calculus was first developed, Etchemendy [1990] has developed a more critical view of Tarski's formal definition of truth and logical consequence. To address some of Etchemendy's concerns, we shall recast our original semantic definitions to make it clear that the various models of our language do not constitute different interpretations of the constants and predicates. This results in a more perspicuous definition of truth and logical truth. To this end, we take the domains required for the interpretation of the terms, quantifiers, and modal operators *out* of the models! The domain of objects, the domain of relations, and the domain of possible worlds shall be grouped together and specified antecedently as part of an interpretation of the language. So an *interpretation* of the language will be distinguished from the *models* that can be defined for that interpretation. This way, the domains of the interpretation will not vary from model to model. The models for an interpretation **I** simply represent the various ways that the objects of **I** can be assigned to the extensions of the relations of **I** (at the various possible worlds of **I**). On this picture, the only thing that varies from model to model is: (a) the function that indicates which objects exemplify which relations at each world, and (b) the function that indicates which objects encode which properties.

On this picture of modal language, the notion of 'truth at a world in a model' is defined for an an *interpreted* sentence. We do *not* evaluate a purely formal sentence at world **w** in model **M** by considering the extension of the terms and predicates at world **w** in **M**. Instead, we consider an interpretation **I** of the sentence and find out what objects and relations the terms and predicates of the sentence denote with respect to **I**. Then once we have these, we may ask, with respect to a world **w** in the interpretation **I**, whether model **M** structures the relations and objects in the appropriate way at **w**. In other words, we interpret formulas first and then discover whether they are true. Thus, truth (and logical truth) will be defined relative to a fixed interpretation of the formulas.[11]

Consider, as an example, the formula 'Pa'. Before we can decide whether it is true, we have to know what it means. So let us fix an interpretation **I** by supposing that 'a' denotes Socrates (a member of the domain of objects in **I**), that 'P' denotes the property of being snub-nosed (a member of the domain of properties in **I**), and that the domain of possible worlds in **I** contains three possible worlds $\mathbf{w}_1$, $\mathbf{w}_2$, and a distinguished actual world $\mathbf{w}_\alpha$. Now a given model **M** for this interpretation will tell us at which worlds Socrates is an element of the exemplification extension of the property of being snub-nosed, for **M** will be defined, in part, by a function that assigns ordered sets of **I**-objects as the exemplification extensions of the **I**-relations at each **I**-world. The formula 'Pa' under interpretation **I** will be true at world **w** in model **M** iff **M** assigns Socrates to the exemplification extension of the property of being snub-nosed at world **w**. The formula 'Pa' under interpretation **I** will be true (*simpliciter*) in a model **M** iff **M** assigns Socrates to the exemplification extension of the property of being snub-nosed at the distinguished world $\mathbf{w}_\alpha$.

Now consider the modal formula '$\Box Pa$' under interpretation **I**. This interpreted formula will be true at world **w** in model **M** iff **M** assigns Socrates to the exemplification extension of the property of being snub-nosed at all the worlds accessible from **w** (suppose that **I** also specifies the accessibility relation on worlds). It will be true (*simpliciter*) in model **M** iff **M** assigns Socrates to the exemplification extension of the property of being snub-nosed at all the worlds accessible from the actual world $\mathbf{w}_\alpha$. Since we have adopted an S5 modal logic, all worlds are accessible to each other and so we shall simplify our semantics by eliminating the accessibility relation altogether.

Since extensions are *not* directly assigned to the predicates of the language, our semantical treatment differs from Tarski [1936, 1944], though we shall still employ

[11]For an extended discussion motivating this conception of modal logic, see [Zalta, 1993b].

his notion of satisfaction (see below). Nor do we assign extensions to predicates *relative* to worlds, and so our treatment also differs from that of [Kripke, 1959, 1963; Hintikka, 1961]. In general, we do not use the method of intension and extension developed by Carnap [1947] and extended by Montague [1974]. That is, we do not recover an 'intension' function from the assignment of extensions to constants, predicates, and sentences at each possible world. Instead, we take relations as primitive, fine-grained entities and in the course of defining models, something analogous to a Montagovian 'intension' function is used to indicate what extensions these primitive relations have at various worlds. This semantic method, therefore, takes up Frege's [1892] idea that a predicate denotes a 'concept' and that an 'extension' is something associated, in the first instance, with a concept rather than with a predicate.

These explanatory remarks should help motivate the following formal definitions of an interpretation, intended interpretation, model, assignments and denotation, satisfaction, truth, and logical truth.

Interpretations

An interpretation of the modal object calculus is a quadruple $\mathbf{I} = \langle \mathbf{W}, \mathbf{D}, \mathbf{R}, \mathbf{F} \rangle$, the members of which are as follows:

1. $\mathbf{W}$ is the nonempty set of *possible worlds* with a distinguished *actual world* $\mathbf{w}_\alpha$ (the S5 modal operators will be construed as quantifiers over $\mathbf{W}$ in the simplest possible way, namely, without an accessibility relation).

2. $\mathbf{D}$ is the nonempty domain of *objects* (which includes both abstract and ordinary objects).

3. $\mathbf{R}$ is the domain of primitive *relations* (defined as the union of a sequence of nonempty, pairwise disjoint sets $\mathbf{R}_0, \mathbf{R}_1, \ldots$) which is closed under the following algebraic logical operations (which are discussed below): $\mathbf{PLUG}_i$ (plug-into-the-i^{th}-place), **NEG** (negate), **NEC** (necessitate), **COND** (conditionalize), $\mathbf{UNIV}_i$ (universalize-on-the-i^{th}-place), $\mathbf{REFL}_{i,j}$ (reflect-the-i^{th}-and-j^{th}-places), $\mathbf{CONV}_{i,j}$ (convert-the-i^{th}-and-j^{th}-places), and $\mathbf{VAC}_i$ (vacuously-expand-the-i^{th}-place).

4. $\mathbf{F}$ is an interpretation function that maps each object constant of the language to an element of $\mathbf{D}$, and maps each n-place relation constant of the language to an element of $\mathbf{R}_n$.[12]

To complete this definition, we must define the algebraic operations in clause 3.

[12]This function is not yet a full-blooded denotation function because it is not defined on the complex λ-expressions. In what follows, a denotation function, relative to $\mathbf{I}$ and a variable assignment $\mathbf{f}$, will *extend* $\mathbf{F}$ to all terms by assigning a denotation to the variables and to the complex terms.

Intuitively, the algebraic operations harness relations into complex and structured relations that serve as the denotations of the λ-expressions. These operations are analogues of Quine's predicate functors in [1960], but whereas Quine's functors operate on predicates to eliminate variables, our operations are defined on the relations denoted by the predicates. Here are some examples:

1. The operation $\mathbf{PLUG}_1$ takes the 3-place relation $\mathbf{r}$ (in $\mathbf{R}_3$) and an object $\mathbf{o}$ (in $\mathbf{D}$) and plugs $\mathbf{o}$ into the first place of $\mathbf{r}$ to produce the complex relation 2-place $\mathbf{PLUG}_1(\mathbf{r}, \mathbf{o})$ (in $\mathbf{R}_2$).[13] If $\mathbf{r}$ is the denotation of the 3-place predicate R, and $\mathbf{o}$ is the denotation of the constant a, then $\mathbf{PLUG}_1(\mathbf{r}, \mathbf{o})$ serves as the denotation of the 2-place λ-expression $[\lambda xy\ Raxy]$. $\mathbf{PLUG}_3(\mathbf{r}, \mathbf{o})$, on the other hand, serves as the denotation of $[\lambda xy\ Rxya]$. $\mathbf{PLUG}_3$ is not defined for relations having fewer than three places, whereas $\mathbf{PLUG}_1$ is defined for all except the 0-place relations.

2. The operation $\mathbf{NEG}$ produces an n-place relation that is the negation of a given an n-place relation. So, building on a previous example, the 2-place relation $\mathbf{NEG}(\mathbf{PLUG}_1(\mathbf{r}, \mathbf{o}))$ serves as the denotation of the predicate $\lambda xy\ \neg Raxy]$.

3. The operation $\mathbf{UNIV}_2$ maps the 3-place relation $\mathbf{r}$ to the 2-place relation $\mathbf{UNIV}_2(\mathbf{r})$, which serves as the denotation of the 2-place expression

$$[\lambda xy\ \forall z Rxzy].$$

These examples should give the reader some indication of how the operations work.

We shall not reproduce the definitions of the algebraic operations here, for they have been characterized precisely in such places as [Zalta, 1983, 1988; Bealer, 1982; Menzel, 1986]. But it should be mentioned that various constraints must be placed on the (exemplification) extension functions and denotation functions for the algebraic operations to work properly. To see why, recall that each model will assign exemplification extensions to the relations (at possible worlds) in different ways. However, in each model, the extensions of the complex relations produced by the algebraic operations must mesh properly with the extensions of the relations they may have as parts, no matter how the latter extensions are assigned. For example (from the previous paragraph), these constraints ensure that the exemplification extension of the complex relation $\mathbf{PLUG}_1(\mathbf{r}, \mathbf{o})$ at world $\mathbf{w}$ consists of all those pairs $\langle \mathbf{o}_1, \mathbf{o}_2 \rangle$ such that the triple $\langle \mathbf{o}, \mathbf{o}_1, \mathbf{o}_2 \rangle$ is in the exemplification extension of the relation $\mathbf{r}$ at $\mathbf{w}$. The constraints also ensure that the exemplification extension of the complex relation $\mathbf{NEG}(\mathbf{PLUG}_1(\mathbf{r}, \mathbf{o}))$ at world $\mathbf{w}$ consists of

[13]Quine has no need for the **PLUG** operation in his [1960] because he eliminates constants in favor of predicates.

all those pairs $\langle \mathbf{o}_1, \mathbf{o}_2 \rangle$ that fail to be in the exemplification extension of the relation $\mathbf{PLUG}_1(\mathbf{r}, \mathbf{o})$ at $\mathbf{w}$. An appropriate constraint is therefore defined for each algebraic function, and as a group, the constraints guarantee that the interpreted instances of the logical axiom λ-Conversion are true in every model, since the exemplification extension functions of every model will be constrained in these ways.

There is one other subtlety to the logic of the algebraic operations. This concerns the fact that they 'overgenerate' denotations. For example, if we let the constant 'b' denote object $\mathbf{o}'$, then both relations $\mathbf{PLUG}_2(\mathbf{PLUG}_3(\mathbf{r}, \mathbf{o}), \mathbf{o}')$ and $\mathbf{PLUG}_2(\mathbf{PLUG}_2(\mathbf{r}, \mathbf{o}'), \mathbf{o})$ could equally well serve as the denotation of $[\lambda x\ Rxba]$. So we need some means of ensuring that the predicate $[\lambda x\ Rxba]$ receives a unique denotation, i.e., that the denotation function is well-defined on the λ-expressions. In our previous work, we have partitioned the λ-expressions into syntactic categories that correspond with the algebraic logical operations. Then the denotation function, which maps each term of the language to an entity in the relevant domain, is defined so that it maps each λ-expression to an appropriately structured relation based on the expression's syntactic category. This yields a mechanical procedure that selects a unique denotation for the λ-expression. We'll say more about this when we turn to the precise definition of denotation in what follows.

The Intended Interpretation

The *intended interpretation* of our calculus will simply be that interpretation such that: (a) the domains $\mathbf{W}$, $\mathbf{D}$, and $\mathbf{R}$ contain, respectively, all the possible worlds there in fact are, all the relations that there are, and all the objects that there are, and (b) whenever any constant (predicate) is treated as an abbreviation of an English name (predicate), the function $\mathbf{F}$ assigns the object (relation) denoted by the English name (predicate) as the denotation of the constant (predicate). In the remainder of this section, we suppose that the intended interpretation is fixed and we shall simply call it '$\mathbf{I}$'. The interpretation function $\mathbf{F}$ of the intended interpretation will be referred to as $\mathbf{F_I}$.

Models

Before we actually state the definition of a model, let us informally anticipate the definition with some examples, to see how the definitions of truth and logical truth for our *interpreted* language will eventually work. A model $\mathbf{M}$ for the intended interpretation $\mathbf{I}$ will be defined so that: (a) all of the relations in $\mathbf{R}$ receive *exemplification* extensions at each possible world, and (b) all of the properties in $\mathbf{R}_1$ receive, in addition, an *encoding* extension. The definition of truth will then tell us that the formula 'Pa' under the interpretation $\mathbf{I}$ is true at world $\mathbf{w}$ relative to $\mathbf{M}$ iff $\mathbf{F_I}(a)$ is an element of the $\mathbf{M}$-exemplification extension at $\mathbf{w}$ of the property

$\mathbf{F_I}(P)$. And the formula 'aP' under **I** is true at **w** relative to **M** iff $\mathbf{F_I}(a)$ is a member of the **M**-encoding extension of $\mathbf{F_I}(P)$. Of course, the definition of truth won't be stated in terms of $\mathbf{F_I}$, since that function only interprets the primitive constants and predicates. We shall need to define a full-blooded denotation function that extends $\mathbf{F_I}$ and assigns a denotation to every term in the language, including the variables and the complex expressions.

The definition of logical truth will be cast in terms of the definition of truth: a formula φ under the interpretation **I** is logically true iff φ is true at the actual world in every model for **I**. For example, '$Pa \vee \neg Pa$' (under **I**) will be a logical truth, *not* because it turns out true under every interpretation of 'a' and 'P', but rather because it is true (at $\mathbf{w}_\alpha$) no matter how models for **I** assign (exemplification) extensions to $\mathbf{F_I}(P)$. To make this more vivid, suppose the object constant 'a' abbreviates the proper name 'Bill Clinton' and that $\mathbf{F_I}(a)$ is the man Bill Clinton (who is an element of **D**). And assume that the primitive relation constant 'P' abbreviates the property-name 'being a U. S. president' and that $\mathbf{F_I}(P)$ is the property of being a U. S. president (this is an element of $\mathbf{R}_1$). Then the sentence 'Pa' under the interpretation **I** *asserts* that Clinton exemplifies being a U. S. president, and the reason '$Pa \vee \neg Pa$' (under **I**) is a logical truth is that it remains true at $\mathbf{w}_\alpha$ no matter how models for **I** assign an exemplification extension to the property of being a U. S. president. As we mentioned earlier, this formulation of the definition of truth and logical truth has been influenced by the ideas in [Etchemendy, 1990], from which one might conclude that a philosophically proper model-theoretic definition of logical truth should not depend on alternative interpretations of the constants 'a' and 'P'.[14]

To make these ideas precise and more general, we have to define models and extend the definition of truth and logical truth to all the formulas of the language, including those involving variables and/or complex terms. Thus, given our intended interpretation **I**, we can define a model **M** for **I** as consisting of two functions:

(a) **ext**, for $n \geq 1$, is binary function that maps each pair $\langle \mathbf{r}^n, \mathbf{w}\rangle$ consisting of an n-place relation $\mathbf{r}^n$ ($\in \mathbf{R}_n$) and world **w** ($\in \mathbf{W}$) to a set of n-tuples of objects drawn from **D**, and, for $n = 0$, maps each pair $\langle \mathbf{r}^0, \mathbf{w}\rangle$ consisting of a 0-place relation $\mathbf{r}^0$ and world **w** to an element of $\{\mathbf{T}, \mathbf{F}\}$. We hereafter index **ext** to its second argument. The function $\mathbf{ext_w}$ must satisfy a separate constraint for each of the algebraic operations mentioned above (see below).

(b) $\mathbf{ext_A}$ is a function that maps each property in $\mathbf{R}_1$ to a subset of **D**.

We call $\mathbf{ext_w}(\mathbf{r^n})$ the *exemplification* extension of $\mathbf{r}^n$ at **w**. The constraints that $\mathbf{ext_w}$ must satisfy ensure that the complex relations generated by the algebraic operations have exemplification extensions at a world **w** which are defined in terms of

[14] I have also been influenced by a different attempt to accommodate Etchemendy's ideas, in [Menzel, 1990].

the exemplification extensions at $\mathbf{w}$ of the simpler relations the complex relations may have as parts. For example, here are three constraints on $\mathbf{ext_w}$, governing the operations $\mathbf{PLUG}_i$, $\mathbf{NEG}$, and $\mathbf{NEC}$, respectively, for any relation $\mathbf{r} \in \mathbf{R}_n$ and objects $\mathbf{o}, \mathbf{o}_1, \ldots, \mathbf{o}_n \in \mathbf{D}$:

1. $n > 1$: $\mathbf{ext_w}(\mathbf{PLUG}_i(\mathbf{r}^n, \mathbf{o})) = \{\langle \mathbf{o}_1, \ldots, \mathbf{o}_{i-1}, \mathbf{o}_{i+1}, \ldots, \mathbf{o}_n \rangle \mid$

   $\langle \mathbf{o}_1, \ldots, \mathbf{o}_{i-1}, \mathbf{o}, \mathbf{o}_{i+1}, \ldots, \mathbf{o}_n \rangle \in \mathbf{ext_w}(\mathbf{r}^n)\}$

   $n = 1$: $\mathbf{ext_w}(\mathbf{PLUG}_1(\mathbf{r}^1, \mathbf{o})) = \mathbf{T}$ iff $\mathbf{o} \in \mathbf{ext_w}(\mathbf{r}^1)$

2. $n \geq 1$: $\mathbf{ext_w}(\mathbf{NEG}(\mathbf{r}^n)) = \{\langle \mathbf{o}_1, \ldots, \mathbf{o}_n \rangle \mid \langle \mathbf{o}_1, \ldots, \mathbf{o}_n \rangle \notin \mathbf{ext_w}(\mathbf{r}^n)\}$

   $n = 0$: $\mathbf{ext_w}(\mathbf{NEG}(\mathbf{r}^0)) = \mathbf{T}$ iff $\mathbf{ext_w}(\mathbf{r}^0) = \mathbf{F}$

3. $n \geq 1$: $\mathbf{ext_w}(\mathbf{NEC}(\mathbf{r}^n)) =$

   $\{\langle \mathbf{o}_1, \ldots, \mathbf{o}_n \rangle \mid \forall \mathbf{w}'(\langle \mathbf{o}_1, \ldots, \mathbf{o}_n \rangle \in \mathbf{ext_{w'}}(\mathbf{r}^n))\}$

   $n = 0$: $\mathbf{ext_w}(\mathbf{NEC}(\mathbf{r}^0)) = \mathbf{T}$ iff $\forall \mathbf{w}'(\mathbf{ext_{w'}}(\mathbf{r}^0) = \mathbf{T})$

The constraints on the other logical functions were developed in [Zalta, 1983, pages 62–63] and won't be repeated here. But we can now redescribe two of the examples discussed above in more formal terms. In these examples, $\mathbf{r}$ is a 3-place relation and $\mathbf{o}$ is some object in $\mathbf{D}$. The constraint on $\mathbf{PLUG}$ ensures that $\mathbf{ext_w}(\mathbf{PLUG}_1(\mathbf{r}, \mathbf{o}))$ consists of all those pairs $\langle \mathbf{o}_1, \mathbf{o}_2 \rangle$ such that the triple $\langle \mathbf{o}, \mathbf{o}_1, \mathbf{o}_2 \rangle$ is an element of $\mathbf{ext_w}(\mathbf{r})$. The constraint on $\mathbf{NEG}$ ensures that

$$\mathbf{ext_w}(\mathbf{NEG}(\mathbf{PLUG}_1(\mathbf{r}, \mathbf{o})))$$

consists of all those pairs $\langle \mathbf{o}_1, \mathbf{o}_2 \rangle$ that fail to elements of $\mathbf{ext_w}(\mathbf{PLUG}_1(\mathbf{r}, \mathbf{o}))$.

We call $\mathbf{ext_A}(\mathbf{r})$ the *encoding* extension of $\mathbf{r}$. Note that $\mathbf{ext_A}$ is not defined relative to a world, and so, in a given model, stays fixed from world to world. If one were to abandon the Axiom of Encoding, one could allow this function to vary with worlds. Notice also that we have not explicitly asserted, as a *logical* axiom, that ordinary objects can't encode properties, and so we need not, at this time, constrain $\mathbf{ext_A}(\mathbf{r})$ to contain only abstract objects. It is an axiom of the proper theory of abstract objects that ordinary objects can't encode properties, and so this constraint will be satisfied by any model of the proper theory.[15]

[15] We should note that not only will ordinary objects fail to encode properties in models of the proper theory, but $\mathbf{ext_A}$ will stay fixed among such models. The reason is that the comprehension principle for abstract objects guarantees there is an abstract object corresponding to each (expressible) set of properties. In the intended interpretation, $\mathbf{R}_1$ is fixed and contains all the properties there are, and so the subdomain of abstract objects and the encoding extension of properties will not vary across models in which the comprehension principle is true. This gives us a two senses in which the properties an abstract object encodes are 'essential' to it. Not only are predications 'xF' necessary if possibly true, but the proper (non-logical) facts of encoding do not vary across different models of the possibluum (i.e., across different ways of assigning exemplification extensions to relations at worlds).

Though models $\mathbf{M}$ (= $\langle \mathbf{ext_w}, \mathbf{ext_A} \rangle$) are defined relative to (the intended) interpretation $\mathbf{I}$, we hereafter omit the index on $\mathbf{M}$. The main thing to remember is that models can vary depending on how they assign extensions to the relations (at worlds).

Assignments and Denotation

Now relative to our fixed interpretation $\mathbf{I}$, we define an assignment function $\mathbf{f}$ to the variables of the language as any function mapping object variables to elements of $\mathbf{D}$ and mapping n-place relation variables to elements of $\mathbf{R}_n$. Then we may define, for *each* term τ of the language, a denotation function $\boldsymbol{\delta}_{\mathbf{I},\mathbf{f}}(\tau)$ relative to interpretation $\mathbf{I}$ and assignment $\mathbf{f}$. Suppressing the indices '$\mathbf{I}$' and '$\mathbf{f}$' for the moment, we can explain how $\boldsymbol{\delta}$ works. The denotation of a constant (predicate) is what the interpretation function $\mathbf{F_I}$ assigns to that constant (predicate). The denotation of a variable is what the assignment function $\mathbf{f}$ assigns to that variable. And the denotation of a λ-expression is defined recursively, depending on the structure of the expression. As we mentioned earlier, the λ-expressions may be partitioned into one of ten mutually exclusive classes (for the complete definition, see [Zalta, 1983, pages 64–65]. Eight of these classes contain expressions having a syntactic structure that corresponds to an algebraic logical operation; the ninth class is the repository of the elementary λ-expressions of the form $[\lambda y_1 \ldots y_n\ F_1^y \ldots y_n]$; the tenth class is the repository of the 0-place λ-expressions of the form $[\lambda\ \varphi]$. The denotation function $\boldsymbol{\delta}$ works as follows:

1. To each λ-expression in the first eight classes, $\boldsymbol{\delta}$ assigns a structured relation involving the algebraic operation that corresponds to the syntactic category of the λ-expression. For example, if $\boldsymbol{\delta}(R)$ and $\boldsymbol{\delta}(a)$ are the denotations of the terms R and a, respectively, then $\boldsymbol{\delta}([\lambda x\ Rxa])$ is the property $\mathbf{PLUG}_2(\boldsymbol{\delta}(R), \boldsymbol{\delta}(a))$, because the λ-expression has been syntactically defined to be the '2nd-plugging' of the elementary expression $[\lambda xy\ Rxy]$ by the term a. Similarly, if $\boldsymbol{\delta}(P)$ is the denotation of the expression P, then $\boldsymbol{\delta}([\lambda x\ \Box Px])$ is the property $\mathbf{NEC}(\boldsymbol{\delta}(P))$, because the λ-expression has been syntactically defined to be the 'necessitation' of the elementary expression $[\lambda x\ Px]$.

2. To each elementary λ-expression of the form $[\lambda y_1 \ldots y_n\ \ F^n y_1 \ldots y_n]$, $\boldsymbol{\delta}$ assigns $\mathbf{F_I}(F^n)$ as its denotation.

3. To each 0-place λ-expression $[\lambda\ \varphi]$, $\boldsymbol{\delta}$ recursively assigns a proposition in $\mathbf{R}_0$—the structure of the proposition it assigns corresponds to the complexity of φ. For example, if $\boldsymbol{\delta}(P)$ and $\boldsymbol{\delta}(a)$ are the denotations of the terms P and a respectively, $\boldsymbol{\delta}([\lambda\ Pa])$ is the proposition $\mathbf{PLUG}_1(\boldsymbol{\delta}(P), \boldsymbol{\delta}(a))$. Similarly, $\boldsymbol{\delta}([\lambda\ \Box Pa])$ is

$$\mathbf{NEC}(\mathbf{PLUG}_1(\boldsymbol{\delta}(P), \boldsymbol{\delta}(a))).$$

The full definition of the denotation function won't be repeated here (see [Zalta, 1983, pages 65–67], but the foregoing discussion should give the reader a good idea of how that definition works. Now that we can specify the denotation of all the terms of the language under **I** and **f**, we can define satisfaction.

Satisfaction

Satisfaction is defined for interpreted formulas and we shall use the notation $[\varphi]_\mathbf{I}$ to indicate the formula φ under the interpretation **I**. If given a model **M** and an assignment **f**, we define **f** *satisfies*$_\mathbf{M}$ $[\varphi]_\mathbf{I}$ *with respect to* world **w** as follows:[16]

1. If φ is an atomic exemplification formula of the form $\rho^n o_1 \ldots o_n$,

$$\mathbf{f}\text{ satisfies}_\mathbf{M}\ [\rho^n o_1 \ldots o_n]_\mathbf{I} \text{ with respect to } \mathbf{w} \text{ iff}$$
$$\langle \boldsymbol{\delta}_{\mathbf{I},\mathbf{f}}(o_1), \ldots, \boldsymbol{\delta}_{\mathbf{I},\mathbf{f}}(o_n)\rangle \in \mathbf{M\text{-}ext_w}(\boldsymbol{\delta}_{\mathbf{I},\mathbf{f}}(\rho^n))$$

2. If φ is a atomic exemplification formula of the form ρ^0,

$$\mathbf{f}\text{ satisfies}_\mathbf{M}\ [\rho^0]_\mathbf{I} \text{ with respect to } \mathbf{w} \text{ iff}$$
$$\mathbf{M\text{-}ext_w}(\boldsymbol{\delta}_{\mathbf{I},\mathbf{f}}(\rho^0)) = \mathbf{T}$$

3. If φ is an atomic encoding formula of the form $o\rho^1$,

$$\mathbf{f}\text{ satisfies}_\mathbf{M}\ [o\rho^1]_\mathbf{I} \text{ with respect to } \mathbf{w} \text{ iff}$$
$$\boldsymbol{\delta}_{\mathbf{I},\mathbf{f}}(o) \in \mathbf{M\text{-}ext_A}(\boldsymbol{\delta}_{\mathbf{I},\mathbf{f}}(\rho^1))$$

4. Satisfaction for molecular and quantified φ is classical

5. If φ is a modal formula of the form $\Box\psi$,

$$\mathbf{f}\text{ satisfies}_\mathbf{M}\ [\Box\psi]_\mathbf{I} \text{ with respect to } \mathbf{w} \text{ iff}$$
$$\text{for all } \mathbf{w}' \in \mathbf{W}, \mathbf{f}\text{ satisfies}_\mathbf{M}\ [\psi]_\mathbf{I} \text{ with respect to } \mathbf{w}'$$

Note that because (a) our algebraic operations assign exemplification extensions to propositions in the correct way, and (b) atomic exemplification formulas φ can be turned into terms $[\lambda\ \varphi]$ that denote propositions, we could have collapsed clauses (1) and (2) as follows:

If φ is any atomic exemplification formula,

$$\mathbf{f}\text{ satisfies}_\mathbf{M}\ [\varphi]_\mathbf{I} \text{ with respect to } \mathbf{w} \text{ iff}$$

[16] Readers familiar with [Zalta, 1983, 1988] should realize that we do not include definite description in the present system, and thus we need not account for terms that may fail to denote. So we can formulate the definition of satisfaction in the usual, simple way. Had there been terms that might fail to denote, we would have to use the more general formulation in the earlier works, in which satisfaction conditions explicitly ensure that each term of the formula has a denotation.

$$\mathbf{M}\text{-}\mathbf{ext}_{\mathbf{w}}(\boldsymbol{\delta}([\lambda\ \varphi])) = \mathbf{T}$$

In languages where every formula φ can be turned into a 0-place term $[\lambda\ \varphi]$, this clause would be the only clause necessary for the definition of satisfaction. But the language of the modal object calculus is not such a language.

Truth and Logical Truth

Finally we turn to truth at a world, truth, and logical truth. We define $[\varphi]_{\mathbf{I}}$ is *true at* world $\mathbf{w}$ *in* model $\mathbf{M}$ (in symbols: $\mathbf{M}, \mathbf{w} \models [\varphi]_{\mathbf{I}}$) as follows:

$$\mathbf{M}, \mathbf{w} \models [\varphi]_{\mathbf{I}} =_{df} \text{ Every } \mathbf{f} \text{ satisfies}_{\mathbf{M}}\ [\varphi]_{\mathbf{I}} \text{ with respect to } \mathbf{w}$$

We define $[\varphi]_{\mathbf{I}}$ is *true in* model $\mathbf{M}$ (in symbols: $\mathbf{M} \models [\varphi]_{\mathbf{I}}$) as follows:

$$\mathbf{M} \models [\varphi]_{\mathbf{I}} =_{df} \mathbf{M}, \mathbf{w}_{\alpha} \models [\varphi]_{\mathbf{I}}$$

And we define $[\varphi]_{\mathbf{I}}$ is logically true (in symbols: $\models [\varphi]_{\mathbf{I}}$) as follows:

$$\models [\varphi]_{\mathbf{I}} =_{df} \text{ For every model } \mathbf{M}, \mathbf{M} \models [\varphi]_{\mathbf{I}}$$

Thus, we have defined truth and logical truth for the language of the modal object calculus. As we have set things up, these notions apply to interpreted formulas. Logical truth is not defined by considering variations in the interpretations of the terms, but rather by considering various ways extensions can be assigned (at possible worlds) to the relations denoted by the predicates. The domains of objects, relations, and worlds do not vary from model to model. Rather, given a fixed interpretation of the language in the domain of objects and relations, and given a fixed set of possible worlds over which the modal operators quantify, the models are distinguished as total ways in which extensions to the relations are distributed at all possible worlds. Moreover, the denotation function is *not* relativized to a world. This means that the truth of a modal sentence such as '$\Box Pa$' is not evaluated by examining the denotation of the terms 'a' and 'P' at other possible worlds! Rather, it is evaluated by examining whether the object 'a' in fact denotes and the extension of the relation 'P' in fact denotes are structured in the right way at all possible worlds.

4 A NEW INTERPRETATION OF THE CALCULUS

Since the modal object calculus has been designed to serve as the background logic for the proper theory of abstract objects, the intended interpretation yields truth conditions for the non-logical axioms of the theory. When the calculus and proper axioms together are used to analyze philosophically puzzling data, explanatory success often depends on the fact that some abstract objects encode the very same properties that they exemplify.[17] Given our semantics, this means that in the models of the proper theory, there are properties $\mathbf{r}$ such that $\mathbf{ext}_{\mathbf{w}_\alpha}(\mathbf{r})$ and $\mathbf{ext}_{\mathbf{A}}(\mathbf{r})$ contain the same abstract objects. But recall that while abstract objects are not sets of properties (not, at least, in the intended interpretation), intuitively, they are correlated with sets of properties, and for the purpose of building a set-theoretic interpretation and set-theoretic model in which the proper axioms are true, it is natural to think of them as such. The problem is, however, in what model-theoretic sense can a (well-founded) set of properties exemplify the very properties it has as members?[18] The intended interpretation gives us no further model-theoretic understanding of the very important primitive fact that an abstract object can exemplify the very same properties that it encodes. The principal logical question concerning the set-theoretic representation of the calculus, therefore, is how to explain this primitive fact.

Peter Aczel has offered the following suggestion for modeling the calculus and proper theory in standard Zermelo-Fraenkel set theory:

> ...start with a somewhat larger domain of concrete individuals, some of them *ordinary* and the others *special*, and assume suitable domains of ordinary n-place relations ($n \geq 0$) over the *concrete* individuals. So an ordinary proposition $Ra_1 \ldots a_n$ can be formed by an exemplification whenever R is an ordinary n-place relation and $a_1, \ldots, a_n$ are concrete, possibly special, individuals. Now form the abstract objects (i.e., sets of ordinary properties) ... . To get over the problem of ordinary exemplification of abstract objects we take the following steps. Choose, in whatever way you wish, an assignment of a concrete individual $|a|$ to each abstract individual a. (It might be best that $|a|$ should always be special, but this does not seem necessary.) On the grounds of cardinality, many abstract objects will be assigned the same concrete object. ... Extend the assignment to all objects by putting $|a| = a$ if a is already concrete. We can now take ordinary exemplification $Ra_1 \ldots a_n$ to stand for the ordinary proposition $R|a_1| \ldots |a_n|$, even when some of the $a_1, \ldots, a_n$ are abstract.[19]

In the remainder of this section, we follow up on Aczel's suggestion and recast it in our modal setting.

Although the 'Aczel-interpretations' are rich enough to demonstrate the consistency of the proper theory of abstract objects as well as of the modal object

[17] For example, all abstract objects exemplify the property of *nonsquareness*, so consider the abstract objects that encode this property. As another example, note that Principle C, for which much of the explanatory success of the overall system depends, asserts that there are objects that encode just the property $\bar{E}!$ (i.e., the negation of the property $E!$), and so by the definition of 'abstract' and λ-Conversion, it follows that any such object exemplifies this property as well.

[18] If you try to represent xF as $F \in x$ and represent Fx as $x \in F$, then the presence of objects such that xF & Fx violates the Foundation Axiom of Zermelo-Fraenkel set theory.

[19] Personal communication; letter of January 10, 1991.

calculus, such interpretations will help us to visualize the idea that abstract objects, though correlated with sets of properties, can exemplify the very same properties that they encode. The reader is cautioned, however, not to think that the *intended* interpretation of the modal object calculus is an Aczel-interpretation. In Aczel-interpretations, abstract objects are identified as sets of properties. But in the intended interpretation, abstract objects are not sets of properties. In fact, sets are not included as elements of any domain, except when the theory is applied to the analysis of mathematical objects as abstract objects of a certain sort.[20] To think that abstract objects are sets is to mistake a metaphysical entity for a mathematical one. The properties that an abstract object encodes, in an important new sense, characterize that object; they are predicable of that object. But the properties that are members of a set of properties do not characterize that set in any way. Encoding predication is introduced to describe a phenomena that sets do *not* exhibit. In the technical study of the formal properties of the object calculus, sets prove useful as 'models' of abstract objects, but it doesn't follow from this that abstract objects just are sets.

Aczel-interpretations

An Aczel-interpretation **I** is formalizable as a 9-tuple $\langle \mathbf{W}, \mathbf{O}^*, \mathcal{P}([\mathbf{O}^*]^n), \mathbf{R}, \mathbf{A}, \|a\|, \mathbf{D}, |\mathbf{o}|, \mathbf{F}\rangle$ the elements of which satisfy the following conditions:

1. **W** is a nonempty set of possible worlds, and contains the distinguished element $\mathbf{w}_\alpha$.

2. $\mathbf{O}^*$ is a nonempty primitive domain of *ordinary** objects consisting of two nonempty, disjoint subsets, a set $\mathcal{O}$ of *ordinary* objects and a set $\mathcal{S}$ of *special* objects.[21]

3. $\mathcal{P}([\mathbf{O}^*]^n)$ is the power set of the nth Cartesian product of $\mathbf{O}^*$, for each $n \geq 1$. That is, for $n \geq 1$,

$$\mathcal{P}([\mathbf{O}^*]^n) = \mathcal{P}(\underbrace{\mathbf{O}^* \times \ldots \times \mathbf{O}^*}_{n \text{ times}})$$

   In the definition of a model, we will take the exemplification extension of an n-place relation $\mathbf{r}^n$ ($n \geq 1$) at world **w** to be a member of this set.

[20] And even then, 'the Zermelo-Fraenkel sets' are objects that encode rather than exemplify the property of being a set, since that is a property attributed to them in the theory we use to conceptualize them.

[21] In the above quotation, Aczel called the members of $\mathbf{O}^*$ the 'concrete' objects. However, we are calling them *ordinary** objects. The ordinary objects (i.e., the members of $\mathcal{O}$) constitute a subset of the ordinary* objects, and intuitively, the 'concrete' objects constitute a subset of the ordinary objects, namely, the ones that exemplify spatio-temporal location at the actual world. So the ordinary objects divide up into the concrete objects and the objects that have spatio-temporal locations at other possible worlds. We should also mention that, intuitively, $\mathcal{S}$ should outstrip $\mathcal{O}$ in size by a couple of orders of magnitude, but for the present purposes, we need not make this explicit.

4. **R** is a nonempty, primitive domain of *relations* satisfying the following two conditions:

   (a) **R** is the union of a sequence of nonempty, pairwise disjoint sets, $\mathbf{R}_0$, $\mathbf{R}_1, \ldots$ (i.e., $\mathbf{R} = \bigcup_{n \geq 0} \mathbf{R}_n$) such that each member $\mathbf{R}_n$ (the set of *n-place relations*), for $n \geq 1$, is greater in size than the set of all functions from **W** into $\mathcal{P}([\mathbf{O}^*]^n)$.

   (b) **R** is closed under the following operations: $\mathbf{PLUG}_i$, **NEG**, **NEC**, **COND**, $\mathbf{UNIV}_i$, $\mathbf{REFL}_{i,j}$, $\mathbf{CONV}_{i,j}$, and $\mathbf{VAC}_i$.

   Condition (a) will guarantee that there are more relations than there are Montagovian intensions [Montague, 1974]). This will ensure that some distinct relations have the same Montagovian intension.

5. $\mathbf{A} = \mathcal{P}(\mathbf{R}_1)$; i.e., the set **A** of *abstract* objects is simply the power set of the set of properties.

6. $\|\mathbf{a}\|$ is a mapping which is defined for $\mathbf{a} \in \mathbf{A}$ and which takes values in $\mathcal{S}$. Recall that $\mathcal{S}$ is the set of special objects and is a subset of $\mathbf{O}^*$. The special object $\|\mathbf{a}\|$ will serve as a proxy for **a**

7. $\mathbf{D} = \mathbf{A} \cup \mathbf{O}^*$; i.e., the domain **D** of *all* objects is the union of **A** and $\mathbf{O}^*$.

8. $|\mathbf{o}|$ is a mapping which is defined for $\mathbf{o} \in \mathbf{D}$ and which takes values in $\mathbf{O}^*$. It must satisfy the following conditions:

   if $\mathbf{o} = \mathbf{a}$, for some $\mathbf{a} \in \mathbf{A}$, $|\mathbf{o}| = \|\mathbf{a}\|$
   if $\mathbf{o} = \mathbf{o}^*$, for some $\mathbf{o}^* \in \mathbf{O}^*$, $|\mathbf{o}| = \mathbf{o}^*$

   This extends the function $\|\mathbf{a}\|$ to a function $|\mathbf{o}|$ defined on all the members of **D**. $|\mathbf{o}|$ agrees with $\|\mathbf{a}\|$ on the abstract objects and maps each ordinary* object to itself.

9. **F** is an interpretation function that maps the object constants of the language to an element of **D** and the n-place relation constants to an element of $\mathbf{R}_n$.

Models for Aczel-interpretations

A model **M** for an Aczel-interpretation **I** consists of two functions:

1. **ext** is a function that meets the following conditions:

   (a) for $n \geq 1$, **ext**: $\mathbf{R}_n \times \mathbf{W} \to \mathcal{P}([\mathbf{O}^*]^n)$.

   (b) for $n = 0$, **ext**: $\mathbf{R}_0 \times \mathbf{W} \to \{\mathbf{T}, \mathbf{F}\}$.

   (c) **ext** satisfies the constraints on the operations $\mathbf{PLUG}_i$, **NEG**, **NEC**, **COND**, $\mathbf{UNIV}_i$, $\mathbf{REFL}_{i,j}$, $\mathbf{CONV}_{i,j}$, and $\mathbf{VAC}_i$.

We henceforth index **ext** to its second argument, and so $\mathbf{ext_w}$ maps each $\mathbf{r}^n \in \mathbf{R}_n$ to its exemplification extension at world **w**.

2. $\mathbf{ext_A}$ is a function defined on the elements of $\mathbf{R}_1$ as follows:

$$\mathbf{ext_A}(\mathbf{r}^1) = \{\mathbf{o} \in \mathbf{D} \mid \mathbf{o} \in \mathbf{A} \text{ and } \mathbf{r}^1 \in \mathbf{o}\}.$$

Two observations are in order here. The first is that the function $\mathbf{ext_w}$ is essentially the same function as its counterpart in the models for the intended interpretation, with the exception that it takes values in $\mathcal{P}([\mathbf{O}^*]^n)$ rather than $\mathcal{P}([\mathbf{D}]^n)$. Since the abstract objects (i.e., sets of properties) in **A** are not elements of $\mathbf{O}^*$, they are not officially in the exemplification extension of any relation. But we will exploit the fact that their *proxies* are in the exemplification extensions of relations: the definition of satisfaction will allow an abstract object **a** to 'exemplify' a relation if its proxy $\|\mathbf{a}\|$ is an element of the exemplification extension of that relation. The second observation is that the function $\mathbf{ext_A}$ is more specific than its counterpart in the models for the intended interpretation. It requires that the encoding extension of a property contain as elements all those abstract objects of which it is a member. Notice that the encoding extension of a property neither varies from world to world nor varies from model to model.[22] The different models for a given Aczel-interpretations can therefore only vary in the way they assign exemplification extensions to relations at the various possible worlds.

The Denotation Function

If given an Aczel-interpretation **I**, we next fix an assignment **f** to the variables of the object language, as before. Then we may define a denotation function $\boldsymbol{\delta}_{\mathbf{I},\mathbf{f}}(\tau)$, relative to interpretation **I** and function **f**, for all the terms τ of the language, as follows (suppressing the subscripts on $\boldsymbol{\delta}$):

(a) where o is an object constant, $\boldsymbol{\delta}(o) = \mathbf{F_I}(o)$

(b) where o is an object variable, $\boldsymbol{\delta}(o) = \mathbf{f}(o)$

(c) where ρ^n is an n-place predicate, $\boldsymbol{\delta}(\rho^n) = \mathbf{F_I}(\rho^n)$

(d) where ρ^n is an n-place predicate variable, $\boldsymbol{\delta}(\rho^n) = \mathbf{f}(\rho^n)$

(e) where μ is a λ-expression $[\lambda \nu_1 \ldots \nu_n\, \varphi]$, $\boldsymbol{\delta}(\mu)$ is defined as on pages 65–66 of [Zalta, 1983], with the following substitution:

[22] The reason is that in Aczel-interpretations, the domain of abstract objects **A** is fixed as the power set of the set properties. Since the set of properties $\mathbf{R}_1$ is an element of the Aczel-interpretation, its power set is independent of models **M**. So whether an object $\mathbf{o} \in \mathbf{D}$ encodes a property **r** is simply a matter of whether **o** is an element of the subdomain **A** and $\mathbf{r} \in \mathbf{o}$. So the $\mathbf{ext_A}$ function is fixed in all models for Aczel-interpretations.

where μ is the ith-plugging of λ-expression ξ by term o,
$\boldsymbol{\delta}(\mu) = \mathbf{PLUG}_i(\boldsymbol{\delta}(\xi), |\boldsymbol{\delta}(o)|)$

(f) where $[\lambda\, \varphi]$ is a 0-place λ-expression, $\boldsymbol{\delta}([\lambda\, \varphi])$ is defined as on p. 67 of [Zalta, 1983], with the following substitution:[23]

if $\varphi = \rho^n o_1 \ldots o_n$, $\boldsymbol{\delta}([\lambda\, \varphi]) =$

$$\mathbf{PLUG}_1(\ldots(\mathbf{PLUG}_n(\boldsymbol{\delta}(\rho^n), |\boldsymbol{\delta}(o_n)|), \ldots), |\boldsymbol{\delta}(o_1)|)$$

The denotation function, therefore, works in the same manner as its counterpart in the previous section, except for the changes described in clauses (e) and (f). If we let $\boldsymbol{\delta}(o)$ be the object **o**, then these clauses tell us that the complex relational properties and propositions denoted by λ-expressions containing the object term o have the object $|\mathbf{o}|$ as a constituent instead of the object **o**. This makes a difference only for the complex relations and propositions denoted by λ-expressions containing terms that denote an abstract object. In such cases, the relations and propositions in question have the proxies rather than the abstract objects themselves as constituents.

Satisfaction for Interpreted Formulas

If given an arbitrary formula φ, we may now define, assignment **f** *satisfies*$_{\mathbf{M}}$ $[\varphi]_{\mathbf{I}}$ *with respect to* world **w**, as follows:

1. If φ is an atomic exemplification formula of the form $\rho^n o_1 \ldots o_n$,

   f satisfies$_{\mathbf{M}}$ $[\rho^n o_1 \ldots o_n]_{\mathbf{I}}$ with respect to **w** iff
   $\langle |\boldsymbol{\delta}_{\mathbf{I,f}}(o_1)|, \ldots, |\boldsymbol{\delta}_{\mathbf{I,f}}(o_n)| \rangle \in \mathbf{M\text{-}ext}_{\mathbf{w}}(\boldsymbol{\delta}_{\mathbf{I,f}}(\rho^n))$

2. If φ is a atomic exemplification formula of the form ρ^0,

   f satisfies$_{\mathbf{M}}$ $[\rho^0]_{\mathbf{I}}$ with respect to **w** iff
   $\mathbf{M\text{-}ext}_{\mathbf{w}}(\boldsymbol{\delta}_{\mathbf{I,f}}(\rho^0)) = \mathbf{T}$

3. If φ is an atomic encoding formula of the form $o\rho^1$,

   f satisfies$_{\mathbf{M}}$ $[o\rho^1]_{\mathbf{I}}$ with respect to **w** iff
   $\boldsymbol{\delta}_{\mathbf{I,f}}(o) \in \mathbf{M\text{-}ext}_{\mathbf{A}}(\boldsymbol{\delta}_{\mathbf{I,f}}(\rho^1))$

4. Satisfaction for molecular and quantified φ is classical

5. If φ is a modal formula of the form $\Box\psi$,

[23] In [Zalta, 1983], we let φ itself go proxy for the 0-place relation term $[\lambda\, \varphi]$. But it now seems more perspicuous to avoid such an abbreviation.

$$\mathbf{f} \text{ satisfies}_{\mathbf{M}} \; [\Box\psi]_{\mathbf{I}} \text{ with respect to } \mathbf{w} \text{ iff}$$
$$\text{for all } \mathbf{w}' \in \mathbf{W}, \mathbf{f} \text{ satisfies}_{\mathbf{M}} \; [\psi]_{\mathbf{I}} \text{ with respect to } \mathbf{w}'$$

The only real difference between the definition of satisfaction for Aczel-interpretations and its counterpart for intended interpretations concerns clause 1. It defines the sense in which an abstract object can exemplify a relation, namely, by proxy. Consider the formula 'Pa' under **I**, and suppose that $\mathbf{F_I}(a) = \mathbf{o}$ and $\mathbf{F_I}(P) = \mathbf{r}$. Then **f** satisfies$_{\mathbf{M}}$ 'Pa' at **w** iff either (1) **o** is ordinary* and an element of $\mathbf{M}$-$\mathbf{ext_w}(\mathbf{r})$ (since $|\mathbf{o}|$ is just **o** itself when **o** is ordinary*), or (2) **o** is some abstract object **a** and $||\mathbf{a}||$ is an element of $\mathbf{M}$-$\mathbf{ext_w}(\mathbf{r})$ (since $|\mathbf{o}|$ is $||\mathbf{o}||$ when **o** is an abstract object). We should also remark that it would have been somewhat more direct to define clause 3 in the definition of satisfaction as:

$$\mathbf{f} \text{ satisfies}_{\mathbf{M}} \; [o\rho^1]_{\mathbf{I}} \text{ with respect to } \mathbf{w} \text{ iff}$$
$$\boldsymbol{\delta}_{\mathbf{I,f}}(o) \in \mathbf{A} \;\text{ and }\; \boldsymbol{\delta}_{\mathbf{I,f}}(\rho^1) \in \boldsymbol{\delta}_{\mathbf{I,f}}(o)$$

As it stands, however, clause 3 demonstrates that the satisfaction of atomic encoding formulas appeals to an *extension* of the property denoted.

Truth and Logical Truth

With satisfaction defined, we may define $[\varphi]_{\mathbf{I}}$ is *true at* world **w** *in* model **M** in the usual way:

$$\mathbf{M}, \mathbf{w} \models [\varphi]_{\mathbf{I}} =_{df} \text{ Every } \mathbf{f} \text{ satisfies}_{\mathbf{M}} \; [\varphi]_{\mathbf{I}} \text{ with respect to } \mathbf{w}$$

And, finally, we define $[\varphi]_{\mathbf{I}}$ is *true in* model **M** and define $[\varphi]_{\mathbf{I}}$ is *logically true*, as follows:

$$\mathbf{M} \models [\varphi]_{\mathbf{I}} =_{df} \mathbf{M}, \mathbf{w}_\alpha \models [\varphi]_{\mathbf{I}}$$

$$\models [\varphi]_{\mathbf{I}} =_{df} \text{ For every model } \mathbf{M}, \mathbf{M} \models [\varphi]_{\mathbf{I}}$$

5 CONCLUSION

Aczel-interpretations demonstrate the consistency of the modal object calculus as well as the consistency of the proper theory of abstract objects. Though we shall not develop the proofs in detail here, it is relatively straightforward to show that, under Aczel-interpretations, the five groups of logical axioms for the calculus described in Section 1 are logically true and the three rules of inference preserve logical truth. In particular, the proof that the axioms of propositional logic, quantificational logic, and S5 modal logic are logically true are essentially classical, as are the proofs that the rules MP, Gen, and RN preserve logical truth. To see that the axiom of encoding (i.e., $\Diamond xF \rightarrow \Box xF$) is logically true, pick an arbitrary model $\mathbf{M}$ and reason from the point of view of the actual world: if at some possible world $\boldsymbol{\delta}(x) \in \mathbf{ext_A}(\boldsymbol{\delta}(F))$, then since the encoding extension of a property does not vary from world to world, $\boldsymbol{\delta}(x) \in \mathbf{ext_A}(\boldsymbol{\delta}(F))$ at all worlds.

To see that the axioms for the logic of relations are logically true, consider first the λ-Conversion principle with respect to an arbitrarily chosen model $\mathbf{M}$. Given the facts that the denotation function assigns each λ-expression an appropriately structured relation and that $\mathbf{ext_w}$ is appropriately constrained for each logical function, it is relatively straightforward to show, by induction on the complexity of φ, that for every world $\mathbf{w}$, every assignment function $\mathbf{f}$ satisfies the formula $[\lambda y_1 \ldots y_n\ \varphi]x_1 \ldots x_n$ iff $\mathbf{f}$ satisfies $\varphi^{x_1,\ldots,x_n}_{y_1,\ldots,y_n}$ with respect to $\mathbf{w}$. For no matter what n-tuple of objects $\langle \mathbf{o}_1, \ldots, \mathbf{o}_n \rangle$ that $\mathbf{f}$ assigns to the variables $x_1, \ldots, x_n$, that n-tuple is an element of $\mathbf{ext_w}(\boldsymbol{\delta}([\lambda y_1 \ldots y_n\ \varphi]))$ iff $\mathbf{f}$ satisfies (at $\mathbf{w}$) the formula that results by replacing the y_i in φ with the x_i, that is, iff the objects $\mathbf{o}_1, \ldots, \mathbf{o}_n$ are in the exemplification extensions of relations at $\mathbf{w}$ in just the way $\varphi^{x_1,\ldots,x_n}_{y_1,\ldots,y_n}$ says.[24] The λ-Identity$_1$ principle ($[\lambda y_1 \ldots y_n\ F^n y_1 \ldots y_n] = F^n$) is logically true, notwithstanding the special definition of relation identity (in which F^n and G^n are said to be 'identical' iff no matter how you 'plug' them up with $n-1$ objects the two resulting properties are encoded by the same objects). This principle holds because the denotation function guarantees that the denotation of $[\lambda y_1 \ldots y_n\ F^n y_1 \ldots y_n]$ is the same relation as the denotation of F^n. Thus, in an arbitrary model, the *defined* conditions for relation identity between $[\lambda y_1 \ldots y_n\ F^n y_1 \ldots y_n]$ and F^n will be trivially true at the actual world, given that these expressions denote the same relation in the semantics. Finally, the λ-Identity$_2$ principle ($[\lambda y_1 \ldots y_n\ \varphi] = [\lambda y'_1 \ldots y'_n\ \varphi']$) is logically true, since mere alphabetic changes don't affect the denotation of the λ-expression. Again, the special definition of relation identity is trivially true at the actual world (in any arbitrarily chosen model) if $[\lambda y_1 \ldots y_n\ \varphi]$ and $[\lambda y'_1 \ldots y'_n\ \varphi']$ denote the same relation.

Finally, readers familiar with the proper theory of abstract objects may also wish to consider the constraints that must be placed on the models for Aczel-interpretations if the proper axioms of the theory of abstract objects are to turn out

[24] As we remarked earlier, a given λ-expression is syntactically categorized either as the plugging of another λ-expression in the ith place by a certain term, or as the conditionalization of two λ-expressions, or as the negation of another λ-expression, or as the conversion of another expression about the ith and j^{th} places, etc. Each syntactic category corresponds to an algebraic logical operation. Complex λ-expressions can therefore be thought of as structural transformations of simpler ones, and the denotation of the whole expression will be built up from the denotations of the simpler expressions in a way that mirrors the transformation process. The constraint on $\mathbf{ext_W}$ of the complex relation denoted at the final stage of transformation can then be decomposed into a variety of constraints on the simpler relations denoted by the simpler expressions involved in the transformation process. So the semantic reason why objects exemplify the complex relation at a world iff they stand in the simpler relations at that world is that the constraints on the exemplification extension of the complex relation decompose into the right constraints on the exemplification extensions of the simpler relations it may have as parts. That is why the λ-conversion principle is logically true.

true. We simply require that models assign an extension to the property denoted by the distinguished 1-place relation constant $E!$ (having a spatio-temporal location) in the right way. Consider those models **M** for an Aczel-interpretation **I** that satisfy the following two conditions:

1. $\mathbf{M}\text{-}\mathbf{ext}_{\mathbf{w}}(\mathbf{F_I}(E!)) \subseteq \mathcal{O}$, for each **w**
2. $\mathcal{O} = \bigcup_{\mathbf{w} \in \mathbf{W}} \mathbf{M}\text{-}\mathbf{ext}_{\mathbf{w}}(\mathbf{F_I}(E!))$

In such models, the property denoted by $E!$ (having spatio-temporal location) has only ordinary objects in its exemplification extension at each world (recall that $\mathcal{O}$ is the set of ordinary objects and constitutes a subset of the ordinary* objects in $\mathbf{O}^*$). Note that these conditions allow the same ordinary object to appear in the exemplification extension of $\mathbf{F_I}(E!)$ at more than one world. Therefore, the property denoted by the defined 1-place relation constant $O!$ (i.e., $[\lambda x \; \Diamond E!x]$), which identifies the 'ordinary' objects, will have an exemplification extension at each world **w** that consists of all the objects that exemplify existence at some world or other.[25] So at each **w** (including $\mathbf{w}_\alpha$), $\mathbf{ext}_{\mathbf{w}}(\delta_{\mathbf{I},\mathbf{f}}(O!)) = \mathcal{O}$, no matter what the assignment **f**. Moreover, the property denoted by the defined 1-place relation constant $A!$ (i.e., $[\lambda x \; \neg\Diamond E!x]$), which identifies the 'abstract' objects, will have an exemplification extension at each world that consists of all the special objects in $\mathcal{S}$. Consequently, each set **a** in the power set of the set of properties (i.e., each $\mathbf{a} \in \mathcal{P}(\mathbf{R}_1)$) will be in the exemplification extension of the property denoted by $A!$ in each world, since $|\mathbf{a}|$ is an element of $\mathcal{S}$.[26] These facts guarantee that Principle B turns out true: the assertion that ordinary objects necessarily fail to encode properties (i.e., $O!x \rightarrow \Box\neg\exists F\, xF$) is true in models satisfying the above constraints, since no ordinary objects encode properties at any world. If $\delta(x) \in \mathcal{O}$, then $\delta(x) \notin \mathbf{A}$, and so (at every world **w**) no property **r** is such that $\mathbf{r} \in \delta(x)$.

In any model **M** for an Aczel-interpretation satisfying the above constraints, the comprehension principle for abstract objects is true. Recall Principle C:

$$\exists x(A!x \;\&\; \forall F(xF \equiv \varphi)), \text{ where } \varphi \text{ has no free } x\text{s.}$$

[25] To see why, suppose for simplicity that $\Diamond$ is a primitive modal operator and that there is a primitive algebraic logical operation **POS** which is the dual of **NEC**. Then the denotation of $[\lambda x \; \Diamond E!x]$ would be defined as: $\mathbf{POS}(\mathbf{F_I}(E!))$. Constraints on $\mathbf{ext_w}$ would guarantee that:

$$\mathbf{ext_w}(\mathbf{POS}(\mathbf{F_I}(E!))) = \{\mathbf{o} \mid \exists \mathbf{w}'(\mathbf{o} \in \mathbf{ext}_{\mathbf{w}'}(\mathbf{F_I}(E!)))\}$$

So at the actual world, the exemplification extension of the property denoted by $O!$ consists of all those objects that exemplify existence at some world or other.

[26] For those readers intimately familiar with the theory, we also need to give an extension to the primitive relation of identity for the ordinary objects. In more precise formulations of the proper theory, we employ the distinguished 2-place relation constant $=_E$ to denote this relation and Principle A (described in Section 1) is cast as an axiom that gives necessary and sufficient conditions for $x =_E y$, namely, that x and y are both ordinary objects which necessarily exemplify the same properties. To validate this axiom, we must constrain models **M** as follows:

$$\mathbf{M}\text{-}\mathbf{ext_w}(\mathbf{F_I}(=_E)) = \{\langle \mathbf{o}, \mathbf{o} \rangle \mid \mathbf{o} \in \mathcal{O}\}, \text{ for each } \mathbf{w}$$

This validates the axiom of identity for ordinary objects.

The reason this axiom is true in a model **M** satisfying the above constraints is that, for any given φ, there is an abstract object (i.e., a set of properties) **a** in **A** such that:

(a) $|\mathbf{a}| \in \mathcal{S}$, and

(b) $\boldsymbol{\delta}_{\mathbf{I},\mathbf{f}}(F) \in \mathbf{a}$ iff **f** satisfies$_{\mathbf{M}}$ φ w.r.t. $\mathbf{w}_\alpha$ (for any assignment **f**).

Since we have the power set of the set of properties to choose from, we know that there is a set of properties having as members precisely those properties that 'satisfy' the formula φ. Notice that the definition for the identity of abstract objects (Principle D) is also justified. The definition says that abstract objects are 'identical' iff necessarily, they encode the same properties. But clearly, in Aczel-interpretations, abstract objects **a** and **b** are sets of properties, and so they are identical iff they have the same properties as elements. And if **a** and **b** have the same properties as elements, this fact is true at every possible world. So, the semantic fact that $\mathbf{a} = \mathbf{b}$ iff $\forall\mathbf{w}\forall\mathbf{r}(\mathbf{r} \in \mathbf{a}$ iff $\mathbf{r} \in \mathbf{b})$ justifies the object-language definition that two abstract objects x and y are 'identical' iff necessarily, they encode the same properties.

Center for the Study of Language and Information
Stanford University

Acknowledgments. I am indebted to the Center for the Study of Language and Information for providing me with office space. I would also like to thank Chris Menzel, Bernard Linsky, and Kees van Deemter for examining some of the logical details described herein.

REFERENCES

[Bealer, 1982] G. Bealer. *Quality and Concept*. Clarendon, Oxford, 1982.

[Carnap, 1947] R. Carnap. *Meaning and Necessity*. University of Chicago Press, Chicago, 1947.

[Castañeda, 1974] H. Castañeda. Thinking and the structure of the world. *Philosophia*, 4:3–40, 1974.

[Došen, 1988] K. Došen. Second-order logic without variables. In W. Buszkowski, W Marciszewski, and J. van Benthem, editors, *Categorial Grammar*, John Benjamins, Amsterdam, 1988.

[Etchemendy, 1990] J. Etchemendy. *The Concept of Logical Consequence*. Harvard University Press, Cambridge, MA, 1990.

[Frege, 1892] G. Frege. On sense and reference. In P. Geach and M. Black, editors, *Translations from the Philosophical Writings of Gottlob Frege*, pages 56–78, Basil Blackwell, Oxford, 1892.

[Hintikka, 1961] J. Hintikka. Modality and quantification. *Theoria*, 27:110–128, 1961.

[Kripke, 1959] S. Kripke. A completeness theorem in modal logic. *Journal of Symbolic Logic*, 24:1–14, 1959.

[Kripke, S] S. Kripke. Semantical considerations on modal logic. *Acta Philosophica Fennica*, 16:83–94, 1963.

[Mally, 1912] E. Mally. *Gegenstandstheoretische Grundlagen der Logik und Logistik*. Barth, Leipzig, 1912.

[Meinong, 1904] A. Meinong. Über Gegenstandstheorie. In A. Meinong, editor, *Untersuchungen zur Gegenstandstheorie und Psychologie*. Barth, Leipzig, 1904. (English translation: On the theory of objects. In R. Chisholm, editor, *Realism and the Background of Phenomenology*, pages 76–117. The Free Press, Glencoe, 1960.)

[Menzel, 1986] C. Menzel. A complete, type-free "second-order" logic and its philosophical foundations. Technical Report No. CSLI–86–40, Center for the Study of Language and Information, Stanford, CA, 1986.

[Menzel, 1990] C. Menzel. Actualism, ontological commitment, and possible world semantics. *Synthese*, 85:355–389, 1990.

[Montague, 1974] R. Montague. *Formal Philosophy: Selected Papers of Richard Montague*, Richmond Thomason, editor, Yale University Press, New Haven, 1974.

[Parsons, 1980] T. Parsons. *Nonexistent Objects*. Yale University Press, New Haven, 1980.

[Quine, 1960] W. Quine. Variables explained away. In *Selected Logical Papers*, pages 227–235. Random House, New York, 1960.

[Rapaport, 1978] W. Rapaport. Meinongian theories and a Russellian paradox. *Nous*, 12:153–180, 1978.

[Routley, 1979] R. Routley. *Exploring Meinong's Jungle and Beyond*, Departmental Monograph #3, Philosophy Program, Research School of the Social Sciences, Australian National University, Canberra, 1979.

[Tarski, 1936] A. Tarski. On the concept of logical consequence, 1936. Translated from the Polish and published in *Logic, Semantics, and Metamathematics*. Clarendon, Oxford, 1956.

[Tarski, 1944] A. Tarski. The semantical conception of truth and the foundations of semantics. *Philosophy and Phenomenological Research*, 4:341–376, 1944.

[Zalta, 1983] E. Zalta. *Abstract Objects: An Introduction to Axiomatic Metaphysics*. D. Reidel, Dordrecht, 1983.

[Zalta, 1987] E. Zalta. On the structural similiarities between worlds and times. *Philosophical Studies*, 51:213–239, 1987.

[Zalta, 1988] E. Zalta. *Intensional Logic and the Metaphysics of Intentionality*. MIT/ Bradford, Cambridge, MA, 1988.

[Zalta, 1993a] E. Zalta. Twenty-five basic theorems in situation and world theory. *Journal of Philosophical Logic*, 22:385–428, 1993.

[Zalta, 1993b] E. Zalta. A philosophical conception of propositional modal logic. *Philosophical Topics*, 20(2):263–281, 1993.

NAME INDEX

SUBJECT INDEX

APPLIED LOGIC SERIES

1. D. Walton: *Fallacies Arising from Ambiguity.* 1996 ISBN 0-7923-4100-7
2. H. Wansing (ed.): *Proof Theory of Modal Logic.* 1996 ISBN 0-7923-4120-1
3. F. Baader and K.U. Schulz (eds.): *Frontiers of Combining Systems.* First International Workshop, Munich, March 1996. 1996 ISBN 0-7923-4271-2
4. M. Marx and Y. Venema: *Multi-Dimensional Modal Logic.* 1996 ISBN 0-7923-4345-X
5. S. Akama (ed.): *Logic, Language and Computation.* 1997 ISBN 0-7923-4376-X
6. J. Goubault-Larrecq and I. Mackie: *Proof Theory and Automated Deduction.* 1997 ISBN 0-7923-4593-2
7. M. de Rijke (ed.): *Advances in Intensional Logic.* 1997 ISBN 0-7923-4711-0

KLUWER ACADEMIC PUBLISHERS – DORDRECHT / BOSTON / LONDON